Wafer Level 3-D ICs Process Technology

Series on Integrated Circuits and Systems

Series Editor: Anantha Chandrakasan
Massachusetts Institute of Technology
Cambridge, Massachusetts

Wafer Level 3-D ICs Process Technology
Chuan Seng Tan, Ronald J. Gutmann, and L. Rafael Reif (Eds.)
ISBN 978-0-387-76532-7

Adaptive Techniques for Dynamic Processor Optimization: Theory and Practice
Alice Wang and Samuel Naffziger (Eds.)
ISBN 978-0-387-76471-9

mm-Wave Silicon Technology: 60 GHz and Beyond
Ali M. Niknejad and Hossein Hashemi (Eds.)
ISBN 978-0-387-76558-7

Ultra Wideband: Circuits, Transceivers, and Systems
Ranjit Gharpurey and Peter Kinget (Eds.)
ISBN 978-0-387-37238-9

Creating Assertion-Based IP
Harry D. Foster and Adam C. Krolnik
ISBN 978-0-387-36641-8

Design for Manufacturability and Statistical Design: A Constructive Approach
Michael Orshansky, Sani R. Nassif, and Duane Boning
ISBN 978-0-387-30928-6

Low Power Methodology Manual: For System-on-Chip Design
Michael Keating, David Flynn, Rob Aitken, Alan Gibbons, and Kaijian Shi
ISBN 978-0-387-71818-7

Modern Circuit Placement: Best Practices and Results
Gi-Joon Nam and Jason Cong
ISBN 978-0-387-36837-5

CMOS Biotechnology
Hakho Lee, Donhee Ham and Robert M. Westervelt
ISBN 978-0-387-36836-8

SAT-Based Scalable Formal Verification Solutions
Malay Ganai and Aarti Gupta
ISBN 978-0-387-69166-4, 2007

Ultra-Low Voltage Nano-Scale Memories
Kiyoo Itoh, Masashi Horiguchi and Hitoshi Tanaka
ISBN 978-0-387-33398-4, 2007

Routing Congestion in VLSI Circuits: Estimation and Optimization
Prashant Saxena, Rupesh S. Shelar, Sachin Sapatnekar
ISBN 978-0-387-30037-5, 2007

Continued after index

Chuan Seng Tan • Ronald J. Gutmann
L. Rafael Reif
Editors

Wafer Level 3-D ICs Process Technology

Foreword by Scott List

Springer

Editors
Chuan Seng Tan
School of Electrical and Electronic Engineering
Nanyang Technological University
Singapore
TanCS@ntu.edu.sg

Ronald J. Gutmann
Center for Integrated Electronics
Rensselaer Polytechnic Institute
Troy, NY
USA
gutmar@rpi.edu

L. Rafael Reif
Department of Electrical Engineering
Massachusetts Institute of Technology
Cambridge, MA
USA
reif@mit.edu

ISSN: 1558-9412
ISBN: 978-0-387-76532-7 e-ISBN: 978-0-387-76534-1
DOI: 10.1007/978-0-387-76534-1

Library of Congress Control Number: 2008920894

Printed on acid-free paper

springer.com

Foreword

Three-dimensional (3D) integration is clearly the simplest answer to most of the semiconductor industry's vexing problems: heterogeneous integration and reductions of power, form factor, delay, and even cost. Conceptually the power, latency, and form factor of a system with a fixed number of transistors all scale roughly linearly with the diameter of the smallest sphere enclosing frequently interacting devices. This clearly provides the fundamental motivation behind 3D technologies which vertically stack several strata of device and interconnect layers with high vertical interconnectivity. In addition, the ability to vertically stack strata with divergent and even incompatible process flows provides for low cost and low parasitic integration of diverse technologies such as sensors, energy scavengers, nonvolatile memory, dense memory, fast memory, processors, and RF layers. These capabilities coupled with today's trends of increasing levels of integrated functionality, lower power, smaller form factor, increasingly divergent process flows, and functional diversification would seem to make 3D technologies a natural choice for most of the semiconductor industry.

Since the concept of vertical integration of different strata has been around for over 20 years, why aren't vertically stacked strata endemic to the semiconductor industry? The simple answer to this question is that in the past, the 3D advantages while interesting were not necessary due to the tremendous opportunities offered by geometric scaling. In addition, even when the global interconnect problem of high-performance single-core processors seemed insurmountable without innovations such as 3D, alternative architectural solutions such as multicores could effectively delay but not eliminate the need for 3D. Cost and risk avoidance are also major factors delaying the implementation of 3D. Geometric scaling has a fundamental $2x$ cost reduction per technology node while 3D from a simple wafer perspective has an additional cost of vertical wafer bonding and interconnection. It is only with recent trends toward divergent process flow integration that 3D offers the potential for substantial cost reduction. The relative immaturity of the novel 3D process flows has also delayed its adoption.

So what is the future of 3D? It appears as if its time has finally come. The increasingly more difficult challenges to continued geometric scaling have made 3D the most attractive option to continue increasing the integrated functionality of chips. The trend for reduced form factor has already resulted in commercial implementation of through-silicon via technologies in stacked memories for cell phones. This innovation has primed the pump for related 3D technologies. The vertical integration of divergent flows with through-silicon vias will be implemented within a couple years on cell phones, and high-performance, low-power applications with higher via density are not much further out. Perhaps the greatest potential for 3D will come when more conventional applications drive the technology to sufficient maturity to enable vastly more aggressive 3D integration. Conceptually new biochips in 100-μm cubes may be introduced into the body, scavenge energy, selectively attract cancer cells, sense the type of cell, turn off if the wrong cell is attracted, zap the correct cells with high current, store the event and periodically transmit a unique RF signal to an outside receiver of their identity and running cancer cell kills per specified category.

Three-dimensional integration can be defined in as many different ways as there are researchers in the field. This book provides the most complete differentiation of the various 3D technologies in the literature. It also provides sufficient detail to fully understand their capabilities, limitations, and targeted applications, and closely couples the reader to a quiet revolution in the making.

Intel/SRC Scott List

Preface

Three-dimensional (3D) integration has emerged as an attractive contender as the semiconductor industry faces serious obstacles with interconnect scaling and as demand for on-chip functionality continues to increase. The advent of 3D integration is a direct result of active research in academia, research laboratories, and industry over the past several years. Today, 3D integration takes many forms depending on the application and it promises to be a viable future technology alternative. At the time of this writing, there are already commercial products featuring chip stacks vertically interconnected by through-silicon vias (TSVs).

The idea of a book on 3D technology dates back to more than a year ago. There were then (and continues to be now) an increasing number of publications and conferences that focused on 3D integration. However, a reference book on this emerging field was lacking. While the initial idea was to author a book, we soon realized that such an endeavor would be extremely challenging given the many varieties of 3D integration technologies. We revisited the plan and decided to edit a book instead with contributions from experts in academia, research laboratories, and industry. After careful planning, we identified and invited chapter contribution from an impressive line-up of highly qualified researchers. It took a full 1 year for planning, writing, editing, and printing.

The objective of this book is to present novel ideas in pre-packaging wafer-level 3D integration technologies. The book covers process technologies from the frontend to the backend of the line. All process technologies are carefully described and potential applications are listed. Technical challenges are also highlighted. This book is particularly beneficial to researchers or engineers who are already working or are beginning to work on 3D technology.

This book would not have been possible without a team of highly qualified and dedicated people. We are particularly grateful to Carl Harris of Springer for initiating this undertaking and for providing his support. We thank Anantha Chandrakasan, the series editor, for his recommendation and view on the contents of this book. Katie Stanne worked alongside with us and provided us with the necessary editorial support. The three co-editors were funded for many years through the MARCO and DARPA funded Interconnect Focus Center (IFC) as well as the DARPA funded 3D IC Program; our 3D technology platform research, and this book, would not have been possible without this extended research support.

C.S. Tan was also partially supported by SRC and an Applied Materials Graduate Fellowship previously. He is currently supported through a Lee Kuan Yew Postdoctoral Fellowship at the Nanyang Technological University. Last but not least, we are extremely thankful to authors who accepted our invitation and contributed chapters to this book.

We hope that the readers will find this book useful in their pursuit of 3D technology. Please do not hesitate to contact us if you have any comments or suggestions.

Singapore — Chuan Seng Tan
Troy, USA — Ronald J. Gutmann
Cambridge, USA — L. Rafael Reif

Contents

Contributors

Sitaram Arkalgud
SEMATECH, Austin, TX, USA, Sitaram.Arkalgud@SEMATECH.Org

Brian Aull
Lincoln Laboratory, Massachusetts Institute of Technology, Lexington, MA, USA, aull@LL.MIT.EDU

Kaustav Banerjee
University of California, Santa Barbara, CA, USA, kaustav@ece.ucsb.edu

P. Benkart
Institute of Electron Devices and Circuits, University of Ulm, Albert-Einstein-Alle 45, 89081 Ulm, Germany, peter.benkart@microgan.com

Robert Berger
Lincoln Laboratory, Massachusetts Institute of Technology, Lexington, MA, USA, bb@LL.MIT.EDU

Eric Beyne
IMEC, Kapeldreef 75, B-3001 Leuven, Belgium, beyne@imec.be

James Burns
Lincoln Laboratory, Massachusetts Institute of Technology, Lexington, MA, USA, jab@LL.MIT.EDU

Timothy S. Cale
Rensselaer Polytechnic Institute, Troy, NY, USA, calet@rpi.edu

Mansun Chan
Hong Kong University of Science and Technology, Hong Kong, mchan@ust.hk

Nisha Checka
Massachusetts Institute of Technology, Cambridge, MA, USA, nchecka@ll.mit.edu

Chang-Lee Chen
Lincoln Laboratory, Massachusetts Institute of Technology, Lexington, MA, USA, clchen@LL.MIT.EDU

Chenson Chen
Lincoln Laboratory, Massachusetts Institute of Technology, Lexington, MA, USA, chen@LL.MIT.EDU

Kuan-Neng Chen
IBM T. J. Watson Research Center, Yorktown Heights, NY, USA, chenku@us.ibm.com

Piet De Moor
IMEC, Kapeldreef 75, B-3001 Leuven, Belgium, demoor@imec.be

Andy Fan
Massachusetts Institute of Technology, Cambridge, MA, USA, fana@mtl.mit.edu

Shari Farrens
SUSS MicroTec, Waterbury Center, VT, USA, sfarrens@suss.com

Pascale Gouker
Lincoln Laboratory, Massachusetts Institute of Technology, Lexington, MA, USA, pgouker@LL.MIT.EDU

Ronald J. Gutmann
Rensselaer Polytechnic Institute, Troy, NY, USA, gutmar@rpi.edu

S. Brad Herner
SanDisk Corporation, Milpitas, CA, USA, Brad.Herner@sandisk.com

Sharath Hosali
SEMATECH, Austin, TX, USA, Sharath.Hosali@SEMATECH.Org

H. Hübner
Qimonda AG, Gustav-Heinemann-Ring 212, 81739 Munich, Germany, Holger.Huebner@qimonda.com

Robert E. Jones
Freescale Semiconductor, Inc., Austin, TX, USA, Robert.E.Jones@freescale.com

Anne Jourdain
IMEC, Kapeldreef 75, B-3001 Leuven, Belgium, jourdain@imec.be

A. Kaiser
Institute of Electron Devices and Circuits, University of Ulm, Albert-Einstein-Alle 45, 89081 Ulm, Germany, a.kaiser@reinhardt-microtech.de

Craig Keast
Lincoln Laboratory, Massachusetts Institute of Technology, Lexington, MA, USA, keast@LL.MIT.EDU

Jeffrey Knecht
Lincoln Laboratory, Massachusetts Institute of Technology, Lexington, MA, USA, jmk@LL.MIT.EDU

Steven J. Koester
IBM T. J. Watson Research Center, Yorktown Heights, NY, USA, skoester@us.ibm.com

E. Kohn
Institute of Electron Devices and Circuits, University of Ulm, Albert-Einstein-Alle 45, 89081 Ulm, Germany, erhard.kohn@uni-ulm.de

Douglas C. La Tulipe
IBM T. J. Watson Research Center, Yorktown Heights, NY, USA, dlip@us.ibm.com

Sheng-Chih Lin
University of California, Santa Barbara, CA, USA, sclin@ece.ucsb.edu

Jian-Qiang Lu
Rensselaer Polytechnic Institute, Troy, NY, USA, luj@rpi.edu

A. Munding
Institute of Electron Devices and Circuits, University of Ulm, Albert-Einstein-Alle 45, 89081 Ulm, Germany, andreas.munding@liebherr.com

Robert S. Patti
CTO, Tezzaron Semiconductor, Naperville, IL, USA, rpatti@tezzaron.com

S. Penka
Infineon Technologies AG, Otto-Hahn-Ring 6, 81739 Munich, Germany, sabine.penka@infineon.com

Christopher Petti
SanDisk Corporation, Milpitas, CA, USA, cjpetti@comcast.net

Scott K. Pozder
Freescale Semiconductor, Inc., Austin, TX, USA, Scott.Pozder@freescale.com

L. Rafael Reif
Massachusetts Institute of Technology, Cambridge, MA, USA, reif@mit.edu

Greg Smith
SEMATECH, Austin, TX, USA, Gregory.Smith@SEMATECH.Org

Larry Smith
SEMATECH, Austin, TX, USA, Larry.Smith@SEMATECH.Org

Antonio Soares
Lincoln Laboratory, Massachusetts Institute of Technology, Lexington, MA, USA, soares@LL.MIT.EDU

Vyshnavi Suntharalingam
Lincoln Laboratory, Massachusetts Institute of Technology, Lexington, MA, USA, vyshi@LL.MIT.EDU

Bart Swinnen
IMEC, Kapeldreef 75, B-3001 Leuven, Belgium, swinnenb@imec.be

Chuan Seng Tan
Nanyang Technological University, Singapore, tancs@ntu.edu.sg

Anna W. Topol
IBM T. J. Watson Research Center, Yorktown Heights, NY, USA, atopol@us.ibm.com

Brian Tyrrell
Lincoln Laboratory, Massachusetts Institute of Technology, Lexington, MA, USA, tyrrell@LL.MIT.EDU

Susan Vitkavage
SEMATECH, Austin, TX, USA, susan.vitkavage@lmco.com

Andrew Walker
Schiltron Corporation, Mountain View, CA, USA, andyjan@aol.com

Keith Warner
Lincoln Laboratory, Massachusetts Institute of Technology, Lexington, MA, USA, warner@LL.MIT.EDU

Bruce Wheeler
Lincoln Laboratory, Massachusetts Institute of Technology, Lexington, MA, USA, wheeler@LL.MIT.EDU

Peter Wyatt
Lincoln Laboratory, Massachusetts Institute of Technology, Lexington, MA, USA, wyatt@LL.MIT.EDU

Donna Yost
Lincoln Laboratory, Massachusetts Institute of Technology, Lexington, MA, USA, yost@LL.MIT.EDU

Albert M. Young
IBM T. J. Watson Research Center, Yorktown Heights, NY, USA, yalbert@us.ibm.com

Chapter 1
Overview of Wafer-Level 3D ICs

Chuan Seng Tan, Ronald J. Gutmann, and L. Rafael Reif

1.1 Background and Introduction

Over the past 40 years, higher computing power was achieved primarily through commensurate performance enhancement of transistors by continuously scaling down the device dimensions as described by Moore's Law. Integrated circuits (ICs) have essentially remained a planar platform throughout this period of rigorous scaling. As performance enhancement through device scaling becomes more challenging and demand for higher functionality increases, there is tremendous potential to explore the third dimension, i.e., the vertical dimension of ICs. This was rightly envisioned and pointed out by Richard Fenyman, physicist and Nobel Laureate, when he delivered a talk on "Computing Machines in the Future" in Japan in 1985; his original text reads: "Another direction of improvement (of computing power) is to make physical machines three dimensional instead of all on a surface of a chip. That can be done in stages instead of all at once – you can have several layers and then add many more layers as time goes on" [1].

While dimensional scaling has consistently improved device performance in terms of gate switching delay, it has a reverse effect on global interconnect latency [2]. The global interconnect resistance–capacitance (RC) delay has increasingly become the circuit performance limiting factor especially in the deep submicron regime. Even though Cu/low-κ multilevel interconnect structures improve interconnect RC delay, they are not a long-term solution since the diffusion barrier required with Cu metallization has a finite thickness that is not readily scaled. The effective resistance of the interconnect is larger than that would be achieved with bulk Cu, and the difference increases with reduced interconnect width. Surface electron scattering further increases the Cu line resistance, and hence the RC delay suffers [3].

When chip size continues to increase to accommodate more functionality, the total interconnect length increases at the same time. This causes a tremendous amount of power to be dissipated unnecessarily in the interconnect and in repeaters

C.S. Tan
Nanyang Technological University, Singapore
e-mail: tancs@ntu.edu.sg

C.S. Tan et al. (eds.), *Wafer-Level 3D ICs Process Technology*,
DOI: 10.1007/978-0-387-76534-1_1, © Springer Science+Business Media, LLC 2008

used to minimize delay and latency. On-chip signals also require more clock cycles to travel across the entire chip as a result of increasing chip size and operating frequency. Implementation of system-on-a-chip (SoC) using a planar IC process will result in larger chip size, longer interconnects, and longer process time as the functional blocks require additional process steps. We are also constrained to use a similar substrate which might not have desirable material and device properties for certain applications. It is clear that as demand for functionalities continues to grow, conventional planar ICs will not be able to accommodate such mounting demand without compromising performance, process complexity, and cost.

Three-dimensional (3D) integration in a system-in-a-package (SiP) implementation (packaging-based 3D) is becoming increasingly used in consumer, computer, and communication applications where form factor is critical. In particular, the hand-held market for a growing myriad of voice, data, messaging, and imaging products is enabled by packaging-based 3D (i.e., postsingulation of wafers into individual chips) integration. The key drivers are for increased memory capacity and for heterogeneous integration of different IC technologies and functions.

Wafer-level 3D integration (i.e., 3D stacking prior to singulation of wafers into individual chips) has become an increasingly active research topic. While wafer-level 3D technology is appreciably more complex than packaging-based 3D, as described in this book, there are significant advantages when compared to packaging-based 3D integration, namely:

- Higher density of interchip interconnects
- Lower electrical parasitics of interchip interconnects (therefore, higher interconnect electrical performance and lower power consumption)
- Lower high-volume manufacturing cost, since monolithic, wafer-level interconnectivity is extended to multiple device levels

In addition, the form factor and heterogeneous integration advantages of packaging-based 3-D integration are maintained. For wafer-level 3D integration, a number of inherent issues need to be either accommodated or solved prior to widescale use. These include:

- Establishment of integration architecture and design tools
- Die yield and impact on wafer-level stacking
- Common die size requirement for full silicon area utilization
- Thermal constraints with high power density
- Wafer-level processing equipment qualified for 24/7 manufacturing

While these technical and infrastructure considerations for wafer-level 3D integration appear daunting, the performance advantages of through-chip micron-sized interchip vias for high-speed multicore processors, high memory capacity with reduced processor-memory latency, heterogeneous integration of mixed-signal ICs with high-performance interconnects, and many other leading-edge products with 3-D-enabled integration (some described in this book) are driving research in academic, government, consortia, and individual-company laboratories. Wafer-level

3-D technology platform alternatives, technology concerns, and 3D enabled applications are the focus of this book.

This chapter discusses motivations behind 3D integration and presents a brief description of alternative technology platforms, and concludes with a detailed description of the organization and focus of the remaining chapters.

1.2 Motivations – A More than Moore Approach

Recently, there has been research interest in advanced 3D ICs in the form of a stack of interconnected active layers which has many performance, integration, and cost advantages [4]. Three-dimensional ICs can be defined as a stack of several device layers (with interconnects) that are electrically interconnected by vertical interlayer vias. It is widely seen as a "More than Moore" approach. Advantages offered by 3D integration will be discussed and potential applications will be highlighted in this section.

1.2.1 Interconnect Bottleneck

Today as the device dimension continues to shrink and the chip area continues to increase, the circuit performance has shifted from being device-dominated to interconnect-dominated. As a result of scaling, global interconnects become slower due to increased resistance and capacitance, and total interconnect length also increases as the complexity of the chip increases; as a result, interconnect latency and power consumption increase.

One solution to the interconnect problem is to partition a large chip into smaller blocks followed by thinning, stacking, and interconnecting them with vertical vias on a common substrate as shown in Fig. 1.1. Instead of having to travel across the entire chip, interblock communication is now through vertical vias which are much shorter. With a 3D implementation, one ends up with shorter global and semiglobal interconnects (for clock, power, etc.). This will directly translate into lower propagation delay and power consumption. This in turn will have a positive effect on overall

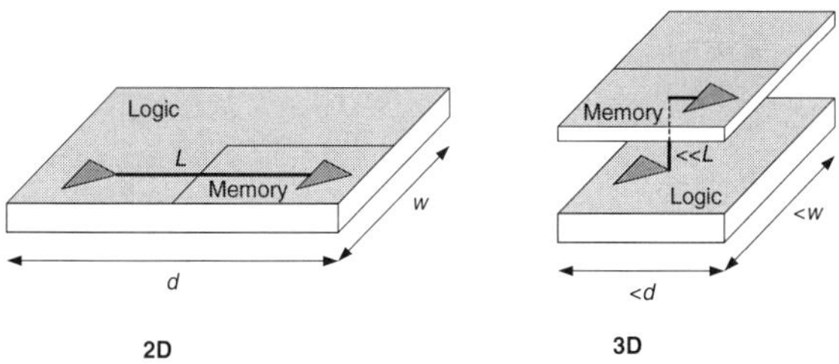

Fig. 1.1 Three-dimensional integration can replace long global and semiglobal wires with shorter vertical interconnects

system performance. To seek a long-term solution to the interconnect bottleneck, the International Technology Roadmap for Semiconductors (ITRS) has outlined 3D interconnects as one of the promising options [5].

1.2.2 Chip Form Factor

By stacking a few device layers in a vertical fashion, more compact ICs can be realized. Packing density, in terms of number of devices or functionalities per unit chip area, will increase and this might have a cost advantage in applications where silicon area is a primary consideration.

1.2.3 Heterogeneous Integration

System-on-a-chip is a potential solution to the mounting demand for multiple functionalities on a single chip. There are several challenges associated with planar implementation of SoC on a single substrate. Each functional block has to be built in sequence, and it is challenging to optimize each functional block on the same substrate. Substrate coupling might cause signal corruption between functional blocks [6]. Three-dimensional integration is an attractive choice for SoC implementation as it allows integration of various functional blocks in a vertical fashion. In this way, each block can be optimized independently and stacked to form a 3D system. Since there is no common substrate in this type of implementation, noise between blocks is expected to improve compared to a planar implementation.

The importance of 3D integration in terms of functional diversification of complementary metal-oxide semiconductor (CMOS) platform was clearly pointed out by Dr. Chang-Gyu Hwang, President and CEO of Samsung Electronics' semiconductor business, when he delivered a keynote speech at the 2006 International Electron Devices Meeting (IEDM) in San Francisco entitled "New Paradigms in the Silicon Industry" [7]. Some important points of his speech are quoted below:

> The approaching era of electronics technology advancement – the **Fusion Era** – will be massive in scope, encompassing the fields of information technology (IT), biotechnology (BT), and nanotechnology (NT) and will create boundless opportunities for new growth to the semiconductor industry.
>
> The core element needed to usher in the new age will be a complex integration of different types of devices such as memory, logic, sensor, processor and software, together with new materials, and advanced die stack technologies, all based on **3-D silicon technology**.

1.2.4 Stacked CMOS

Another attractive advantage to the stacking of active device layers in a vertical fashion is the implementation of stacked CMOS. While the n-metal-oxide semiconductor field-effect transistors (n-MOSFETs) and p-MOSFETs in CMOS inverters have remained largely identical in terms of materials selection in the past, MOSFETs in state-of-the-art CMOS inverters have increasingly diverged; for example, a tensile-strained channel is required for n-MOS while a compressive-strained channel is required for p-MOS [8], and the (1 0 0) orientation sees higher electron mobility while the (1 1 0) orientation sees higher hole mobility [9]. The integration of different high-κ materials and metal gate stacks in 45-nm technology node and beyond has also complicated the process flow. As CMOS becomes "hybrid," a single-substrate implementation and processing can be highly complex. There is opportunity for 3D vertical integration in this area: one can build and optimize n-MOS and p-MOS on two different substrates, bond, and thin back to form stacked CMOS.

1.3 Classification of 3D ICs

There are a number of technology options to arrange ICs in a vertical stack. It is possible to stack ICs in a vertical fashion at various stages of processing: (1) postsingulation 3D packaging (chip-to-chip), and (2) presingulation wafer-level 3D integration (chip-to-wafer, wafer-to-wafer, and monolithic approaches). Active layers can be vertically interconnected using bond wires or through silicon vias (TSVs). Each technology option varies in terms of vertical interconnect density. Since there is substantial overlap between the two, classification of 3D ICs technology is often not straightforward. Table 1.1 lists a few characteristic features of both platforms. The emphasis in this book is on wafer-level 3D integration.

In the case of chip-to-wafer (or pre-bottom wafer singulation 3D integration), postsingulated thinned chips with TSVs that are pretested can be mounted on a wafer using pick-and-place equipment. Only known-good-die (KGD) are aligned and mounted on good die in the substrate wafer, thereby alleviating one major disadvantage of wafer-to-wafer 3D (i.e., the yield penalty of having good die with bad die in a nonrepairable stack). The key disadvantage of this approach is that the die-to-wafer alignment with pick-and-place equipment is limited to 10–20 μm, better than die-to-die assembly but still two orders of magnitude worse in areal density than with wafer-to-wafer alignment.

Wafer-level 3D integration can be further classified as either front-end-of-the-line (FEOL)-based monolithic approaches or back-end-of-the-line (BEOL)-based assembly approaches as the technologies pursued are appreciably different.

Table 1.1 Characteristic features of 3D packaging and 3D integration

	3D Packaging	3D Integration
Infrastructure	Packaging	Foundry
3D Interconnect	Bond wires	Through silicon vias
Active layer thickness (μm)	>50	50 and less
I/O Density (cm^{-2})	10^4–10^5	10^5–10^8

1.3.1 Monolithic Approaches

In these approaches, devices in each active layer are processed sequentially starting from the bottommost layer. Devices are built on a substrate wafer by mainstream process technology. After proper isolation, a second device layer is formed and devices are processed by conventional means on the second layer. This sequence of isolation, layer formation, and device processing can be repeated to build a multilayer structure.

The key technology in this approach is forming a high-quality active layer isolated from the bottom substrate. This bottom-up approach has the advantage that precision alignment between layers can be accomplished. However, it suffers from a number of drawbacks. The crystallinity of upper layers is usually low and imperfect. As a result, high-performance devices cannot be built in the upper layers. Thermal cycling during upper layer crystallization and device processing can degrade underlying devices and therefore a tight thermal budget must be imposed. Due to the sequential nature of this method, manufacturing throughput is low. A simpler FEOL process flow is feasible if polycrystalline silicon can be used for active devices; however, a major difficulty is to obtain high-quality electrical devices and interconnects. While obtaining single-crystal device layers in a generic IC technology is still in the research stage, polycrystalline devices suitable for nonvolatile memory (NVM) have not only been demonstrated but have been commercialized (for example by SanDisk). A key advantage of FEOL-based 3D integration is that IC BEOL and packaging technologies are unchanged; all the innovation occurs in 3D stacking of active layers.

A number of FEOL techniques exist which include laser beam recrystallization [10, 11], seeding-assisted recrystallization [12, 13], selective epitaxy and overgrowth [14], and graphoepitaxy [15].

1.3.2 Assembly Approaches

Most of the research and development (R&D) on wafer-level 3D integration is BEOL-based, where fully processed and interconnected wafers are aligned, bonded, thinned, and interconnected. An interesting approach with the interconnectivity of FEOL-based 3D but with the processing approach and constraints of BEOL-based 3D is to bond two fully processed wafers with oxide-to-oxide bonding after full FEOL processing and local interconnect processing of each wafer. This local interconnect-based 3D approach limits the high aspect ratio (HAR) requirements to form a high density of interwafer vias in BEOL-based 3D platforms, thereby providing close to the active layer interconnectivity obtained with FEOL-based 3D. One difficulty with this approach is the limited die-yield mapping capability prior to wafer-to-wafer bonding; only active devices and test structures can be mapped prior to bonding. The numerous BEOL-based wafer-level 3D integration technology platforms are not as easy to classify. The key differentiators are as follows:

- type of wafer bonding, either metal-to-metal, dielectric-to-dielectric (oxide, adhesive, etc.), or hybrid bondings;
- type of interwafer interconnection, with Cu, tungsten (W), or heavily doped polysilicon (poly) the most dominant;
- via-first or via-last process flow;
- with or without a handling wafer (i.e., face-up or face-down stacking).

The selection of the optimum technology platform is subject to ongoing development and debate. Copper-to-copper bonding with a via-first process flow has significant advantages for highest interwafer interconnectivity, comparable to oxide-to-oxide bonding after local interconnectivity within each wafer as described in the previous paragraph; as a result, this approach is desirable for microprocessors and digitally based SoC technologies. Polymer-to-polymer with a via-last process flow is attractive when heterogeneous integration of diverse technologies is the driver and the interwafer interconnect density is more relaxed; benzocyclobute (BCB) is the polymer most widely investigated as described in Chapter 10.

The types of wafer bonding potentially suitable for wafer-level 3D integration are depicted in Fig. 1.2. Both the oxide-to-oxide bonding and the polymer-to-polymer bonding are inherently via-last process flows; the interwafer vias are formed after wafer-to-wafer alignment and bonding (by HAR etching, metal fill, and chemical–mechanical planarization (CMP)). Any metal-to-metal bonding can be used in a via-first process flow, where the interwafer vias are formed in the bonding process; note that appropriate interconnect processing within each wafer is required to enable 3D interconnectivity.

Oxide-to-oxide bonding of fully processed IC wafers is not considered feasible since the required atomic-scale smoothness is difficult to achieve; in addition, wafer distortions introduced by FEOL and BEOL processing introduce sufficient wafer bowing and warping to prevent sufficient bonding strength. While oxide-to-oxide bonding after FEOL and local interconnect processing has been shown to be promising, the increased wafer distortion and oxide roughness after multilevel interconnect processing does not allow a robust process window.

Both Cu-to-Cu bonding using the via-first process flow described earlier and polymer-to-polymer bonding with a via-last process flow are the most established

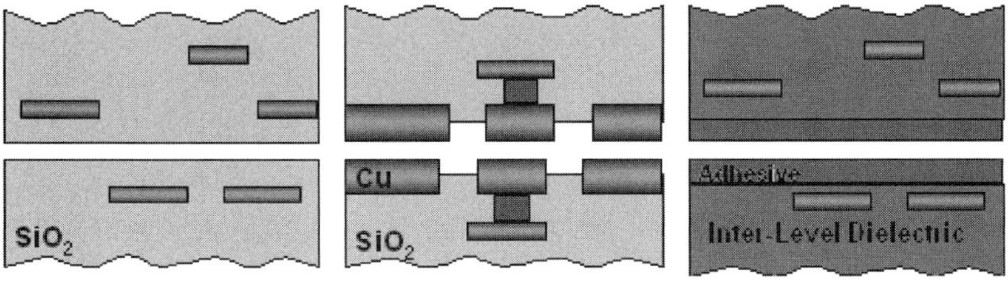

Fig. 1.2 Wafer bonding techniques for wafer-level 3D integration: (**a**) oxide-to-oxide; (**b**) metal-to-metal; (**c**) polymer-to-polymer. Reproduced from [16] with permission from J.-Q. Lu of RPI

and viable approaches with fully interconnected wafers (after FEOL and BEOL processing). In both cases ("face-down" stacking), all process conditions are BEOL-compatible (e.g., $T < 400^{\circ}$C and with limited bonding pressure).

Both of these wafer-level 3D platforms do not utilize handling wafers, but use face-to-face alignment and bonding with all required I/Os brought to the thinned backside of the top wafer (which becomes the face of the two-wafer stack). Another approach is to bond the top wafer to a handling wafer, after which the device wafer is thinned (to expose interwafer interconnects for metal-to-metal via-first bonding or a silicon or oxide surface for subsequent polymer-to-polymer bonding) and bonded to the full-thickness bottom wafer; after this bonding the handling wafer is removed with a compatible energy source such as ultraviolet (UV) light. This is also called a "face-up" stacking.

1.3.3 Wafer-Level 3D Design Opportunities

A key driver for wafer-level 3D technology is high-capacity memory with short access time. With the redundancy and reprogrammability of cache memory, 100% memory cell yield across a die is not required; as a result, die yield issues inherent in wafer-level stacking are alleviated. In addition, memory technology dissipates relatively little power; therefore, power dissipation issues are relatively minor. The initial products introduced by Tezzaron have high-capacity memory, principally due to these two factors.

Another strong technology driver for wafer-level 3D is the promise of incorporating more active silicon within a clock cycle for leading-edge digital processors. Wafer-level 3D for such high-speed, multicore processors are significantly more complex, partially because conventional 2D IC design and physical implementation are less established and partially because the layer partitioning and vertical interconnectivity is more difficult to design and fabricate. High-density interwafer interconnectivity is needed for design flexibility in such products. Die-yield impacts can only be limited by keeping die size small and/or introducing redundancy.

Other opportunities including analog-mixed signal design, software radios, and prime power delivery are also possible. Three-dimensional integration also enables multitechnology design such as Microelectromechanical Systems (MEMS) and sensors, imagers with pixel-by-pixel processing, optical and electro-optical applications, and the micro–nano interface (use of 3D to interface nanotechnology-based IC components with scaled microelectronics-based IC components).

Imagers with pixel-by-pixel processing is the first multitechnology design product that challenges technology capabilities of wafer-level 3D technology. A possible implementation can include signal-processing electronics that are contained in a 3D stack "behind" each pixel element. As the signal-processing electronics shrink with each IC generation, either improved pixel resolution and/or additional signal processing can be introduced. Such imagers will be a wafer-level 3D technology driver, definitely for defence and security applications and probably commercially as well.

1.4 Organization of the Book

The remainder of the book is divided into four principal topic areas, namely:

1. *Front-end-based monolithic approaches*: Front-end approaches of creating multiple layers of single-crystal silicon such as agent-assisted recrystallization and stacked 3D CMOS are described (Chapters 2 and 3);
2. *Back-end-based assembly approaches*: Methods of forming a multilayer ICs stack using wafer bonding and thinning are included in this part of the book. Various options using dielectric, metallic, and hybrid bonding are presented. New enabling process steps such as precision alignment, handle wafer processing, and vertical via formation are described (Chapters 4–11);
3. *Applications*: 3D-enabled applications such as high-density memory and heterogeneous integration are discussed (Chapters 12 and 13);
4. *Challenges and outlook*: Thermal challenges of 3D ICs are discussed. The book concludes with a survey chapter on current status of 3D ICs (Fall 2007) and future outlook (Chapters 14 and 15).

These four components are written in a self-explanatory manner and summarized below. There exists a wide variety of monolithic 3D ICs, and methods of fabricating them. Chapter 2 discusses work based on laser recrystallization of amorphous or polycrystalline silicon, as well as various solid-phase crystallization (SPC) methods, both seeded and unseeded. The higher-quality thin silicon films have been used to realize quite complex 3D logic circuits, while the smaller-grained films have been used to fabricate memory, especially static random access memory (SRAM).

Sequential 3D ICs process often suffers from limited thermal budget which results in poor crystalline quality in the upper silicon layers. One solution is to create high-quality silicon layers prior to device fabrication. This method is presented in Chapter 3. In this chapter, approaches of fabricating stacked FinCMOS are described. Process sequences are described, and characterization results are presented.

Chapter 4 focuses on two important unit processes that are inseparable from the assembly approaches, namely wafer alignment and wafer bonding. Techniques for precision wafer-to-wafer alignment are discussed and compared. An overview of different flavors of wafer bonding is presented: metallic, oxide fusion, dielectric adhesive, etc. The working principles of wafer bonders and aligners are included.

Chapter 5 discusses enabling process steps such as TSVs formation and wafer-thinning techniques. It also presents the concept of handle wafer processing including techniques for temporary bonding and handle wafer release.

Low-temperature Cu-to-Cu wafer bonding is described in detail in Chapter 6. The process flow developed at MIT and subsequently at IBM is included.

Cu–Sn solid–liquid interdiffusion (SLID) bonding is based upon alloying of Cu and Sn at low temperature to form a phase that is stable at high temperature. In Chapter 7, the mechanism behind this technique is described in detail. The application of the SLID process in 3D ICs stacking is presented.

The MIT Lincoln Laboratory's authors will present a 3D flow that is based on low-temperature oxide fusion bonding of SOI wafers in Chapter 8. This is a face-to-face bonding method and no handle wafer is required. Progression to multiple layers is achieved. Specific applications related to image sensors are presented.

Chapter 9 covers 3D flow developed by IBM for high-performance application, which includes low-temperature oxide fusion bonding, glass handle wafer and laser ablation for handle release. Technical challenges are discussed and solutions are highlighted.

Chapter 10 begins with an overview of adhesive bonding options. Authors emphasize the fundamentals of dielectric adhesive bonding for wafer-level 3D integration in this chapter. Hybrid bonding of Cu and BCB useful for redistribution layer bonding is also covered.

In Chapter 11, researchers from IMEC present their approach in 3D stacked IC (3D-SIC) integration based on hybrid bonding of Cu and BCB.

Application of 3D in high-density memory stack forms the core of Chapter 12. The author explains the density gain from 3D stacking of memory. Examples specific to Tezzaron are included.

Circuit architectures for 3D integration will be discussed in Chapter 13. The discussion will cover applications such as SOC, digital systems, and nanoscale/microscale integration. A comparison of 3D integration to planar packaging will be given.

In Chapter 14, authors focus on thermal challenges of 3D ICs and their implications on reliability issues. An analytical die-temperature model of a multilayer IC will be described and accurate chip-level leakage-aware methodology for 3D ICs thermal profile estimation is illustrated.

This book concludes in Chapter 15 with a survey on the current status of wafer-level 3D ICs (Fall 2007) such as technology options and their maturity for production. A projection of the future of 3D integration is provided.

References

1. Feynman RP (2000) The pleasure of finding things out. Perseus Publishing, Cambridge, p 28
2. Sylvester D, Hu C (2001) Analytical modeling and characterization of deep-submicrometer interconnect. Proc IEEE 89(5):634–664
3. Kapur P, McVittie JP, Saraswat KC (2001) Realistic copper interconnect performance with technological constraints. In: Proceedings of the IEEE Interconnect Technology Conference, pp 233–235
4. Banerjee K, Souri SJ, Kapur P, Saraswat KC (2001) 3-D ICs: A novel chip design for improving deep-submicrometer interconnect performance and systems-on-chip integration. Proc IEEE 89(5):602–633
5. Semiconductor Industry Association (2001) International Technology Roadmap for Semiconductors
6. Su DK, Loinaz MJ, Masui S, Wooley BA (1993) Experimental results and modeling techniques for substrate noise in mixed-signal integrated circuits. IEEE J. Solid State Circuits 28(4):420–430
7. Hwang CG (2006) New paradigms in the silicon industry. In: Keynote Speech, IEDM

8. Yang HS, et al. (2004) Dual stress liner for high performance sub-45 nm gate length SOI CMOS manufacturing. In: IEDM Technical Digest, pp 1075–1077
9. Yang M et al (2003) High performance CMOS fabricated on hybrid substrate with different crystal orientations. In: IEDM Technical Digest, pp 453–456
10. Kawamura S, Sasaki N, Iwai T, Nakano M, Takagi M (1983) Three-dimensional CMOS ICs fabricated by using beal recrystallization. IEEE Electron Device Lett 4(10):366–368
11. Kunio T, Oyama K, Hayashi Y, Morimoto M (1989) Three dimensional ICs, having four stacked active device layers. In: IEDM Technical Digest, pp 837–840
12. Subramanian V, Toita M, Ibrahim NR, Souri SJ, Saraswat KC (1999) Low-leakage germanium-seeded laterally-crystallized single-grain 100-nm TFTs for vertical integration applications. IEEE Electron Device Lett 20(7):341–343
13. Chan VWC, Chan PCH, Chan M (2001) Three-dimensional CMOS SOI integrated circuit using high-temperature metal-induced lateral crystallization. IEEE Trans Electron Devices 48(7):1394–1399
14. Pae S, Su T, Denton JP, Neudeck GW (1999) Multiple layers of silicon-on-insulator islands fabrication by selective epitaxial growth. IEEE Electron Device Lett 20(5):194–196
15. Rajendran B, Shenoy RS, Witte DJ, Chokshi NS, DeLeon RL, Tompa GS, Pease RFW (2007) Low temperature budget processing for sequential 3-D IC fabircation. IEEE Trans Electron Devices 54(4):707–714
16. Lu J-Q, McMahon JJ, Gutmann RJ (2006) Via-first inter-wafer vertical interconnects utilizing wafer-bonding of damascene-patterned metal/adhesive redistribution layers. In: 3D Packaging Workshop at IMAPS Device Packaging Conference, Scottdale, AZ

Chapter 2
Monolithic 3D Integrated Circuits

Christopher Petti, S. Brad Herner and Andrew Walker

2.1 Introduction

Monolithic three-dimensional integrated circuits (3D ICs) – defined here to be ICs in which circuit elements are fabricated on a substrate, and at least one layer above this substrate, in a single linear process flow with no material bonding required – were first touted in the literature in the early 1980s as a way to get around what were then perceived as scaling limits in silicon complementary metal-oxide semiconductor (CMOS) devices. Moreover, monolithic 3D ICs were envisioned as one way to reduce interconnect delay bottlenecks in 2D ICs [1].

However, conventional 2D CMOS devices have consistently been scaled beyond all these perceived limits; thus, simply scaling CMOS circuits has been more cost-effective than building-in the third dimension. There have been exceptions: polyload static random access memories (SRAMs), for example, where the elements placed in the third dimension are as simple (and as cheaply made) as possible.

Recently, the amount of interest in monolithic 3D ICs has been increasing, again driven by the increasing difficulty and expense of scaling silicon devices and technologies. It is thus worth reviewing some of the previously introduced stacking methods, as well as examining in detail the current technologies and their applications.

In this chapter, we will begin by discussing the device-stacking schemes using large-grained polysilicon ($>\sim 5\,\mu m$), which are applicable especially for CMOS logic circuits. We will review the different crystallization techniques for the upper layers of these stacked circuits, as well as the process integration for these schemes. We will then consider techniques using small-grained polysilicon for the upper devices; these are applied mostly for memory devices. Finally, we will provide a short synopsis on monolithic techniques for non-silicon devices.

C. Petti
SanDisk Corporation, Milpitas, CA, USA
e-mail: cjpetti@comcast.net

C.S. Tan et al. (eds.), *Wafer-Level 3D ICs Process Technology*,
DOI: 10.1007/978-0-387-76534-1_2, © Springer Science+Business Media, LLC 2008

2.2 Three-Dimensional Circuits Using Large-Grain Upper Layers

The CMOS inverter lends itself readily to 3D integration, as shown in Fig. 2.1. Typically, the p-type metal-oxide semiconductor (PMOS) device is stacked on top of the n-type metal-oxide semiconductor (NMOS) device, or vice versa; a single-gate feature can be used for both devices, to optimize cost at the expense of layout flexibility.

Since device performance and matching in logic circuits is critical, monolithic 3D technologies used for logic applications tend to require device layers with large grain sizes, typically >5 μm. Techniques for achieving these grain sizes are summarized first, followed by a resume of 3D logic process architectures.

2.2.1 Upper-Layer Recrystallization Techniques

2.2.1.1 Laser Recrystallization

Early work on laser recrystallization of polysilicon formed by chemical vapor deposition (CVD), spawned some of the first works on monolithic 3D ICs. This included cw-Argon (cw-Ar) laser annealing [2], wherein a 500-nm CVD polysilicon layer was annealed with a 14-W cw-Ar laser, with a ~40-μm spot size and a scan rate of 12.5 cm s^{-1}, which resulted in a grain size of $2\times 25\ \mu m^2$. Thin-film NMOS transistors (with channel length $L = 10\ \mu m$) fabricated in such Ar-laser-recrystallized polysilicon layers had channel mobilities 60%–70% that of bulk silicon [3].

In a technique known as graphoepitaxy [4], patterning of a 3.8-μm pitch grating on a silica substrate is used to influence the direction of the larger grains produced by Ar-laser annealing. Other variations on the basic laser annealing technique include a planarized polysilicon heat-sink [5] to protect underlying bulk-Si devices from the laser irradiation, and adding a second scanning laser pulse, which trails the first by 0.25 mm [6], which reduces the across-wafer variation of the device mobilities.

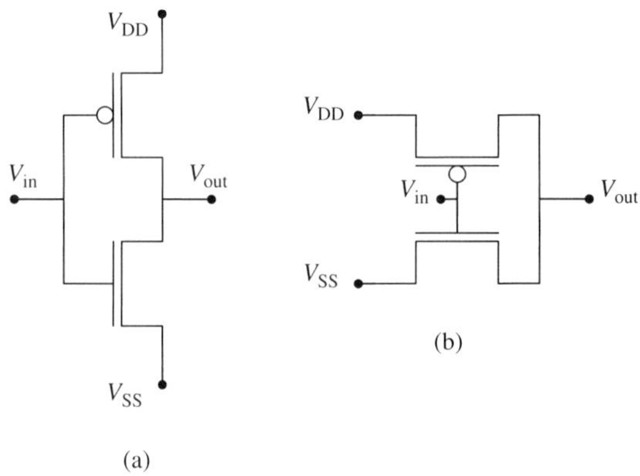

Fig. 2.1 CMOS inverter schematic drawn in (**a**) a conventional 2D style and (**b**) a style suggesting a 3D implementation

To obtain infinitely long grains, stripes of silicon nitride are used as an antireflective layer [7]; devices are built inside the stripes. Each stripe, however, will have a random grain orientation, and thus the variation of device parameters across a wafer will still be high. A solution is to introduce a seed grown from the bulk, which can be done by a selective epitaxial growth inside a via [8].

A recent variation on graphoepitaxy, known as nanographoepitaxy [9], uses nanoimprint technology to form a 190-nm pitch grating on a SiO_2 substrate. Thin (100 nm) films of a-Si are then deposited and annealed with a laser pulse of 15 μs. Since this is much smaller than the ~50-ms dwell time of the scanned laser used in the earlier work, much less energy is imparted into the substrate. This small pitch results in long, narrow, unidirectional grains, in which aggressively scaled transistors can be fabricated. In this way, upper-layer devices can be fabricated without excessive heating of the devices on the substrate, allowing small devices to be fabricated on all layers.

2.2.1.2 Epitaxial Overgrowth and Solid-Phase Crystallization

Epitaxial overgrowth alone (without laser annealing) has also been used to form the upper device layers. A simple method entails opening holes in the gate oxide overlying the bulk device drains, then performing an epitaxial growth followed by a planarization [10].

Solid-phase crystallization (SPC) of amorphous (as-deposited) silicon via a long, low-temperature anneal is another common technique for obtaining large-grained polysilicon [11]. In this technique, nucleation sites in the amorphous film are generated slowly at low temperature; these sites then grow into grains. The size of the resulting grains is limited by the density of nucleation sites. Low temperature (~600°C) is preferred, so that the nucleation rate is low and thus the density of nucleation sites is also low. A 12-h, 600°C anneal can result in an average grain size of 3.5 μm [12]. Note that no crystallization happens in the first ~5 h of such an anneal, as nucleation sites are being generated.

A similar SPC technique uses germanium as a seeding material [13]. Germanium is deposited (through windows in a masking oxide) to contact the surface of an amorphous silicon film. An inert anneal at 500–550°C is performed in order to crystallize the silicon film. During this anneal, a SiGe alloy is formed in the seeding windows, and these alloyed areas act as nucleation sites for Si crystallization. Compared to the standard SPC method, the nucleation site generation time can be reduced or eliminated, thus reducing the total thermal budget needed for the process.

Another SPC technique is metal-induced lateral crystallization (MILC), with a high-temperature anneal [14]. In this technique, amorphous Si is reacted with a metal (usually Ni). The resulting silicide front sweeps laterally through the thin film, leaving long, narrow (typical width <1 μm) grains in its wake. A subsequent high-temperature anneal (e.g., 900°C for 30 min) enlarges some grains and subsumes smaller grains, such that grains in the resulting film are ~10 μm in both directions.

2.2.2 Three-Dimensional Logic Process Architectures

2.2.2.1 Common-Gate Processes

A simple 3D architecture has a thin-film device stacked on top of a bulk device, sharing one gate between them in the arrangement suggested by Fig. 2.1b [15]. This structure, called junction MOS (JMOS) (Fig. 2.2), uses a bottom-gated NMOS on the upper layer; the channel was recrystallized using the simple Ar-laser scheme. A similar approach was used to form a fully connected CMOS inverter [16, 17]. In this scheme, PMOS was placed on the top (see Fig. 2.3) so as to be compatible with a standard NMOS process. The inverter's output node was formed via a silicon-to-silicon buried contact. This common-gate architecture has the advantage of using a single processing sequence for the device gates; however, the top device no longer has self-aligned source-drain regions and thus an extra masking step must be added.

Nevertheless, the common-gate architecture has been studied for its potential to realize high-density logic. It was estimated that this architecture could achieve density increases of $\sim 4X$ [18]. Using the lateral epitaxial overgrowth technique and dual-gated top-layer devices, actual logic circuits were realized with density

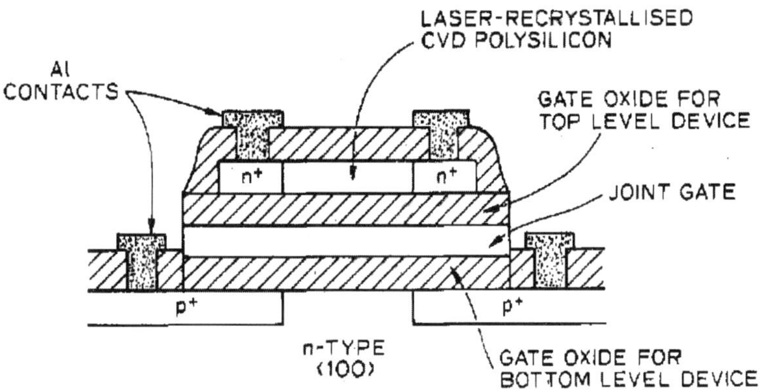

Fig. 2.2 JMOS structure (NMOS stacked on top of PMOS). From [15] © 1980 IEEE

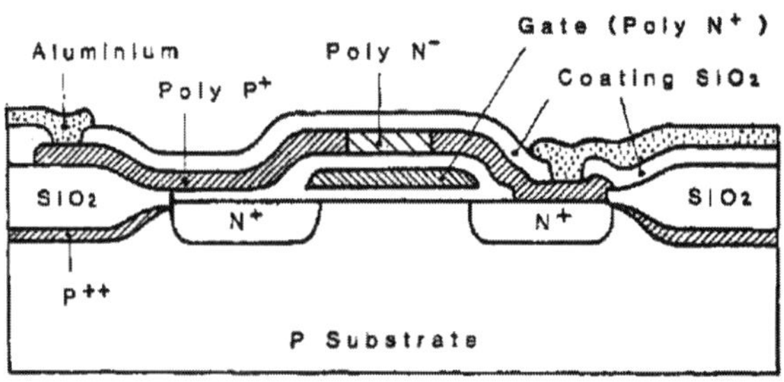

Fig. 2.3 Stacked CMOS inverter with a single gate layer. From [16] © 1981 IEEE

increases of 1.6–3X, depending on the circuit [19]. This process, with NMOS on the bottom and PMOS on the top, uses the number of masking steps (10) as a bulk CMOS process. Like earlier common-gate processes, it uses a silicon-to-silicon contact to connect NMOS and PMOS drains in logic gates, and NMOS and PMOS channel regions must be placed directly on top of each other.

2.2.2.2 Independent Gate Processes

To allow greater layout flexibility, and to avoid the silicon-to-silicon buried contact, the top-layer device can use an independent top gate, as in Fig. 2.4 [20]. Connections between layers are made with metallic contacts; this eliminates the buried contact, which degraded the inverter performance, and allows the structure to be extended to multiple layers above the substrate.

The development of advanced technologies used in 2D circuits – tungsten local interconnect, planarization, filled contacts – have also been used on 3D circuits based on the structure shown in Fig. 2.4. For example, tungsten (W) plug technology is introduced to directly connect PMOS to NMOS devices in an inverter (see Fig. 2.5), reducing cell area by 50% over a 2D layout [21]. The W plug connects to the edges of the silicon-on-insulator (SOI) layers only, and, unlike in conventional 2D logic technology, a single contact is used to connect more than just two conducting layers together. This concept – minimizing the number of separate contact patterning steps – helps to make 3D technologies more cost-effective.

This W-plug edge contact was also used in a 512-Mb SRAM chip with two SOI layers above an NMOS substrate [22]. The SOI layers are single-crystal silicon formed by a seeded epitaxial technique. The NMOS pass transistors are placed on

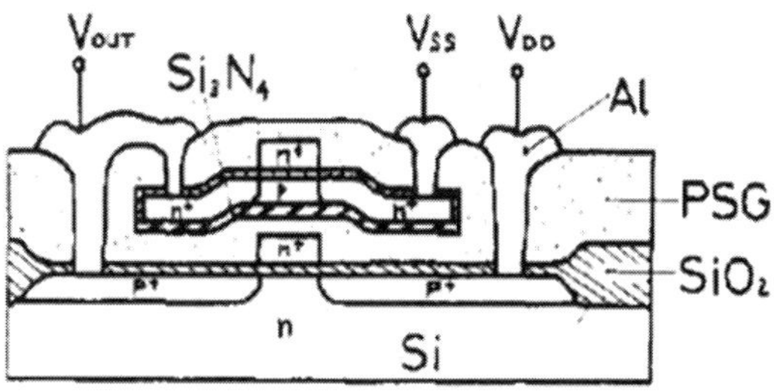

Fig. 2.4 Stacked CMOS inverter with two gate layers. From [20] © 1983 IEEE

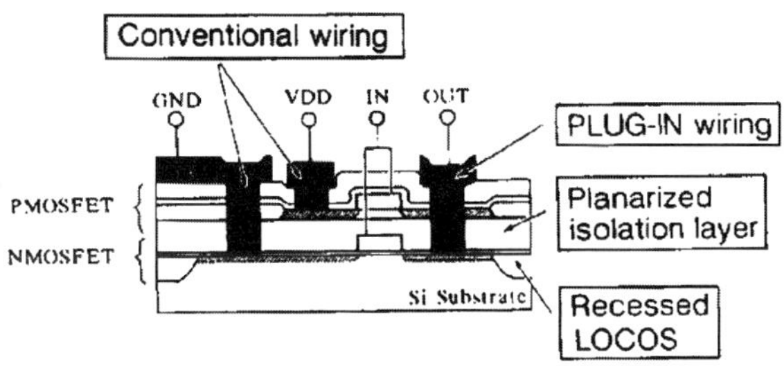

Fig. 2.5 CMOS inverter with W-plug edge contacts. From [21] © 1990 IEEE

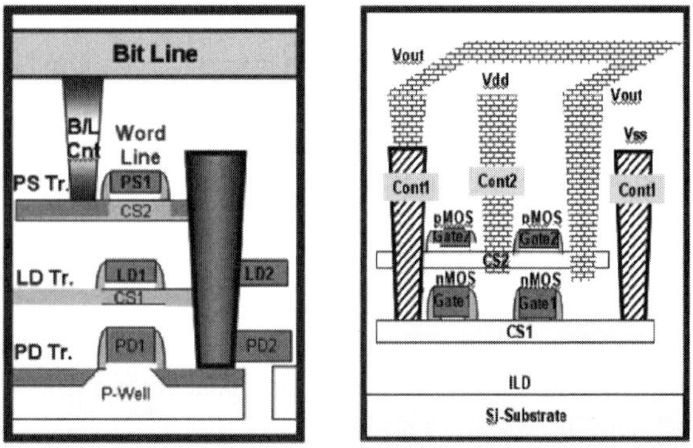

(a) (b)

Fig. 2.6 Cell (**a**) and peripheral circuits (**b**) for a 512 Mb 3D SRAM. From [22] © 2004 IEEE

the top SOI layer, the PMOS load transistors are placed in the second SOI layer, and the NMOS pull-down transistors are in the bulk silicon (see Fig. 2.6a). Peripheral circuits were also fabricated in the SOI layers; NAND gates were realized entirely in the SOI layers with no devices in the bulk (Fig. 2.6b).

The same SOI technology was used to fabricate a 3D NAND-type nonvolatile memory [23], with one set of NAND strings in the substrate and one in the single upper layer. This is discussed further in section 2.3.3.

NMOS and PMOS devices have also been placed together in upper layers. Examples include an 8 K CMOS SRAM, with 4-k bits on the bulk-Si layer and 4-k bits on the upper layer [6], and a 25-pixel, three-layer image processor [8] consisting of photodiodes on the top layer, CMOS A/D converters in the middle layer, and NMOS ALU and support circuitry in the substrate.

A summary of monolithic 3D circuit applications in large-grained polycrystalline silicon is shown in Table 2.1. Circuit density increases have averaged ~50% (i.e.,

Table 2.1 Comparison of 3D logic architectures

Type	Reference	Size reduction (%)	Comment
• N/PMOS on separate layers • Gate shared between layers • Silicon–silicon contact	[17, 19]	40	Fewest process steps
• N/PMOS on separate layers • Separate gate for each layer • Metal layer–layer contact	[20]	50	More layout flexibility, extendable to multiple layers
• N/PMOS on separate layers • Separate gate for each layer • Plug edge contacts	[21, 23]	30	Very dense layout
• N/PMOS together • Separate gate for each layer	[6]	60	Most modular

equivalent to one scaling node), and circuit performance increases average ~10% over 2D bulk-Si designs.

2.3 Three-Dimensional Circuits Using Small-Grained Polysilicon Layers

The preceding discussion focused on monolithic 3D circuit applications in which the upper layers of silicon were processed in such a way as to leave them with large grains, so that the upper-layer devices match the bulk devices as closely as possible, in terms of performance and uniformity. However, the processes used for this transformation were often expensive (e.g., laser annealing) or not proven in manufacturing. This has therefore spurred interest in applications that do not require devices in the upper layers whose quality approached that of bulk devices, in particular where the upper-layer devices can be made from small-grained (<~200 nm) polysilicon.

2.3.1 SRAM

Stand-alone SRAM products have been commonly made in a four transistors and two resistors (4T–2R) configuration, where the load resistors are made of thin-film polysilicon resistors [24] placed above the n-channel bulk transistors. This form of 3D circuit does not require an extensive upper-layer recrystallization.

With technology scaling, the 4T–2R SRAM became increasingly susceptible to data errors because of the weak coupling of the small storage-node capacitance to the voltage supply. The next logical step was to add a gate layer to the polysilicon resistor, to form a polysilicon thin-film PMOS load field-effect transistor (FET) [25]. Since these devices were replacing load resistors, they did not need the characteristics of a bulk-Si PMOS device, and very little crystallization was needed.

As technologies scaled even further, the poor quality PMOS thin-film transistors (TFTs) no longer supplied enough current to keep the cell stable. As a result, a long-anneal SPC process was introduced to increase grain size to ~0.5–1 μm [26]. As described in section 2.2.1, this required depositing silicon in the amorphous state (typically 530°C–550°C), followed by a long (>=6 h), low-temperature (~600°C) grain-growth anneal in an inert ambient.

Other techniques were used to increase the storage-node capacitance by adding various structures to the 3D stack. This included the "fin" capacitors shown in Fig. 2.7 [27], which was combined with a double-gate TFT for better device characteristics.

Thus, 3D SRAM technologies became increasingly complex. However, with the introduction of retrograde-well and shallow-trench isolation processes, using bulk PMOS load devices in SRAMs became more cost effective [28]. The three-layer, large-grained polysilicon technology described above (Fig. 2.6) is the next stage in the evolution of 3D SRAMs.

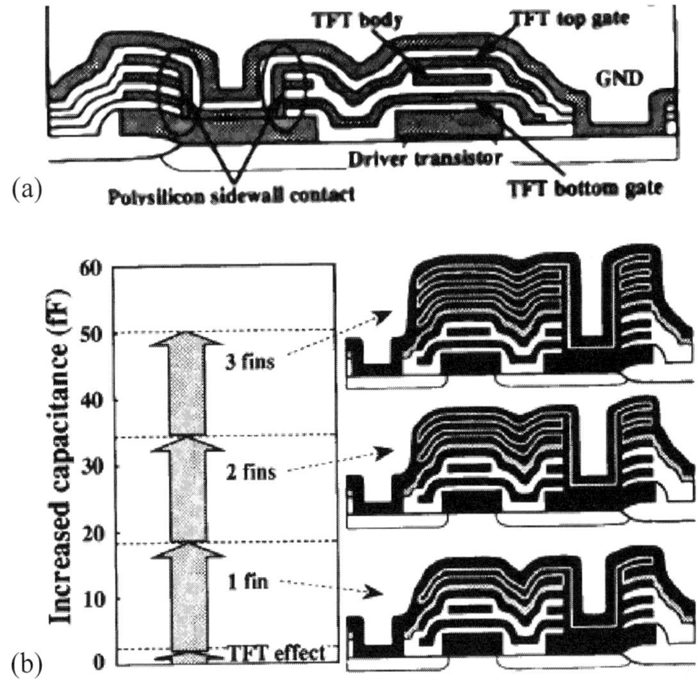

Fig. 2.7 TFT SRAM structures (**a**) with no extra node capacitance (**b**) with "fin" capacitors added. From [27] © 1991 IEEE

2.3.2 Nonvolatile Memory: Cross-Point Memories

A cross-point memory element is a device wherein data storage occurs at the intersection of two perpendicular conductors. Memories based on arrays of cross-point elements can achieve very high densities, since the memory cell size can be as small as $4F^2$, where F is the minimum lithographic feature size. Because of the potential of high density and process simplicity, cross-point elements have been prime candidates for 3D stacking.

2.3.2.1 Cross-Point Memory Architectures

One candidate proposed for a stackable cross-point memory element is a simple resistive cell, wherein a material with an electrically switchable resistance is placed between two arrays of perpendicular conductors, as shown in Fig. 2.8a [29, 30]. However, it is very difficult to make large arrays from this simple cell; the entire array presents a current-flow path in parallel to a selected cell, thus impeding the ability to distinguish between the low-resistance and high-resistance states of the selected cell.

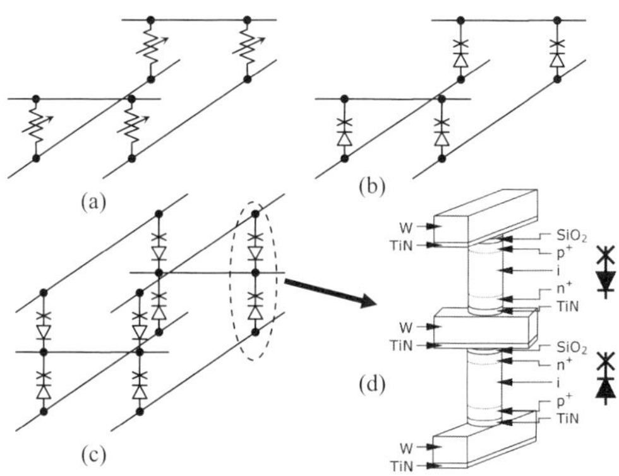

Fig. 2.8 Cross-point memory array types: (**a**) resistor cell array, (**b**) diode/antifuse cell array, (**c**) diode/antifuse array (two layers) and (**d**) diode antifuse cell stack. From [32] © 2004 IEEE

A solution is to place a diode in series with the resistive memory element. In a diode cross-point array, any current path parallel to the selected cell will include at least one reverse-biased (i.e., current-blocking) diode; thus the parallel-path current will be low, and the different states of the selected memory element can be detected.

The memory element in series with the diode can be a simple dielectric antifuse (Fig. 2.8b); in this case, the two memory states are the intact and ruptured antifuse, and the memory is programmable only once. Arrays of these cross-point diode-antifuse elements are very dense and easy to fabricate, and thus are good candidates for 3D stacking. Such 3D one-time-programmable (OTP) memory products, with up to eight layers of memory cells, have been mass-produced [31].

2.3.2.2 Diode/Antifuse Cross-Point Memory: Technology

In these devices, memory layers consist of polysilicon diodes and antifuses formed in pillars, sandwiched between two conducting metal layers. The conducting layers may be shared between one layer and the next layer above it (see Fig. 2.8c, d); in this case, the diodes in succeeding layers will point in opposite directions. This ensures that the shared conducting lines do not need to be biased differently depending on which memory layer is accessed.

Tungsten-filled contact plugs are used to make electrical connection to the memory layers. As in other monolithic 3D memory processes, a given contact can connect three or more conductor levels, thus reducing overall process costs.

The polysilicon diode is formed by CVD deposition of silicon at 540°C [32]. The initial few nanometers of this deposition is formed in a dopant-containing (boron or phosphorus) ambient in order to form the bottom p+ or n+ diode electrode. The rest of the diode is deposited undoped. The top contact (n+ or p+) can be formed by ion implantation, or can be formed in situ. An SiO_2 antifuse, 2–3-nm thick, is grown on top of the diode by rapid thermal oxidation.

Thus, the diode layers are amorphous or microcrystalline as deposited. After all layers are formed, a rapid thermal anneal at 750°C grows grains to a size of ~100 nm in the silicon pillars [33]. The metal layers are elemental refractory metals (e.g., tungsten) or refractory metal silicides, so as to withstand the temperatures needed to grow the antifuses and crystallize the silicon pillars.

When no antifuse is grown on these pillars (i.e., the pillar is contacted directly by the top conductor), the current–voltage curve of the pillar diode is relatively poor, as one would expect from a polysilicon diode with grains ~100 nm [34]. However, when a pulse of 8–10 V is applied for ~1 μs, the diode conducts current much more strongly (Fig. 2.9); in fact, the post-pulsing current–voltage curve is very similar to a single-crystal positive intrinsic negative (PIN) diode. Mobilities and activation energies derived from these characteristics give results very close to single-crystal silicon. Thus, although the 3D memory elements are formed in small-grained polysilicon, during operation these elements are transformed into large-grained polysilicon.

2.3.2.3 Diode/Antifuse Cross-Point Memory: Die Size Considerations

Whereas in the 3D SRAM and logic schemes described previously, the bulk silicon is utilized for part of the memory cell or basic logic gate, in the 3D cross-point memory scheme the bulk silicon underneath the memory array is freed up for support circuitry. This advantage can be expressed in terms of array efficiency α, which can be defined as the ratio of the "footprint" of the user-accessible memory area to the total die area:

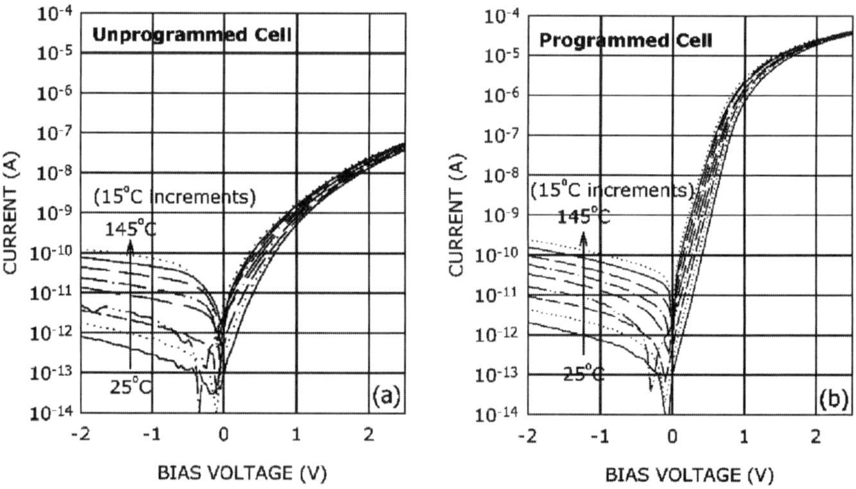

Fig. 2.9 Current-voltage–temperature characteristics of a polysilicon diode, with no antifuse, (**a**) before a voltage pulse is applied, and (**b**) after a voltage pulse is applied. From [33] © 2006 IEEE

$$\alpha = \frac{N_{\mathrm{ubl}} A_{\mathrm{c}}}{A_{\mathrm{D}}} = \frac{N_{\mathrm{ubl}} \beta F^2}{A_{\mathrm{D}}}, \tag{2.1}$$

where N_{ubl} is the number of user-accessible bits on each memory layer of the chip, N_{L} is the number of memory layers, A_{D} is the die area, and A_{C} is memory cell footprint, which can be expressed as βF^2, where F is the technology's minimum feature size. Note that this definition of α is roughly independent of the number of memory layers.

Typically, this metric will be lower for smaller chips, since there is a portion of the chip (bond pads, I/O circuits, charge pumps, page registers, etc.) that remains the same size regardless of the total chip area. However, 3D cross-point memories offer an advantage in array efficiency for smaller die sizes, because of the ability to tuck circuitry underneath the array. This is illustrated in Fig. 2.10, where die area per bit footprint, and array efficiency as defined in Eq. 2.1, are calculated and plotted against the number of bit footprints in a die (for 3D memory, this is the number of memory cells per layer). In this figure, it is assumed that, in the 3D arrangement, 50% of the nonscalable layout can be placed underneath the array. It is seen that simply raising the memory array above the substrate results in a 20–30% increase in array efficiency for smaller memories.

Another measure of the efficiency of memory bit packing is the bit area including overhead, or BAIO, defined as the chip area in square feature sizes, divided by the total number of bits of user-accessible memory [35]. This is similar to the bit footprint area shown in Fig. 2.10; however, it takes into account all of the layers of memory. Figure 2.11 shows the BAIO for an eight-layer 3D cross-point memory, as well as the BAIO for several other 2D memory technologies, plotted against minimum feature size (F). This figure shows that this 3D memory offers a 3X advantage in BAIO, compared to a FLASH memory made with a 40% smaller feature size.

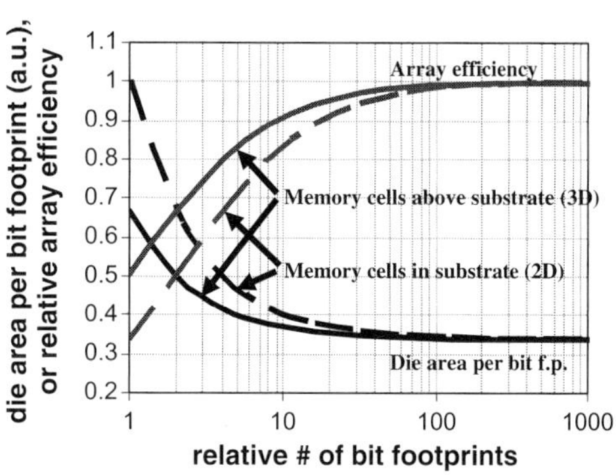

Fig. 2.10 Array efficiency and die area per bit footprint versus number of bit footprints for 2D and 3D cross-point memories

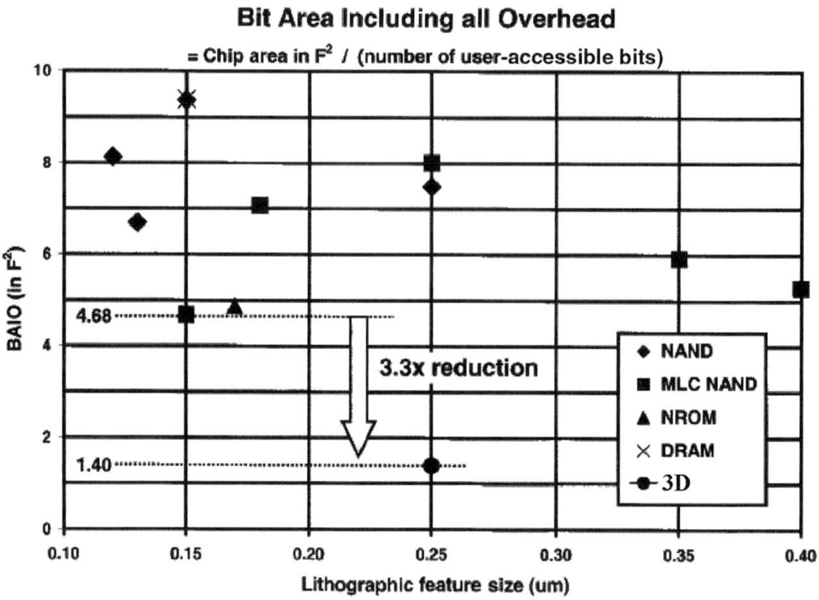

Fig. 2.11 BAIO comparison between various 2D memory types and 3D diode/antifuse cross-point memory. From [35]. © 2003 IEEE

2.3.3 *Nonvolatile Memory: TFT-SONOS*

The use of silicon nitride as a charge-trapping layer in a silicon nonvolatile memory element has a long history [36]. The initial MNOS structure consisted of a silicon nitride layer separated from the channel of a FET by a thin tunnel oxide with the gate electrode lying directly on the nitride layer. To improve retention properties and help reduce program and erase voltages, the SONOS structure was introduced where the silicon nitride layer could be thinned by placing a blocking oxide between the gate electrode and the nitride layer [37].

The first implementation of SONOS in a deep submicron TFT [38] showed that nonvolatile characteristics such as retention, endurance, and program and erase voltages could rival those normally seen in the crystalline version. In this case, the ONO consisted of 2.5-nm tunnel oxide, 7-nm silicon nitride, and 5-nm silicon dioxide. The gate electrode was boron-doped polysilicon. Electrical characterization showed no threshold window closure even after 10^6 program and erase cycles (Fig. 2.12a). Retention after 10^6 cycles extrapolated to a threshold voltage window of around 1.6 V after 10 years.

A NAND string implementation of TFT-SONOS [39] included the use of a multistack ONONO in place of a simple ONO on TFT devices placed in a 16-cell series string. The main challenge in using SONOS in a NAND string approach is the charge disturb associated with accessing a cell for read or program. To be able to read a cell, all other cells in the string must be placed in the conducting state no

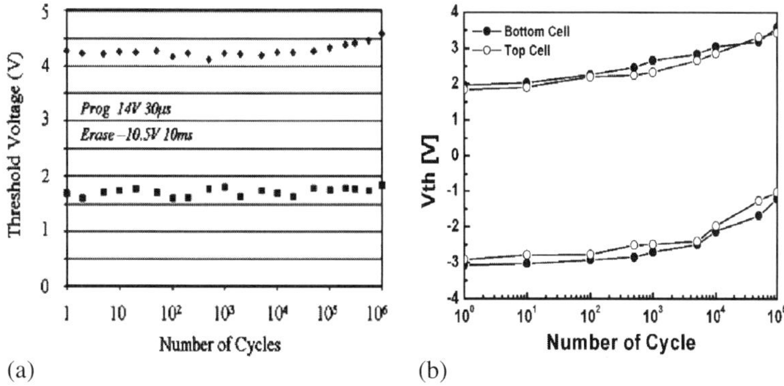

Fig. 2.12 3D SONOS write/erase endurance characteristics: (**a**) polycrystalline TFT, (**b**) stacked single-crystal TFT and substrate cell. (**a**) From [38] © 2003 IEEE. (**b**)From [23] © 2006 IEEE

matter what their programmed state happens to be. The read-pass voltage applied to the gates of these series devices must be larger than the highest threshold voltage of any programmed state. This then slowly programs these pass cells. A critical parameter for any NAND Flash string is the worst-case string current which is the read current out of a string when the cell to be read is in the erased (low threshold voltage) state while all other cells in the string are in the programmed state. This current must be high enough to sense. While this is already a challenge in regular crystalline NAND, the low carrier mobilities in polycrystalline and nanocrystalline silicon make this even more of a daunting prospect to be able to productize such an approach. One approach to increase this string current is to increase the read-pass voltage on all other cells. This however severely increases the disturb in those cells.

The disturb mechanisms described above apply to both NAND strings in the silicon wafer bulk and in TFT form. A challenge unique to the TFT-SONOS NAND is the time to erase a cell in a string. In bulk NAND, direct connection is made to the p-type well in which the NAND strings are placed. Erase consists of applying a high positive potential to this well while all the wordline gates are close to ground. This erases all cells in a block within a few milliseconds. The problem with TFT NAND is that no connection is made to the TFT body. This then results in an elaborate procedure to erase cells individually by connecting them up through conducting cells within the string to either the bitline or the source of the string.

The importance of the NAND Flash market combined with the increasing difficulty in scaling floating gate NAND Flash has resulted in great interest in both charge trap Flash (CTF), such as silicon nitride, and in 3D solutions [23, 40]. The disturb challenges inherent in SONOS NAND are being tackled by increasing the thickness of the tunnel oxide from under 3 nm to between 4 nm and 6 nm [23]. To avoid erase saturation (electron injection from the gate to the silicon nitride), thick aluminum oxide is used as the top blocking oxide and high-workfunction tantalum nitride is used as gate metal. The 3D version of this TANOS approach used single-crystal channel silicon achieved through solid-phase epitaxy. The

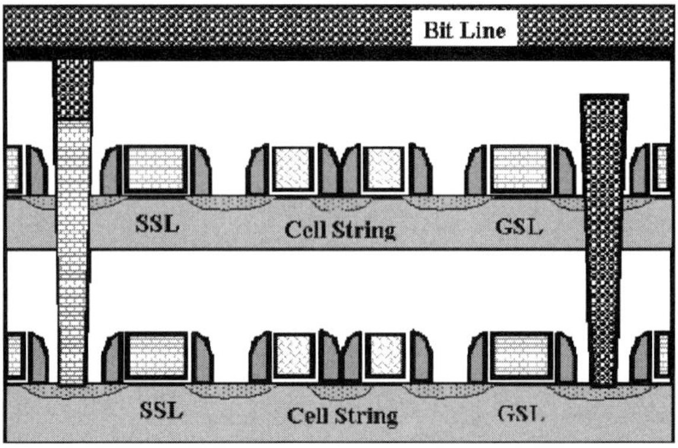

Fig. 2.13 3D (two layer) NAND memory cell structure. From [23] © 2006 IEEE

troublesome slow erase was tackled by implementing a special edge contact plug that shorted the source of the string to its body (Fig. 2.13), allowing a bulk-like erase approach to be used. Identical characteristics between bulk and upper-layer devices are demonstrated, including endurance up to 10^5 cycles (Fig. 2.12b).

The 3D implementation of high-density Flash promises to be especially interesting in the next few years as floating-gate NAND Flash scaling reaches fundamental limits.

2.4 Non-Silicon Monolithic 3D Integrated Circuits

Many of the 3D circuit architectures described in the previous sections can be applied to non-silicon technologies. This can give advantages of low-temperature processing on a variety of substrates, for applications where flexibility or transparency is required, or for combination with optically active materials for hybrid systems.

One example is to use germanium instead of silicon [41], since Ge has a lower melting and crystallization temperature. Ge can be crystallized from a Si seed hole, using a microsecond laser pulse while keeping the Si substrate at $<400^\circ$C.

Other examples use the diode-isolated cross-point memory architecture (Section 2.3.2). In one case, the diodes could be made by using NiO as a p-type semiconductor and TiO_2 as an n-type semiconductor [30]. Process temperatures for such structures are $<=300^\circ$C.

Molecular-based switches embedded into cross-point arrays have also been proposed [42]. There are various candidates for these switches, such as Catenane-based molecules that exhibit hysteretic current–voltage characteristics, switching at +/–2 V [43]. Another molecular-based cross-point fabrication method proposed uses a dielectric support separating two perpendicular wires; this support is then slightly undercut to support self-assembly of a tunnel-barrier molecule [44]. Each end of the long-chain molecules can anchor to the exposed surfaces of the electrodes in the undercut area; the thickness of the dielectric support can be tuned to match the length of the molecule.

Lastly, a more general 3D IC scheme uses stamping methods to apply each layer onto a substrate [45]. For example, transistor channel regions (Si or GaN) are patterned on separate wafers, separated from these source wafers, and "inked" onto an elastomeric stamp, where they are held by van der Waals forces. The stamp can then be used to transfer them onto a substrate. Other layers (transistor gates, metallic interconnect) can be processes with conventional patterning. Although this does not strictly meet the definition of "monolithic" 3D IC as specified in section 2.1, it nevertheless represents an interesting method of stacking active devices of different materials (Si, GaN, carbon nanotubes) onto a single IC; for example, a CMOS inverter consisting of a single-walled nanotube (SWNT) PMOS device and a silicon NMOS device.

2.5 Summary

There exists a wide variety of monolithic 3D ICs, and methods of fabricating them. Much work has been based on laser recrystallization of amorphous or polycrystalline silicon, as well as various SPC methods, both seeded and unseeded. The higher-quality thin silicon films have been used to realize quite complex 3D logic circuits, while the smaller-grained films have been used to fabricate memory, especially SRAM.

The cross-point memory architecture is well suited to stacking into the third dimension; this has been exploited to mass-produce a polysilicon-diode-isolated nonvolatile cross-point memory with up to eight layers. There is much work currently on other cross-point 3D methods, including memory elements based on metal oxide and molecular switches.

References

1. Saraswat KC, Souri SJ, Subramanian V, Joshi AR, Wang AW (1999) Novel 3-D structures [ICs]. In: Proceedings of the 1999 IEEE International SOI Conference, pp 54–55
2. Gat A, Gerzberg L, Gibbons JF, Magee TJ, Peng J, Hong JD (1979) cw laser anneal of polycrystalline silicon: Crystalline structure, electrical properties. Appl Phys Lett 33:775–778
3. Lee KF, Gibbons JF, Saraswat KC (1979) Thin film MOSFET's fabricated in laser-annealed polycrystalline silicon. Appl Phys Lett 35:173–175
4. Geis MW, Flanders DC, Antoniadis DA, Smith HI (1979) Crystalline Silicon on insulators by graphoepitaxy. In: 1979 IEDM Technical Digest, pp 210–212
5. Akiyama S, Ogawa S, Yoneda M, Yoshii N, Terui Y (1983) Multilayer CMOS device fabricated on laser recrystallized Silicon islands. In: 1983 IEDM Technical Digest, pp 352–355
6. Yamazaki K, Yoneda M, Ogawa S, Ueda M, Akiyama S, Terui Y (1986) Fabrication technologies for dual 4-Kbit stacked SRAM. In: 1986 IEDM Technical Digest, pp 435–438
7. Kunio T, Oyama K, Hayashi Y, Morimoto M (1989) Three dimensional ICs, having four stacked active device layers. In: 1989 IEDM Technical Digest, pp 837–890
8. Nishimura T, Inoue Y, Sugahara K, Kusunoki S, Kunamoto T, Nakagawa S, Makaya M, Horiba Y, Akasaka Y (1987) Three dimensional IC for high performance image signal processor. In: 1987 IEDM Technical Digest, pp 111–114

9. Crnogorac F, Witte DJ, Xia Q, Rajendran B, Pickard DS, Liu Z, Mehta A, Sharma S, Yasseri A, Kamins TI, Chou SY, Pease RFW (2007) Nano-graphoepitaxy of semiconductors for 3D integration. Microelectron Eng 84:891–894
10. Zingg RP, Friedrich JA, Neudeck GW, Höfflinger B (1990) Three-dimensional stacked MOS transistors by localized silicon epitaxial overgrowth. IEEE Trans Electron Devices 37:\INTbreak; 1452–1461
11. Zellama K, Germain P, Squelard S, Bourgoin JC, Thomas PA (1979) Crystallization in amorphous silicon. J Appl Phys 50:6995–700
12. Katoh T (1988) Characteristics of MOSFET's on large-grain polysilicon films. IEEE Trans Electron Devices 35:923–928
13. Subramanian V, Saraswat KC (1998) High-performance Germanium-seeded laterally crystallized TFT's for vertical device integration. IEEE Trans Electron Devices 45:1934–1939
14. Wang H, Chan M, Jagar S, Wang Y, Ko P (2000) Submicron super TFTs for 3-D VLSI applications. IEEE Electron Device Lett 21:439–441
15. Gibbons JF, Lee KF (1980) One-gate-wide CMOS inverter on laser-recrystallized polysilicon. IEEE Electron Device Lett EDL-1:117–118
16. Colinge J-P, Demoulin E (1981) A high density CMOS inverter with stacked transistors. IEEE Elecctron Device Lett EDL-2:250–251
17. Colinge J-P, Demoulin E, Lobet M (1982) Stacked transistors CMOS (ST-MOS), an NMOS technology modified to CMOS. IEEE Trans Electron Devices ED-29:585–589
18. Höfflinger B, Liu ST, Vajdic B (1984) A three-dimensional CMOS design methodology. IEEE Trans Electron Devices ED-31:171–173
19. Roos G, Höfflinger B (1992) Complex 3D CMOS circuits based on a triple-decker cell. IEEE J Solid-State Circuits 27:1067–1072
20. Kawamura S, Sasaki N, Iwai T, Nakano M, Takagi M (1983) Three-dimensional CMOS ICs fabricated by using beam recrystallization. IEEE Electron Device Lett EDL-4:366–368
21. Oyama K, Kunio T, Koh R, Hayashi Y, Kajiyana K, Tsunenari K (1990) High density dial-active-device-layer (DUAL)-CMOS structure with vertical tungsten plug-in wirings. In: 1990 IEDM Technical Digest, pp 59–62
22. Jung SM, Lim H, Cho W, Cho H, Yeo C, Kang Y, Bae D, Na J, Kwak B, Kim S, Jeong J, Chang Y, Jang J, Kim J, Kim K, Ryu B (2004) Highly area efficient and cost effective double stacked S^3 (Stacked Single-crystal Si) peripheral CMOS SSTFT and SRAM cell technology for 512 M bit density SRAM. In: 2004 IEDM Technical Digest, pp 265–268
23. Jung SM, Jang J, Cho W, Cho H, Jeong J, Chang Y, Kim J, Rah Y, Son Y, Park J, Song M, Kim K, Lim J, Kim K (2006) Three dimensionally stacked NAND flash memory technology using stacking single crystal Si layers on ILD and TANOS structure for beyond 30 nm Nod. In: 2006 IEDM Technical Digest, pp 1–4
24. Minato O, Masuhara T, Sasaki T, Sakai Y, Kubo M, Uchibori K, Yasui T (1979) A high-speed, low-power Hi-CMOS 4 K static RAM. IEEE Trans Electron Devices ED-26:882–885
25. Yamanaka T, Hashimoto T, Hashimoto N, Nishida T, Shimizu A, Ishibashi K, Sakai Y, Shimohigashi K, Takeda E (1988) A 25 μm^2, new poly-Si PMOS load (PPL) SRAM cell having excellent soft error immunity. In: 1988 IEDM Technical Digest, pp 48–51
26. Yoshida T, Kinugawa M, Kanabayashi S, Onga S, Ishihara M, Mikata Y (1991) Crystallization technology for low voltage operated TFT. In: 1991 IEDM Technical Digest, pp 843–846
27. Itabashi K, Mizutani K, Koga T, Masato M, Kawashima S, Sakata M, Ema T, Yabu T, Yoyoda K, Suzuki N, Shimada H (1991) A split wordline cell for 16 Mb SRAM using polysilicon sidewall contacts. In: 1991 IEDM Technical Digest, pp 477–480
28. Helm M, Kavanaugh W, Liew BK, Petti C, Stolmeijer A, Ben-Tzur M, Bornstein J, Lilygren J, Ting W, Trammel P, Allan J, Gray G, Hartranft M, Radigan S, Shanmugan JK, Shrivastava R (1993) A low-cost, microprocessor compatible, 18.4 μm^2 6-T bulk cell technology for high speed SRAMS. In: Symposium on VLSI Technology, 1993, Digest of Technical Papers, pp 65–66

29. Chen Y-C, Chen CF, Chen CT, Yu JY, Wu S, Lung SL, Liu R, Lu C-Y (2003) An access-transistor-free (0/1R) non-volatile resistance random access memory (RRAM) using a novel threshold switching self-rectifying chalcogenide device. In: 2003 IEDM Technical Digest, pp 905–908
30. Baek IG, Kim DC, Lee MJ, Kim HJ, Yim EK, Lee MS, Lee JE, Ahn SE, Seo S, Lee JH, Park JC, Cha YK, Park SO, Kim HS, Yoo IK, Chung U-I, Moon JT, Ryu BI (2005) Multi-layer cross-point binary oxide resistive memory (OxRRAM) for post-NAND storage application. In: 2005 IEDM Technical Digest, pp 750–753
31. Crowley M, Al-Shamma A, Bosch D, Farmwald M, Fasoli L, Ilkbahar A, Johnson M, Kleveland B, Lee T, Liu T-Y, Nguyen Q, Scheuerlein R, So K, Thorp T (2003) 512 Mb PROM with 8 layers of antifuse/diode cells. In: 2003 IEEE International Solid State Circuits Conference Digest of Technical Papers, p 284
32. Herner SB, Bandyopadhyay A, Dunton SV, Eckert V, Gu J, Hsia KJ, Hu S, Jahm C, Kidwell D, Konevecki M, Mahajani M, Park K, Petti C, Radigan SR, Raghuram U, Vienna J, Vyvoda MA (2004) Vertical p-i-n polysilicon diode with antifuse for stackable field-programmable ROM. IEEE Electron Device Lett. 25:271–273
33. Herner SB, Bandyopadhyay A, Jahn C, Kidwell D, Petti CJ, Walker AJ (2006) Polycrystalline memory switching: Electrothermal-induced order. IEEE Trans Electron Devices 53: 2320–2326
34. Kroon MA, van Swaajj RACMM (2001) Spatial effects on ideality factor of amorphous silicon pin diodes. J Appl Phys 90:994–1000
35. Johnson M, Al-Shamma A, Bosch D, Crowley M, Farmwald M, Fasoli L, Ilkbahar A, Kleveland B, Lee T, Liu T-Y, Nguyen Q, Scheuerlein R, So K, Thorp T (2003) 512 Mb PROM with a three-dimensional array of diode/antifuse cells. IEEE J Solid-State Circuits 38:1920–1927
36. Wegener H, Lincoln A, Pao H, O'Connell M, Oleksiak R (1967) The variable threshold transistor, a new electrically-alterable, non-destructive read-only storage device. In: 1967 IEDM Technical Digest, p 70
37. White M, Adams D, Bu J (2000) On the go with SONOS. In: 2000 IEEE Circuits and Devices Magazine 16, pp 22–31
38. Walker A, Nallamothu S, Chen E, Mahajani M, Herner SB, Clark M, Cleeves J, Dunton V, Eckert V, Gu J, Hu S, Knall J, Konevecki M, Petti C, Radigan S, Raghuram U, Vienna J, Vyvoda M (2003) 3D TFT-SONOS memory cell for ultra-high density file storage applications. In: 2003 Symposium on VLSI Technology, pp 29–30
39. Lai E, Lue H, Hsiao Y, Hsieh J, Lee S, Lu C, Wang S, Yang L, Chen K, Gong J, Hsieh K, Ku J, Liu R, Lu C (2006) A highly stackable thin-film transistor (TFT) NAND-type flash memory. In: 2006 Symposium on VLSI Technology, pp 46–47
40. Prall K (2007) Scaling non-volatile memory below 30 nm. In: 2007 22 nd IEEE Non-Volatile Semiconductor Memory Workshop, pp 5–10
41. Wong S, El-Gamal A, Griffin P, Nishi Y, Pease F, Plummer J (2007) Monolithic 3D integrated circuits. In: 2007 International Symposium on VLSI Technology, Systems and Applications, pp 1–4
42. Stan MR, Rose GS, Zielger M (2006) Hybrid CMOS/molecular circuits. In: Proceedings of the 19th International Conference on VLSI Design, pp 703–708
43. Collier CP, Mattersteig G, Wong EW, Luo Y, Beverly K, Sarnpaio J, Raymo F, Stoddart JF, Heath JR (2000) A [2]Catenane-based solid state electronically reconfigurable switch. Science 289:1172–1179
44. Chen W, Liu X, Tan Z, Likharev KK, Lukens JE, Mayr A (2006) Fabrication and characterization of novel cross point structures for molecular electronic integrated circuits. J Vac Sci Technol B. 24:3217–3220
45. Ahn JH, Kim HS, Lee KJ, Jeon S, Kang SJ, Sun Y, Nuzzo RG, Rogers JA (2006) Heterogeneous three dimensional electronics by use of printed semiconductor nanomaterials. Science 314:1754–1757

Chapter 3
Stacked CMOS Technologies

Mansun Chan

3.1 Stacked Complementary Metal-Oxide Semiconductor Topology

Stacking metal-oxide semiconductor field-effect transistors (MOSFETs) on top of each other to form multilayer integrated circuits (ICs) is an effective way to reduce footprint and interconnect distance. In term of topology, stacked complementary metal-oxide semiconductor (CMOS) technologies can be classified into two major categories:

- One that has the active devices and interconnects interleaving each other
- One that has the active device layers and interconnects grouped separately

A schematic diagram illustrating stacked CMOS technology with the interconnect metals interleaving the active device layers is shown in Fig. 3.1. This approach provides flexibility in performing vertical and horizontal routing. However, layer-by-layer fabrication becomes impractical because the allowable thermal budget is significantly reduced after the formation of the interconnect metals. In most cases, the active devices in a single layer and the interconnects for that particular layer are first fabricated individually. Subsequently, the upper active device layer, together with its intralayer interconnects, is transferred to the top of the intralayer interconnects of the lower active device layer to form the stacked structure. The circuit density for this type of CMOS stacking is limited by the via size and spacing that can be achieved. As a result, it becomes less attractive for applications that require very densely packed active devices like semiconductor memory.

A different form of multilayer stacked CMOS IC has all the active device layers formed before the interconnect metal as shown in Fig. 3.2. The active devices can be fabricated layer by layer in a sequential manner in this structure. In addition, this approach allows the active devices to be packed very closely together and has been

M. Chan
Hong Kong University of Science and Technology, Hong Kong
e-mail: mchan@ust.hk

C.S. Tan et al. (eds.), *Wafer-Level 3D ICs Process Technology*,
DOI: 10.1007/978-0-387-76534-1_3, © Springer Science+Business Media, LLC 2008

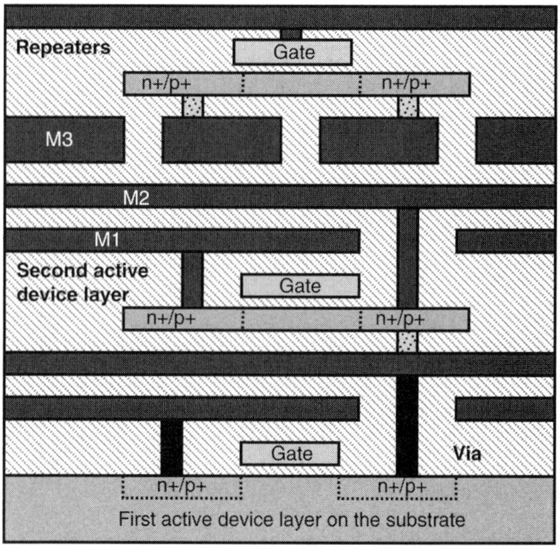

Fig. 3.1 Stacked CMOS IC with active devices and their intralayer interconnects fabricated before forming multiple stacking layers

applied to fabricate high-density static random access memory (SRAM) cells in the early 1980s [1–4].

The major challenge for this type of technology is the formation of high-quality silicon films beyond the first device layer. To achieve high circuit density, wafer bonding after active device fabrication is less preferred due to wafer-to-wafer misalignment between layers. Sequential fabrication processes require the formation of high-quality upper silicon layers on the insulating dielectrics, and some techniques discussed in Chapter 2 may be necessary. Recently, some processes have been developed to utilize multiple layers of single-crystal silicon films resulting from silicon-on-insulator (SOI) technology, and these processes will be described in Sections 3.3 and 3.4.

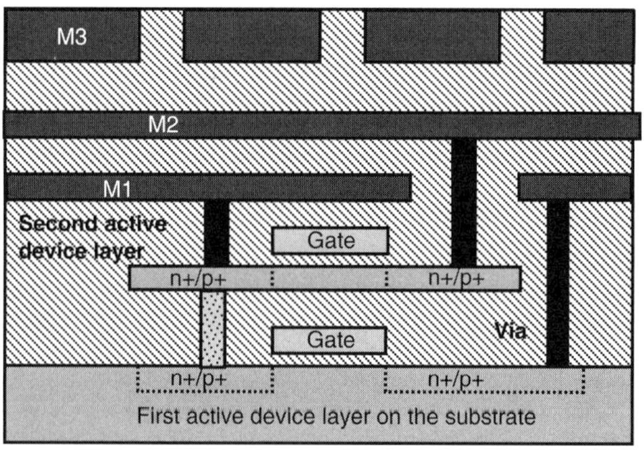

Fig. 3.2 Stacked CMOS IC with all active device layers stacked on top of each other before forming the interconnect layers

Another consideration in the topology of stacked CMOS circuits is whether to have complementary devices on every individual layer or each layer with one type of device (either n-metal-oxide semiconductor field effect transistor (NMOSFET) or PMOSFET) only. The tradeoffs mainly come from the process complexity. For stacked CMOS circuits formed by layer transfer and wafer bonding, the process complexity is usually acceptable. But layer-by-layer fabrication techniques with complementary devices on the same layer complicate the fabrication process too much, and most reported processes have only one type of FET in each stacked layer.

3.2 Stacked CMOS Processes and Device Design

3.2.1 Layer-by-Layer Process

Compared with the conventional CMOS process, more factors have to be considered when fabricating stacked CMOS circuits. A generic layer-by-layer stacked CMOS process, commonly used to fabricate early stacked SRAM cell, is shown in Fig. 3.3.

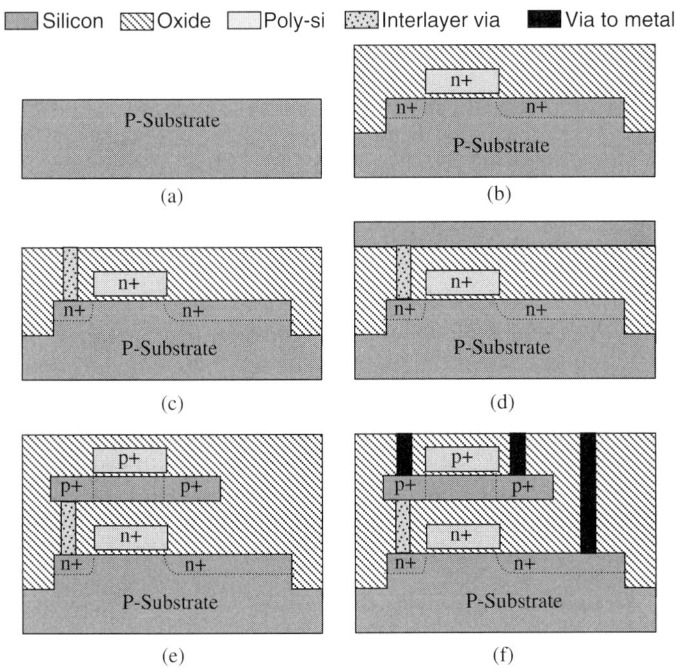

Fig. 3.3 Generic layer-by-layer stack CMOS process: (**a**) Starting with a silicon wafer; (**b**) Following regular NMOS process with shallow trend isolation to form a NMOSFET up to oxide passivation and planarization steps; (**c**) Opening of via for interlayer connection; (**d**) Formation of upper-layer silicon film for PMOSFET fabrication; (**e**) Follows the conventional SOI process to form a PMOSFET; (**f**) Opening of vias to both the top and bottom layers for metal interconnect

In this process, the devices in the bottom active layer are fabricated first before processing the top layer. One major concern in this fabrication approach is the thermal budget experienced by the lower layer when processing the upper devices which may result in serious dopant diffusion in the source, drain, and gate regions. This issue not only limits the dopant activation process for the upper-layer transistors but also the formation of high-quality silicon film for active device fabrication since most techniques (like those described in Chapter 2) require high-temperature treatment. Thus, using layer-by-layer fabrication techniques alone can only produce stacked CMOS circuits with a few layers (typically 2–3). Considering the thermal budget, it is preferred to have NMOSFETs fabricated in the lower layer(s) and the PMOSFETs fabricated in upper layers due to the more serious diffusion at high temperature of p-type dopants (mainly boron) compared with n-type dopants.

Another complication in stacked CMOS fabrication is in the contact opening. With a via-last process flow, the first metal layer has to contact devices on different active layers through vias with different depths. It presents a challenge to the etching process to create small openings through dielectrics with different thickness and stop at the corresponding silicon. In addition, filling up the vias (especially the deep one) without forming internal voids also requires significant effort to optimize the process conditions.

3.2.2 Simultaneous Multilayer Processing

Recently, some simultaneous multilayer processing techniques have been developed in which all transistors on different active layers are being fabricated at the same time. A generic simultaneous multilayer processing technique is illustrated in Fig. 3.4. The figure only depicts the processing sequence, with fabrication methods described later.

This method resolved the thermal budget problems in the layer-by-layer processing. As all the transistors are fabricated after the formation of multilayer single-crystal silicon films, the constraints to anneal the material using low-temperature processes are relieved. More mature material-handling processes such as layer transfer with wafer bonding and double separation by implantation of oxygen (SIMOX) [5] can be used to form the multiple single-crystal silicon device layers. However, the processing of simultaneous layers presents different challenges including (1) gate definition of the lower layer; (2) doping of the bottom devices; and (3) interconnecting different layers.

3.2.2.1 Gate Definition of Lower Layers

As in most gate-all-around or surrounding gate MOSFETs [6], it is difficult to define the gate below an active layer with lithography. The presence of the upper active layers also prevents the etching of the gates underneath; thus, special processing is required to form the gates. The most common process is to use wet-etch to undercut the bottom of the active layers, but the process is very difficult to control. To ensure

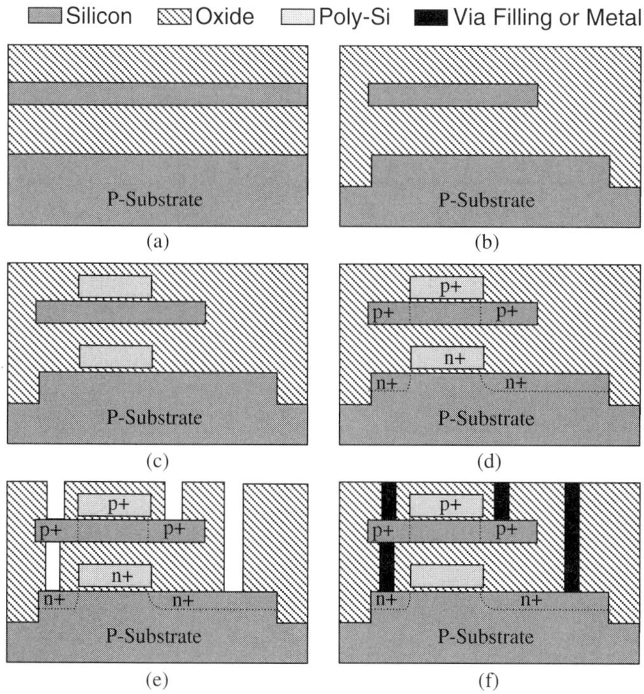

Fig. 3.4 Simultaneous multilayer stacked CMOS process: (**a**) Starts with a wafer with two or more active silicon layers; (**b**) Active area formation by etching and refill; (**c**) Gate formation for devices in all layers; (**d**) Gate and source/drain doping for all active devices; (**e**) Contact opening to all active layers; (**f**) Contact filling for metal interconnect

complete exposure of the silicon region below an active layer, an oversized gate is usually formed causing large gate-to-drain and gate-to-source overlap capacitances. Recently, some new processes have been introduced to allow the formation of the gate at the lower layers after the formation of the upper layers, as described later in this chapter.

3.2.2.2 Doping the Bottom Layer

With the upper silicon layers formed, doping the lower silicon layers becomes difficult. The most commonly used method is to apply a high-energy implant to introduce dopant through the upper layers and target the implant peak at the location of the desired lower active layers [7] as shown in Fig. 3.5. After the implant, dopant activation of all layers can be performed at the same time to relieve the thermal budget issues. However, the heavy implant through the upper layers can potentially create damage, and it is not easy to perform deep implants without a large lateral dopant spread. This limits the maximum implant depth and dictates the need to have only a small number of layers, minimal layer-to-layer separation, and minimum

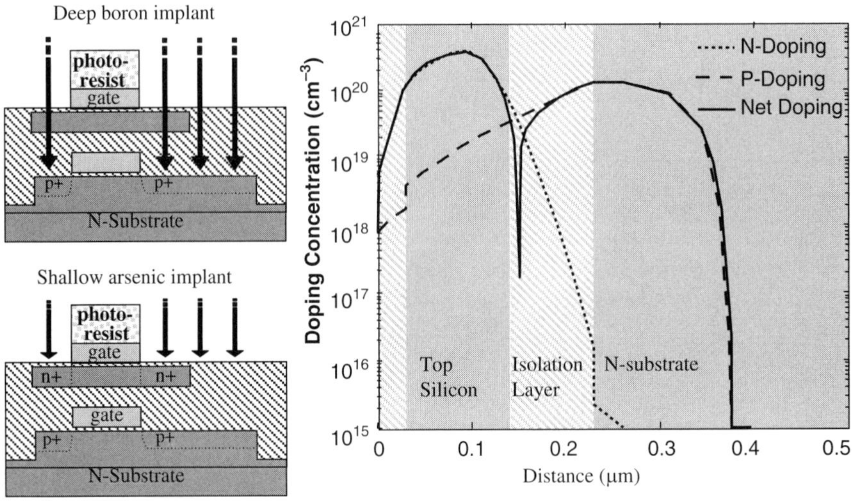

Fig. 3.5 Method to introduce dopants into the top and bottom active layers by implants with different energy. Figure on the right shows the resulting simulated doping profile for boron and arsenic reported in [12] showing the possibility to introduce dopants to a lower layer through the top layer of silicon. © 2005 IEEE

layer thickness. As boron has deeper implant depth relative to phosphorus or arsenic, it is generally preferred to have PMOSFETs at the lower layers and NMOSFETs at the upper layer, which is the opposite of that in layer-by-layer technologies.

3.2.2.3 Making Contact Between Two Active Layers

Contact between the top and bottom layers can be achieved if the top and bottom contacts are at different locations, by using the metal wires above the device layers. This approach consumes silicon area and reduces circuit density. Some better approaches to connect these layers are shown in Fig. 3.6. In Fig. 3.6a, two lithography steps are performed with the first one open up a deep contact through the top

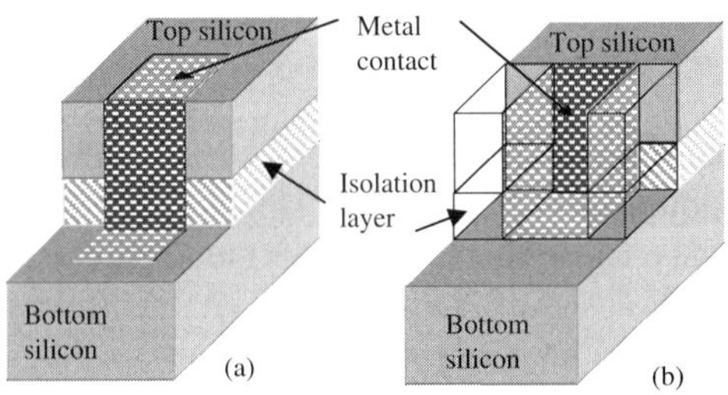

Fig. 3.6 Method to connect the top and bottom silicon layers through (**a**) combined contact to the top and bottom silicon layers; and (**b**) contact to connect the top and bottom silicon layers using the inside sidewall surface of the top silicon layer and the top surface of the bottom silicon layer [12]. © 2005 IEEE

layer and stop at an etch-stop layer. The isolation layer may serve the purpose if it is sufficiently thin. In the second lithography step, a larger contact covering the deep contact is opened at the top silicon layer. Without removing the photoresist, another etch is performed to remove the etch-stop material covering the bottom silicon at the deep contact. The via-filling then connects the top and bottom layers using the combined contact. Figure 3.6b illustrates another method to contact the top and bottom layers using a deep via formed by etching through the top layer directly down to the bottom layer. The via-filling then connects the inside sidewall of the top silicon layer to the top surface of the bottom silicon layer at the same contact. The second method provides an effective way to contact both top and bottom layers without increasing contact dimension. But the contact resistance can be very high for very thin top-layer silicon films.

3.2.3 Layout Issues

As mentioned in previous sections, most reported stacked technologies have either PMOSFETs or NMOSFETs (but not both) on the same layer. To fully exploit the advantage of small footprint offered by stacked CMOS technology, the layout of transistors in standard circuit components must be designed carefully. For example, an inverter, a basic unit of CMOS circuits, requires the PMOSFET to have twice the width of the NMOSFET to provide symmetrical rise and fall time. When stacking the NMOSFET and PMOSFET together, the footprint is limited by the larger device as shown in Fig. 3.7. The area-saving resulting from stacking the CMOS devices is thus lower than expected.

In more advanced nontraditional CMOS technology, double-gate MOSFETs have been introduced. In this case, the PMOSFETs can be implemented with the double-gate structure while the NMOSFETs are designed with a planar structure to reduce the footprint. If NMOSFETs and PMOSFETs can be fabricated in the same layer, the footprint can also be reduced by placing part of the PMOSFETs

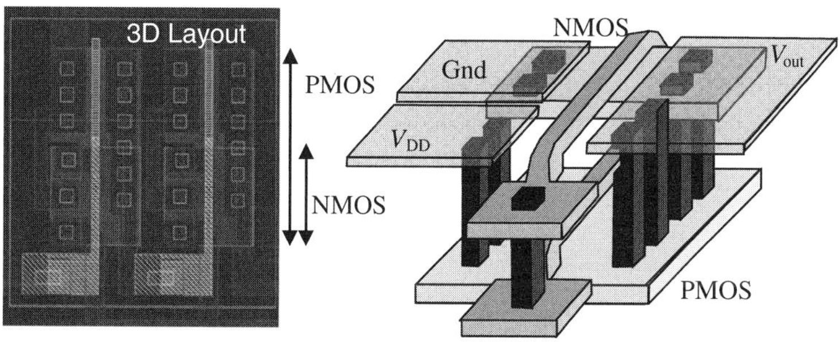

Fig. 3.7 An inverter layout together with its footprint in three-dimensional (3D) view using the stacked CMOS technology

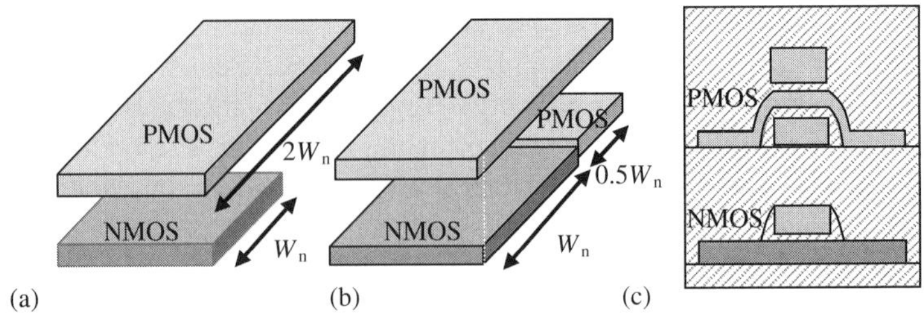

Fig. 3.8 Footprint of a CMOS inverter with (**a**) conventional stacked layout; (**b**) layout with part of the PMOSFET stacking on itself next to the NMOSFET (provided PMOSFETs and NMOSFETs can be formed at the same layer); and (**c**) using double-gate PMOSFETs to reduce the footprint of the PMOSFETs

underneath itself next to the NMOSFETs. Both approaches to reduce the footprint are shown in Fig. 3.8.

3.3 Stacked CMOS on the Thin Film and the Substrate of a SOI Wafer

Despite the difficulty to form stacked CMOS circuits, a few innovative stacked CMOS technologies have been proposed recently. One approach is to utilize both the thin silicon film and the bulk substrate of a SOI wafer to form the stacking transistors [8]. The mature SOI process ensures the high-quality single crystal to be used to form the active device layers. The process follows the simultaneous layer processing approach and the key processing steps are shown in Fig. 3.9.

The process starts with a SOI wafer with initial oxidation and nitride deposition. A shallow trench process is used to define the active area of both the SOI film and the substrate silicon (Fig. 3.9a). The trench is refilled by a thick low-temperature oxide (LTO) film, followed by planarization with chemical–mechanical polishing (CMP) using the nitride as polish stop. The nitride is then removed and another thick (200 nm) LTO is deposited (Fig. 3.9b). The gate region definition step is performed followed by a thin layer of nitride deposition as a spacer for the source/drain region of the bulk MOSFET. Arsenic implant is then performed to form the source/drain region of the nMOSFET on the substrate of the SOI wafer (Fig. 3.9c). A via-hole is opened at regions where connections between the upper- and lower-layer MOSFETs are desired. A thick layer (500 nm) of polysilicon (poly-Si) is then deposited and planarized using CMP with the top nitride as stop layer. The poly-Si is then thinned to about 100 nm by timed wet-etch and the exposed nitride is removed by wet etch (Fig. 3.9d). A 150-nm a-Si is then deposited and the part on top of the gate region is removed by CMP (Fig. 3.9e). Boron is then implanted to dope the source/drain regions of the upper PMOSFET. The active area of the PMOSFET is then defined by removing the excess amorphous and poly-Si. The LTOs occupying the gate regions are removed by wet etch. The exposed oxide covering the channel of the PMOSFET is then removed by wet etch (Fig. 3.9f). A trench on the top of the silicon film and a

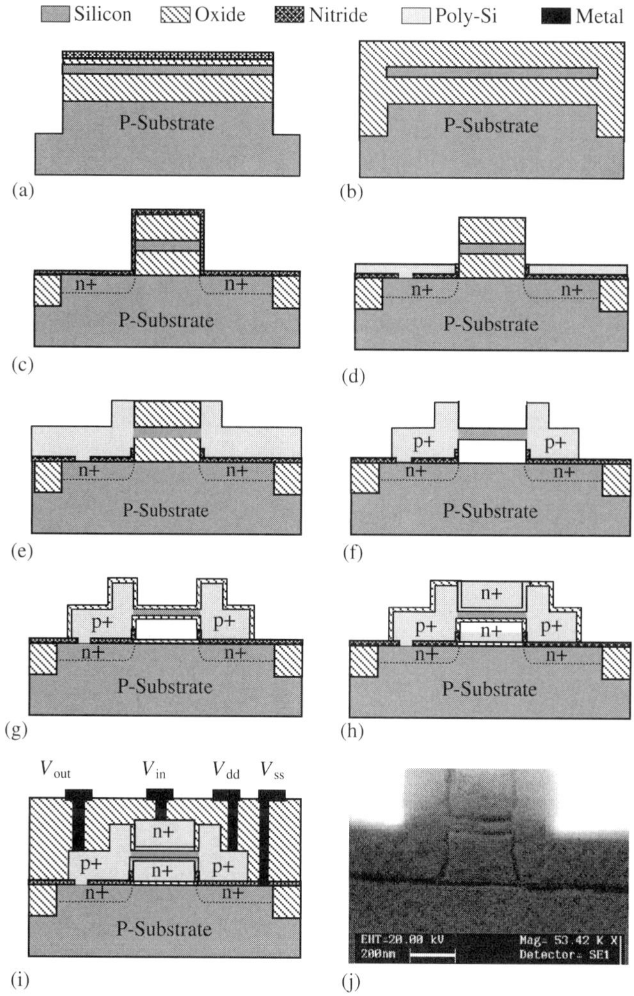

Fig. 3.9 Key processing steps to form the stacked CMOS circuit with the silicon thin-film PMOSFET and substrate NMOSFET from a SOI wafer. The final structure obtained from SEM is also included. From [8] © 2004 IEEE

tunnel beneath it that self-aligned to each other are formed to define the dimension of the three channels (two for the PMOSFET and one for the NMOSFET). The gate oxides of both the NMOSFET and the PMOSFET are grown together with thermal oxidation (Fig. 3.9g) followed by a conformal deposition in-situ-doped poly-Si and patterning by CMP to form the gate electrodes (Fig. 3.9h). The fabrication is completed with conventional back-end processes, including passive layer deposition, contact opening, and metallization (Fig. 3.9i). The final structure resulting from the process is shown in the scanning electron microscopy (SEM) picture (Fig. 3.9j).

The described process produced a very compact stacked CMOS structure with many desirable features including

- double-gate PMOSFET with top and bottom gates self-aligned to each other;
- PMOSFET with thin body and thick source/drain region self-aligned to the gate;
- perfectly stacked PMOSFET and NMOSFET for symmetrical rise and fall time;
- short interconnect distance between the top and bottom devices.

The layout of an inverter formed using the stacked CMOS process is shown in Fig. 3.10. A large amount of area-saving compared with conventional layout can be achieved by stacking the transistors. Very nice and symmetrical *I–V* characteristics for a PMOSFET and a MOSFET having the same footprint resulting from the described process have been achieved (Fig. 3.11). Besides CMOS inverter, other circuits such as 2-input NAND gate, 2-input NOR gate, and 4-input NAND gate are also demonstrated with significant area reduction [9].

Even though the described process provides a breakthrough in forming high-density stacked CMOS circuits, there are still some limitations in the process that require further refinement. Even though the PMOSFET and NMOSFET are fabri-

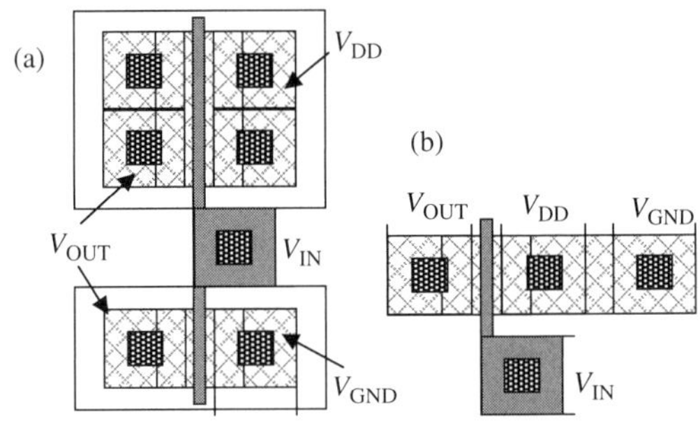

Fig. 3.10 Layout of (**a**) conventional 2D inverter and (**b**) stacked CMOS inverter. From [12] © 2005 IEEE

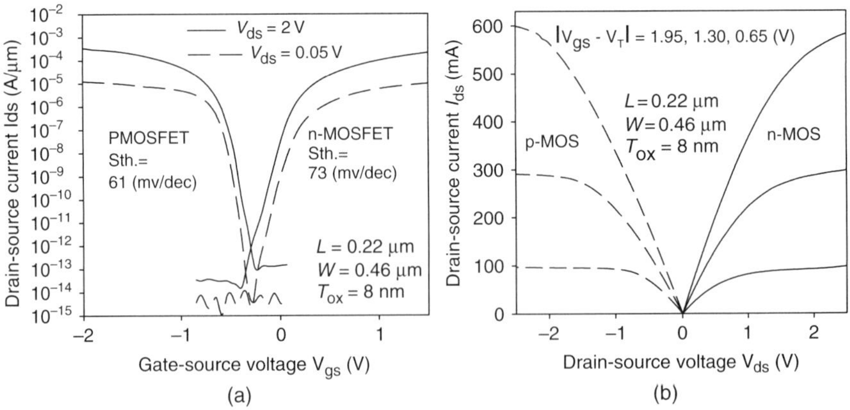

Fig. 3.11 Measured (**a**) I_{ds}–V_{gs} and (**b**) I_{ds}–V_{ds} characteristics from the stacked CMOS transistors on silicon film and substrate of SOI wafer reported in [8]. © 2004 IEEE

cated together, the formation of the source/drain region of the bottom NMOSFET occurs in the early stage of the process. Subsequent high-temperature processes are experienced by the bottom transistors and lead to increase in dopant diffusion. Despite the attempt to reduce the thermal cycle by performing the dopant activation near the end of the process, other unavoidable thermal steps such as gate oxidation still contribute to the extra dopant diffusion at the bottom transistors.

The demonstrated process uses n+ poly-Si gate for both the NMOSFET and PMOSFET. This configuration results in an appropriate threshold voltage for the NMOSFET, but a relatively high threshold voltage for the PMOSFET. To achieve a symmetrical threshold voltage for both NMOSFET and PMOSFET, mid-gap materials have to be used as the gate material. However, the requirement to use more advanced gate material to refill the area beneath the upper channel requires some innovation in process design which has not been demonstrated yet. Despite the complexity, the process demonstrates the feasibility of fabricating very compact stacked CMOS circuits.

3.4 Stacked FinCMOS Technology

Recent advances in nano-CMOS technology have promoted the development of nonplanar CMOS structures to overcome the short channel effects. In particular, the FinFET [10] has been considered to be the most promising structure for manufacturing double-gate MOSFETs using conventional CMOS process with minimal modifications. Based on the FinFET process, a new approach to form stacked CMOS process has been developed [11, 12]. The structure of the stacked FinFET is illustrated in Fig. 3.12. It consists of π-shaped gate straddles on a stacked fin that serves

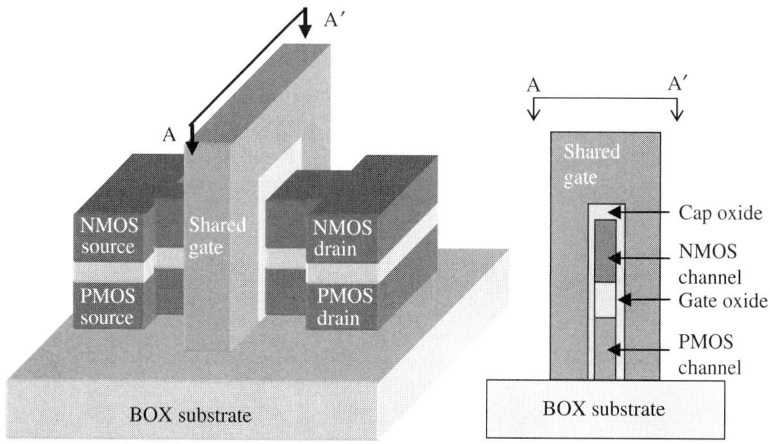

(a) (b)

Fig. 3.12 (**a**) Three-dimensional schematic view and (**b**) cross-section of the stacked FinFET structure reported in [12]. © 2005 IEEE

as the shared gate for both the NMOSFET and PMOSFET. The shared gate is used as an implant mask to achieve self-alignment for both NMOSFET and PMOSFET source/drain regions. Between the two Si active layers, there is an insulation layer to isolate the top and bottom devices.

The stacked FinFET technology is a straightforward extension of the simultaneous multilayer processing approach, with the multilayer single-crystal silicon films being formed prior to active device fabrication. While SOI wafers with conventional thin active-layer silicon films and a passive bulk substrate can be used, FinFETs formed on the bulk substrates can be implemented as well [13]; however, the buried oxide (BOX) thickness should be minimized to reduce the height of the stacked fins.

The process demonstrated in [11, 12] uses a double SIMOX process to form the starting material as shown in Fig. 3.13. Starting with a SIMOX wafer, a second low-energy oxygen implant was performed to separate the top silicon film into two layers with an additional BOX layer. The film thicknesses resulting from the process described in [12] are 120 nm for the top silicon, 90 nm for the insulating oxide, and 80 nm for the bottom silicon layer. A transmission electron microscopy (TEM) micrograph shown in Fig. 3.13 illustrates the structure of the starting material used to demonstrate the process.

The key fabrication steps to form the stacked FinFET are shown in Fig. 3.14. Starting with the double SIMOX wafer, a layer of LTO is deposited to serve as the hard mask for subsequent etching (Fig. 3.14a). The multilayer films are to form the stacked fins (Fig. 3.14b). The gate oxide is then grown followed by in-situ-doped n+ gate poly-Si deposition (Fig. 3.14c). The gate is then defined by a lithography resulting in the structure shown in Fig. 3.14d. Doping of the source/drain region of the top and bottom FinFETs is performed by ion implantation. The doping of the bottom is achieved by high-energy implant through the top layer as described in section 3.2 (Fig. 3.14e). In order to connect to the bottom layer, the unwanted regions of the top silicon layer are etched to expose the bottom layer. The contacts where the top transistors are to be connected to the bottom ones are also opened in this step (Fig. 3.14f). It is followed by phosphosilicate glass (PSG) passivation and contact opening to various layers. The contact vias are filled to form the interconnect (Fig. 3.14g). A SEM image of the cross-section and a top view of the stacked FinCMOS

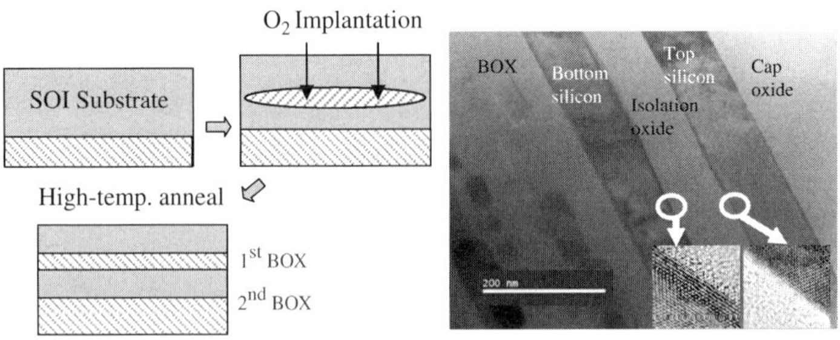

Fig. 3.13 Method to form the double silicon film as described in [12] and the TEM view of the material structure to form the double FinFET. © 2005 IEEE

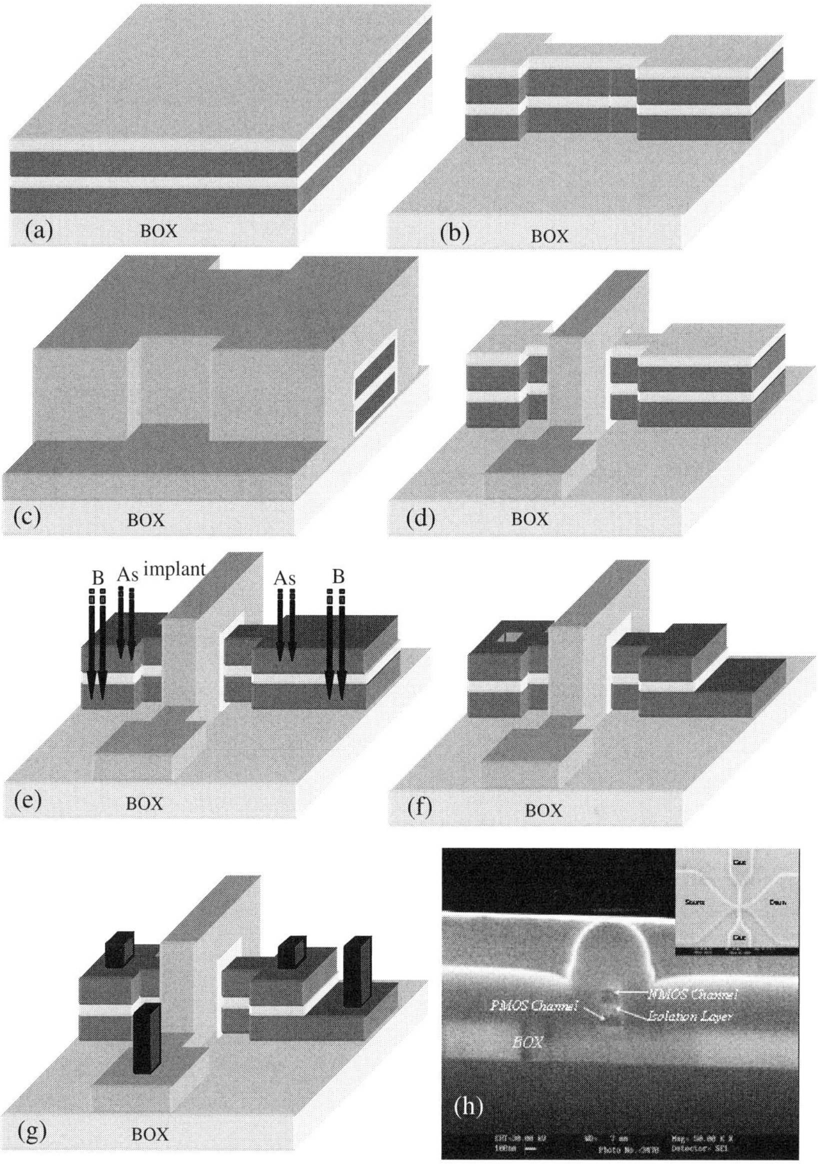

Fig. 3.14 Key processing steps to form the stacked FinCMOS inverter. The SEM image of the top view and cross-section of the fabricated FinCMOS inverter is also included. From [12]. © 2005 IEEE

inverter are shown in Fig. 3.14h. The fabricated device characteristics are shown in Fig. 3.15.

Compared with other reported processes, the stacked FinCMOS process is relatively simple and has a number of advantages, including

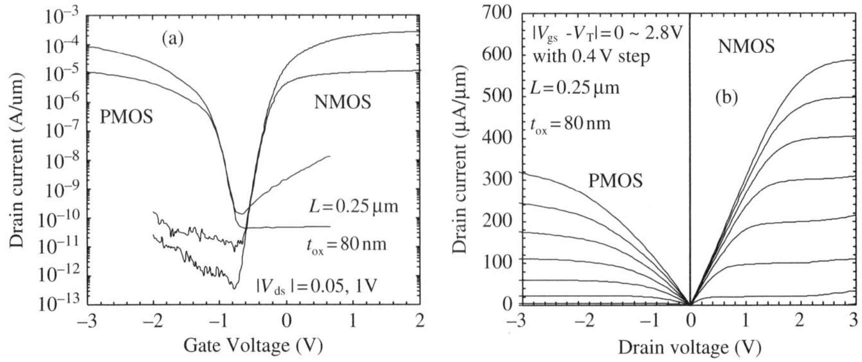

Fig. 3.15 Measured (**a**) I_{ds}–V_{gs} and (**b**) I_{ds}–V_{ds} characteristics from the stacked FinCMOS transistors as reported in [12]. © 2005 IEEE

- compatibility with existing planar CMOS technology,
- perfectly stacked NMOSFET and PMOSFET devices,
- simultaneous formation of top and bottom MOSFETs and thus no accumulation of thermal cycles,
- ratio of NMOSFET and PMOSFET currents can be adjusted using different fin heights.

Despite the numerous advantages and promises offered by the stacked FinCMOS technology, a number of improvements are still required in addition to the reported technology. First of all, similar to the stacked CMOS on the thin film and the substrate of a SOI wafer described in section 3.3, the demonstrated stacked FinCMOS process used n+ poly-Si gate for both NMOSFETs and PMOSFETs. As a result, the threshold voltages of the NMOSFETs and PMOSFETs are not symmetrical, about $V_{gs} = 0$ V, with the threshold voltage of the NMOSFET too low and the threshold voltage of the PMOSFET too high as shown in Fig. 3.15a. To produce symmetrical threshold voltage, mid-gap materials have to be used to adjust the threshold voltage.

Secondly, the fabricated PMOSFETs and NMOSFETs do not have the same current drive. The current drive of the PMOSFETs and NMOSFETs can be adjusted by using the fin height of the top and bottom silicon layers. However, it is restricted by the starting material (unless it is acceptable to use very short top fin height).

Thirdly, due to the need to dope the bottom layer using high-energy implant through the top layer as described in Section 3.2, only NMOSFETs on top of PMOSFETs have been demonstrated. Fabricating PMOSFETs on top of NMOSFETs becomes relatively difficult and more careful process optimization is required.

Finally, a processing challenge occurs to etch the tall fins composed of the top silicon fin, bottom silicon fin, and the isolation layer with vertical sidewall and high aspect ratio (HAR). By using inductive coupled plasma (ICP) etching, HAR fins with vertical sidewall angle of 89.5°–90° can be formed as shown in Fig. 3.16, but the sidewall quality has not been carefully studied.

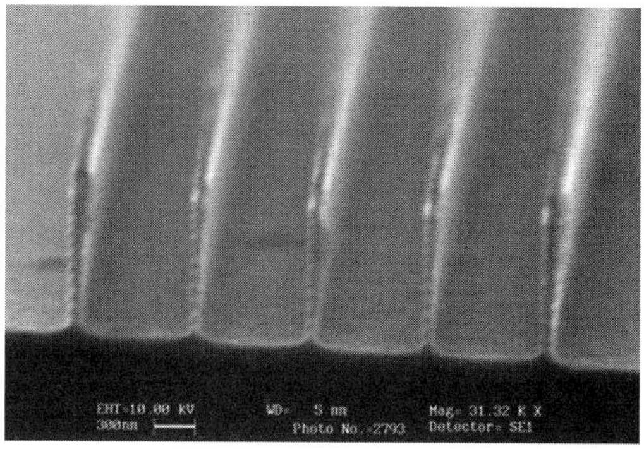

Fig. 3.16 Scanning electron microscopy image of tall silicon film with HAR and vertical side wall (89.5°–90°) achieved by ICP etching

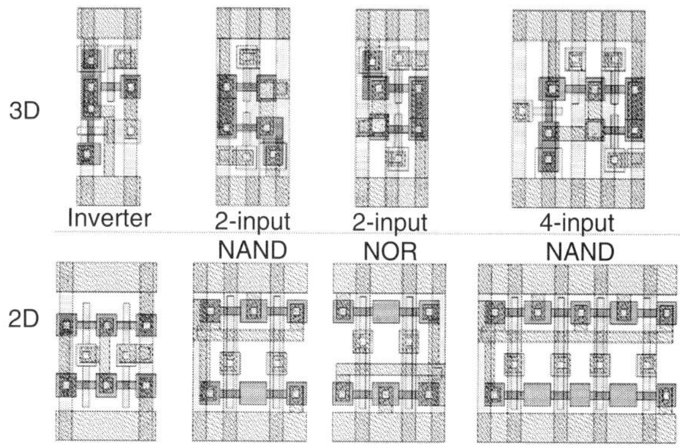

Fig. 3.17 Layouts of some of the 3D FinCMOS standard cells compared with the conventional 2D layout

Despite the challenges to improve the stacked FinCMOS process, a number of standard cells have been demonstrated and the layouts of the inverter, 2-input NAND, 2-input NOR, 4-input NAND are shown in Fig. 3.17 with the standard 2D layout for comparison. In all cases, area and capacitive loading reduction from 40%

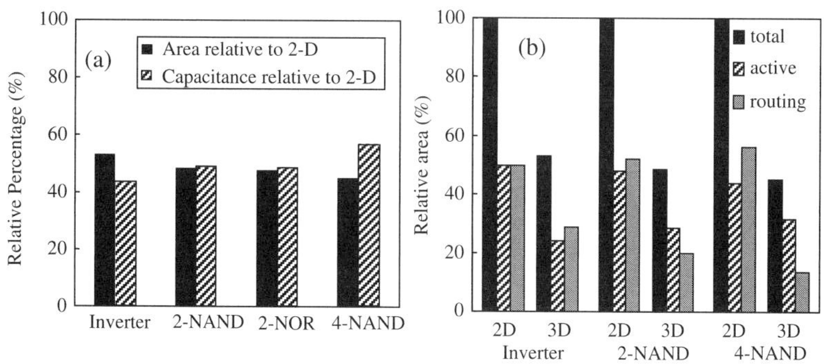

Fig. 3.18 Comparison of area and capacitance between 2D planar gates and 3D stacked FinCMOS gates. (**a**) Percentage of area and capacitance of the gates constructed by the FinCMOS structure relative to the 2D case. (**b**) Relative reduction of the area coming from the active devices and the intragate interconnects. From [12]. © 2005 IEEE

to 60% can be achieved as shown in Fig. 3.18. The reduction comes from both more compact active devices placement and shorter interconnect distance between active devices within the cell. The standard cell approach is very useful in developing 3D IC because it is compatible with the current planar IC design methodology and computer-aided design (CAD) Tools. Once the standard cells have been reconfigured using the 3D process, 3D IC can be designed using high-level hardware description languages (e.g., very high speed IC hardware description language (VHDL) and Verilog) and automatic placement and routing. This approach allows the benefit provided by 3D IC to be quickly incorporated into the current technology.

Among the different circuits, memory array has the most regular structure with repeating units. Redesigning the individual unit in a memory array with stacked CMOS technology is expected to produce the most significant area saving and performance improvement among all circuits. In particular, SRAM cells are composed of mainly CMOS transistors and can quickly adopt the stacked CMOS technologies. Figure 3.19 shows the layout of SRAM cell in 2D [13] and using the stacked FinCMOS technology, together with the output characteristics of the fabricated circuits. Although each active layer contains only one type (either n-type of p-type) of transistor, the advantage in terms of compactness is very obvious.

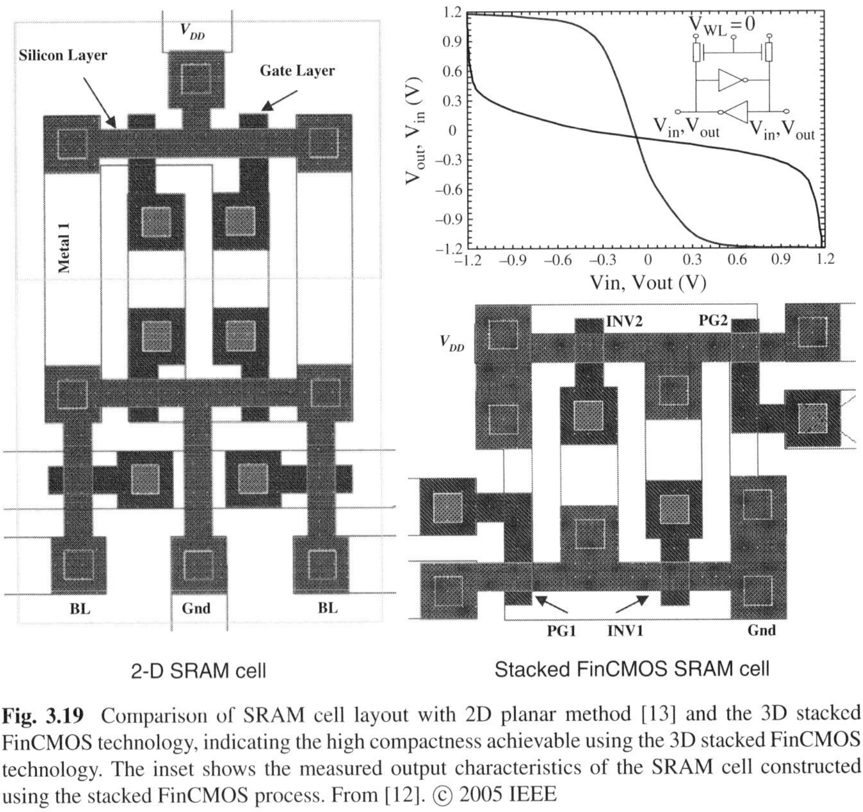

Fig. 3.19 Comparison of SRAM cell layout with 2D planar method [13] and the 3D stacked FinCMOS technology, indicating the high compactness achievable using the 3D stacked FinCMOS technology. The inset shows the measured output characteristics of the SRAM cell constructed using the stacked FinCMOS process. From [12]. © 2005 IEEE

3.5 Summary

Stacked CMOS process has been an effective approach to create very compact circuits. However, there are many challenges to form stacked CMOS circuits. In addition to material issues in forming high-quality silicon film beyond the first device layer, other processing issues such as thermal budget and method to contact various active device layers require careful consideration. With the advancement of CMOS technology to the nanoregime, more complex processing techniques become more acceptable to the technology communities and a number of new approaches to form stacked CMOS circuits have been proposed. In this chapter, two of such approaches have been introduced, and the stacked FinCMOS seems to be promising for further exploration.

References

1. Gibbons JF, Lee KF (1980) One-gate-wide CMOS inverter on laser-recrystallized polysilicon. IEEE Electron Device Lett 1:117–118
2. Chen CE, Lam HW, Smlhi SDS, Pinizzotto RF (1983) Stacked CMOS SRAM cell. IEEE Electron Device Lett 4:272–274
3. Colinge JP, Demoulin E, Lobet M (1982) Stacked transistors CMOS (ST-MOS), and nMOS technology modified to CMOS. IEEE Trans Electron Devices 29:585–589
4. Kunio T, Oyama K, Hayashi Y, Morimoto M (1989) Three-dimensional ICs, having four stacked active device layers. IEDM Tech Dig:837–840
5. Matsamura A, Kawamura K, Hamaguchi I, Takayama S, Yano T, Nagatake Y (1999) Recent progress in low-dose SIMOX wafers fabricated with internal-thermal-oxidation (ITOX) process. Mtg Abs Electrochem Soc:391
6. Colinge J, Gao M, Romano-Rodriguez A, Maes H, Claeys C (1990) Silicon-on-insulator gate-all-around device. IEEE IEDM Tech Dig:595–598
7. Current MI, Pramanik D (1984) MeV implant for silicon device fabrication. Solid State Technol 27:211–216
8. Zhang S, Han R, Lin X, Wu X, Chan M (2004) A stacked CMOS technology on SOI substrate. IEEE Electron Device Lett 25:661–663
9. Lin X, Zhang S, Wu X, Chan M (2006) Local clustering 3-D stacked CMOS technology for interconnect loading reduction. IEEE Trans on Electron Devices 53:1405–1410
10. Hisamoto D, Lee WC, Kedzierski J, Takeuchi H, Asano K, Kuo C, Anderson E, King TJ, Bokor J, Hu C (2000) FinFET – A self-aligned double-gate MOSFET scalable to 20 nm. IEEE Trans Electron Device 47:2320–2325
11. Wu X, Chan PCH, Zhang S, Feng C, Chan M (2005) Stacked 3-D fin-CMOS technology. IEEE Electron Device Lett 26:416–418
12. Wu X, Chan PCH, Zhang S, Ceng C, Chan M (2005) A three-dimensional stacked fin-CMOS technology for high-density ULSI circuits. IEEE Trans Electron Devices 52:1998–2003
13. Nowak EJ, Rainey BA, Fried DM, Kedzierski J, Ieong M, Leipold W, Wrignt J, Breitwisch M (1990) A functional FinFET-DGCMOS SRAM cell. IEEE IEDM Tech Dig:411–414

Chapter 4
Wafer-Bonding Technologies and Strategies for 3D ICs

Shari Farrens

4.1 Introduction

Wafer bonding has often been described as a key enabling step for three-dimensional (3D) integration. This is the processing step in which the individual wafers are aligned and bonded so that the benefits of the layer-to-layer interconnects can be realized. Many of the alignment techniques and bonding methods have evolved from microelectromechanical system (MEMS) fabrication methods. The fundamental difference between MEMS and 3D integration is that the alignment accuracy had to be improved by 5–10 times. This has led to an evolution in the equipment platforms which are now able to achieve micron and submicron alignment accuracies in the final products. The following sections of this chapter will describe the process flow (align, fixture, and bond) and describe the options available, with enough details to determine the appropriate methods for a specific device application.

4.2 Wafer-Bonding Equipment Overview

Wafer-bonding equipment is available in manual and automated platforms. The equipment must perform several operations, including surface preparation and cleaning, wafer alignment, wafer bonding, and post-bond metrology. For throughput improvements, a chill plate is usually included so that wafers can be unloaded from the bonder above room temperature. Not every bonding method requires all of these steps, with the layout of the production system utilizing only those modules necessary to the process flow.

Figure 4.1 shows each of the module options available to support process flow options in a production system. The front-end interface will depend on the production line wafer size. Most of the work to date has been done on 200-mm wafers, with four cassette stations enclosed in the ULPA-filtered minienvironment of the cluster

S. Farrens
SUSS MicroTec, Waterbury Center, VT, USA
e-mail: sfarrens@suss.com

C.S. Tan et al. (eds.), *Wafer-Level 3D ICs Process Technology*,
DOI: 10.1007/978-0-387-76534-1_4, © Springer Science+Business Media, LLC 2008

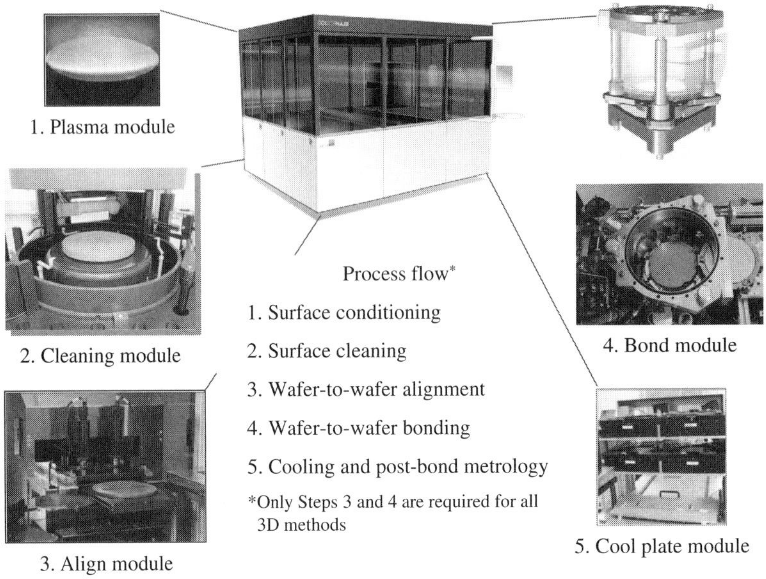

Fig. 4.1 Typical production of 3D integrated circuit (IC) bond cluster showing the variety of process modules available

tool providing the interface to the robotic handling system. As the research and development (R&D) processes have been transitioned to preproduction on 300 mm substrates, and the equipment suppliers have migrated to front-opening unified pod (FOUP) interfaces and equipment front-end automation modules (EFAMs). The FOUP and EFAM interfaces provide additional isolation of the wafers and process modules from the mechanics of the material-handling unit and robotics, and provides a cleaner system.

The leading bond methods for 3D integration are based on MEMS bonding techniques that have existed for over a decade in production. The perceived or perhaps true lack of standardization within the MEMS industry has led the equipment suppliers to settle on standardized process modules that can populate cluster bonding system to provide the "customized" process flow needed for each MEMS device or product. With integrated device manufacturers (IDMs) and silicon foundries, the stability and reliability of the equipment is essential and standardization is highly valued. Thus the cluster approach with standard modules is an excellent platform to address 3D IC wafer-to-wafer bonding needs.

A generic 3D process would start with wafers loaded into FOUPs, and a centralized robot would retrieve the wafers one at time, record lot and wafer identification information, and then proceed with the processing. Some process flows such as the "Ziptronix method", and "SAB method" will require plasma surface modification or chemical treatments [1–3]. Nearly all wafers will receive some manner of surface preparation, even if it is only a megasonic deionized (DI) water rinse to remove particles. Next the wafers must be aligned, very precisely, and placed in a transport fixture for processing in the bonder. The bonder module is used to provide the appropriate temperature ramp, down force, and atmosphere for bonding. Finally, the bonded pair should be properly cooled and checked for accuracy before additional processing occurs. The remaining sections in this chapter will examine each of the possible wafer-to-wafer 3D bonding steps in detail to delineate the advantages and disadvantages provided by the various options.

4.3 Surface Preparation Treatments

Surface preparation treatments can be used for surface modification or surface cleaning. Dry processing generally refers to plasma treatments and can be done in closed chambers using vacuum plasma technologies or with atmospheric pressure plasma sources. Wet processing involves a cleaning arm that moves back and forth over a rotating wafer or wafer pair dispensing chemicals and water. Acid vapor cleaning refers to a process whereby the substrates are exposed to the vapor phase of an oxide-reducing acid to remove surface scale. This process can be done offline in a batch process or incorporated onto the bond cluster tool. It should be noted that while users may refer to these stations as cleaning stations this is not precisely the correct description of these modules. Since these are single modules and the chemistries used are dilute and at room temperature, true cleaning procedures (e.g., RCA 1/RCA 2) should be done in batch process in traditional wetbench cleaning systems.

4.3.1 Surface Preparation Treatments – Wet Chemistry

One method of 3D integration is silicon direct or fusion bonding. Fusion bonding involves the creation of hydrophilic surfaces (note that there are also hydrophobic direct wafer-bonding methods but these do not apply to 3D integration methods at this time, refer to [21] for hydrophobic bonding) that attract one another when brought into contact. The theory of a direct bond is that adsorbed [OH] groups attach to the dangling bonds on the wafer surface. These [OH] groups attract water molecules on the surface and the angular water groups have inherent dipole moments. When two treated wafers are brought into proximity the dipole moments will attract one another and bond due to van der Waals forces [4]. As the wafers are pulled together by the attractive forces, a ring structure called a cyclic tetramer forms at the interface at room temperature. This is the origin of the so-called "bond wave" which can be viewed with infrared (IR) transmission cameras. The wafers will hover over one another until a point contact initiates the bond and penetrates the near-surface electrostatic repulsive forces. Then as the wafers bond the air is forced out from the interface and appears as a wave transgressing across the interface. As the wafers are pulled together by the attractive forces, a ring structure called a cyclic tetramer forms at the interface at room temperature.

Wet chemistries can be used to create the hydrophilic surfaces that lead to direct bonds. A typical process would be a (1:1:10)-NH_4OH:H_2O_2:H_2O at 55°C for 3 min in a traditional wetbench. Alternatively, the cleaning module in the bond cluster can be used, with a 1–4% NH_4OH solution at room temperature, to create the hydrophilic surface. Typical processing in the cleaning module for activation will involve two strokes of the cleaning arm in which the NH_4OH solution is dispensed followed by the megasonic DI water rinse and spin dry. The advantage of the point-of-use activation in the bond cluster is that once the surfaces become hydrophilic the surface is not only attractive to other silicon wafers but also the van der Waals attraction will cause any particles in the environment to adhere to the surface as well. Therefore, wafers being queued up for bonding may stay in a cleaner state if they are not activated in batch but rather immediately before alignment in the cluster platform. The tradeoff is the serial processing in the bond cluster versus the batch processing in the wetbench, the quality of the environment, and duration of the queue time. Throughput modeling is used to determine if point-of-use activation or cleaning will become a bottleneck to the process flow. In the event that throughput is adversely affected additional cleaning stations can be added or the process step can be completed offline.

With both the offline and point-of-use activation the wafers will move to the alignment step after activation and then be heated to full strength usually in a batch process in a quartz oven. If the activation is a chemistry-only process the bonds will not reach full strength until annealed above 900°C. If the activation involves plasma surface treatments or combined plasma and chemical treatments then it is possible to reduce the thermal annealing to below 400°C. Plasma treatments will be covered in later sections.

Both the leading manufacturers of wafer-bonding equipment provide integrated cleaning modules in the production tools for 3D wafer-to-wafer bonding [5, 6]. Both vendors supply systems with the ability to rinse the wafer surfaces with hot or cold DI water, or use megasonic DI water to create cavitation events at the particle–surface interface to assist in the removal of adsorbed particulates. In addition, the tools are equipped to dispense dilute chemistries (~4% NH_4OH or 4% HCl) onto the surface. The EVG tool can also be equipped with a scrubbing brush identical to those found on post-chemical–mechanical planarization (post-CMP) cleaning systems. The polyvinyl alcohol (PVA) brush is wetted with DI water or the same diluted chemistries as dispensed by the cleaning arm. It is rare that wafers intended for 3D integration require brush cleaning but the option exists. All available systems allow for spin drying but the SUSS systems have the capability to use IR heat lamps to increase the temperature during the spin dry as necessary. This is used to manipulate the thickness of the adsorbed water layers. The other differentiator is that the EVG tool is a one-wafer-at-a-time system. The SUSS surface preparation module can process one or two wafers simultaneously, which substantially improves throughput and cost of ownership (COO).

After the wafer-bonding process the surfaces become buried interfaces and any particle or contamination layer is permanently entombed, potentially affecting both die yield and device performance. Hence, particle performance of the cleaning station is the critical metric used for maintaining the cleaning modules. Testing of these stations involves measuring the number of particles on an incoming wafer and

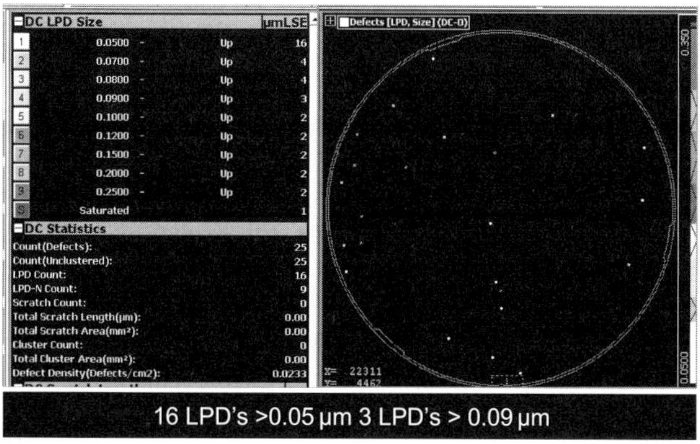

Fig. 4.2 Cleaning results for a 300-mm silicon wafer with thermal oxide. Typical vendor specification is zero added particles above 0.12 µm. Data measurements taken with a KLA Tencor SP2 particle measurement system

comparing that to the number of particles on the wafer after the process recipe is complete. Figure 4.2 shows a typical tool qualification result on 300-mm wafers taken from a KLA Tencor SP2 particle measurement tool after cleaning on a SUSS MicroTec CL300 module. Most silicon on insulator (SOI) facilities will run daily particle checks on the equipment since even one particle is not allowed. In 3D preproduction, most users are checking on a less regular schedule. If a particle problem does arise, the recipe steps are broken down and wafers are counted before and after each recipe step to identify the source of the problem. Typical specification is zero particles added above 0.10–0.12 µm, although this varies between locations and users.

4.3.2 *Surface Preparation Treatments – Plasma Activation*

Plasma treatments can be used for two very different purposes in wafer-to-wafer bonding. If the ultimate goal is a silicon direct fusion bond then the plasma treatment is used to "activate the surface". On the other hand, the etching properties of the plasma have also been used to clean metal surfaces in preparation for copper-to-copper (Cu-to-Cu) diffusion bonds; in the Ziptronix process plasma etching is a means to enhance a subsequent chemical activation and surface reaction [1].

Plasma activation has its origins in SOI fabrication and it rapidly gained popularity in other application areas such as MEMS and 3D integration. This bonding technique dramatically reduces the annealing temperatures needed to achieve full-strength silicon direct bonds and increases initial attractive forces between the wafers. There have been several implementations of plasma activation, including extremely low-vaccum applications, capacitive-coupled reactive ion etch (RIE)-style systems, atmospheric plasma sources, and downstream reactors to name a few [2, 3, 7, 8]. SUSS MicroTec, EVGroup, Silicon Genesis (SiGen), and Mitsubishi all provide plasma treatment systems intended for wafer bonding.

All plasma systems using any type of gas will render a wafer hydrophilic because the plasma will either create dangling bonds or deposit charged ions onto the surface. The surface then immediately attracts water vapor from the treatment chamber and the polar water molecule creates a hydrophilic surface. Therefore, gas selection for the plasma system can include reducing gases, such as H_2/He, H_2/Ar, or reactive gases, such as O_2, N_2, synthetic air, or water.

Plasma activation is a dry process that is substituted for wet chemical processes described in the previous section. The plasma COO is less because the amount of gas used is minimal and considerably less expensive than chemical costs and associated infrastructure and disposal costs. The treatment time ranges from a few seconds to 45 s and does not affect other materials on the wafer.

Van der Waals forces are weak dipole interactions and while they are strong enough to facilitate transfer of the bonded wafers between modules it is possible to separate the wafers with a razor or wedge. Plasma activation will demonstrate its first advantage by increasing the strength of the van der Waals forces by a factor of 2 to as much as an order of magnitude. While the mechanisms are not fully understood or agreed upon, it is generally accepted that the benefits are related to an increased number of potential bonding sites, surface cleanliness and removal of hydrocarbons and other volatiles, as well as modified surface charge [8, 9]. The net result is that when the initial attractive forces are increased the surfaces will bond in spite of slightly elevated surface roughness or bow-and-warp issues.

Whenever a bond can spontaneously initiate (as is the case with the direct bond method) at room temperature without the use of vacuum or applied force the bond can be initiated in the aligner and the accuracy of wafer-to-wafer alignment will increase. A strong initial attraction due to plasma treatment extends the variety of applications that are suitable for bonding in the aligner by relaxing the stringent requirements for surface quality and wafer shape.

The second and perhaps other major advantage of plasma activation is that the kinetics of the van der Waals bond transformation to covalent Si–O–Si silanol bonds is increased dramatically: not only is the annealing time shortened, but the overall temperature of the anneal is reduced to below 400°C. This appreciable reduction in temperature makes plasma activation an indispensable process step if silicon direct bonding is desired; in comparison, wet chemical activation of wafers for silicon fusion bonding requires thermal treatments over 900°C. Without the plasma treatment, this bond technology is not compatible with any 3D integration solution, since interconnected complementary metal-oxide semiconductor (CMOS) technology limits the thermal budget to ~400°C.

4.3.3 Surface Preparation Treatments – Combined Plasma and Wet Chemistry

In the Ziptronix method plasma is used to create bonding sites for [NH] ammine groups and to increase the reaction kinetics. While the mechanism explaining why the ammine groups behave so differently from other [OH]-based reactions has not yet been fully divulged, the bonding results available in the literature are remarkable [1]. Three-dimensional integration schemes have been developed using this silicon direct bonding in conjunction with a proprietary metal referred to as the direct bond interconnect (DBI) metal. The DBI metal is recessed relative to the dielectric isolation layer with both material layers exposed on the surfaces. The plasma treatment is used to achieve very strong direct bonding between the dielectric layers on each substrate. Subsequently, during the annealing process the silicon direct bonds form Si–O–Si bonds almost immediately and as the temperature is increased the metal layers expand and either diffusion-bond or cold-weld at the interface under compression.

The tool qualification for plasma systems usually involves particle qualifications with and without postexposure rinsing; in addition, for vacuum plasma sources the system is checked for vacuum leaks that can affect the plasma chemistry. Vendors supply recommended periodic maintenance and process checks specific to their designs.

4.3.4 Surface Preparation Treatments – Vapor Cleaning

Copper-to-copper diffusion bonding is used extensively in 3D integration whenever the electrical connections will be completed during the bonding process; this process flow is referred to as "via first". However, Cu readily oxidizes at room temperature and is not self-limiting. Figure 4.3 shows the oxidation free energy for Cu as a function of temperature and partial pressure of oxygen. This Ellingham diagram shows that under the standard processing conditions used for Cu-to-Cu wafer-level bonding the avoidance of oxidation is not possible without surface treatment. Copper self-diffusion rates can decrease several orders of magnitude depending upon the thickness of the surface contamination preventing bond formation altogether.

The removal of the oxide layer can be achieved by various oxide etch chemistries, but the wet chemical (acid-based) chemistry can damage other layers in the CMOS wafers. Dry methods such as acid vapor treatments are advantageous because the wafers do not have to be immersed in fluids, and etching of underlying passivation, adhesion, and dielectric layers can be minimized or excluded. One chemistry that has worked well is formic acid (H_2CO_2) vapor cleaning.

Formic acid vapor cleaning is used extensively in the SUSS MicroTec C4NP mold transfer equipment and die-bonding equipment [12, 13]. Chemical and vapor delivery systems have been developed and safety qualified for IDM high-volume production. For wafer-level bonding the process implementation can be internal or external to the cluster bonder. The most simple approach is to pre-clean the wafer prior to alignment in a batch process by purging a cleaning chamber with the formic acid vapor. The wafers do not need to be heated but low temperature ($\sim$100°C) accelerates the oxide removal rate. It is also possible to clean the wafers at point of use in the cluster tool before wafer-to-wafer alignment. The wafers are rinsed and dried after vapor cleaning and proceed to the aligner.

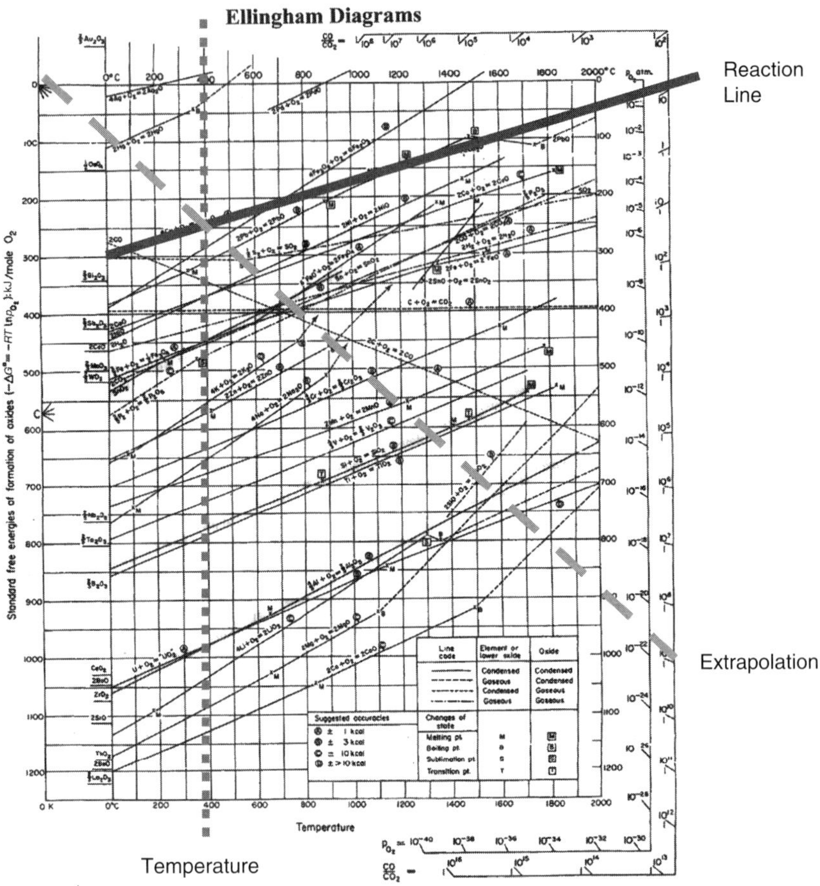

Fig. 4.3 Ellingham plot for Cu_2O. By extrapolation from the intersection of the reaction line and the processing temperature line, the limit for the partial pressure of oxygen necessary for surface reduction is determined. From [10] © 1961 Wiley-VCH Verlag GmbH & Co. KGaA. Reproduced with permission. Originally from [11] © 1953 McGraw Hill

To illustrate the efficacy of formic acid vapor cleaning, a comparison was made between wafers cleaned with rinse-only techniques and wafers cleaned with formic acid vapor. The wafers were annealed face-up at 400°C in vacuum for 20 min. They were not bonded with another wafer, but simply exposed to the thermal and environmental controls of a standard Cu thermal diffusion bonding recipe. Figure 4.4 compares the before and after condition of the wafer surface for the rinse-only sample. The oxidation is appreciable and observation of the metal traces and alignment keys is difficult.

A second wafer was cleaned with the formic acid vapor and rinsed with megasonic DI water and spun dry. Figure 4.5 shows that although an adsorbed water layer remains after the rinse-and-dry step the formic acid passivates the surface against

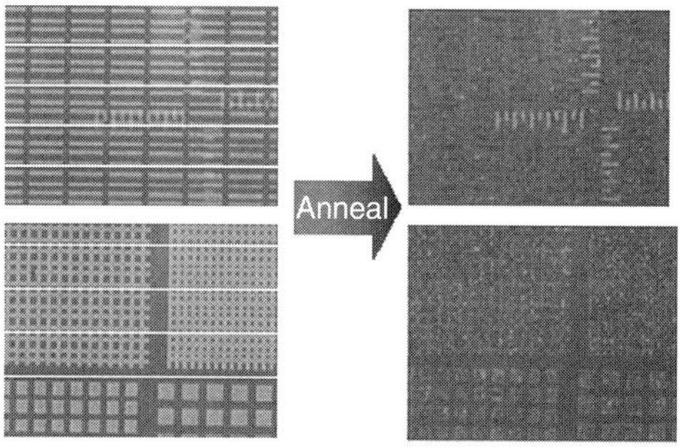

Fig. 4.4 Unpassivated Cu films readily oxidize if heated to 400°C even in vacuum

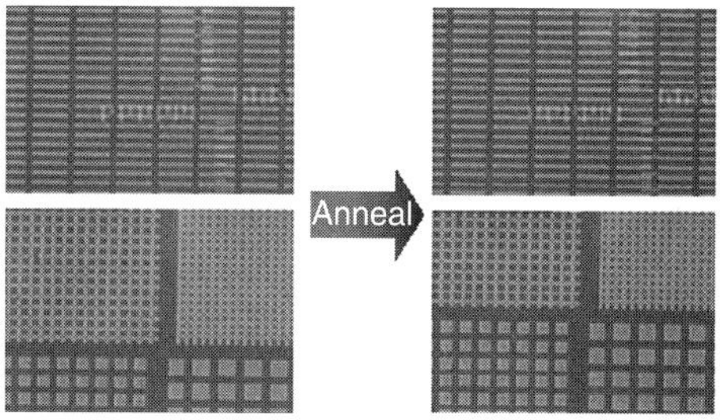

Fig. 4.5 Copper films cleaned and passivated with formic acid vapor do not oxidize if heated to 400°C

additional oxidation. Comparison of the before and after images of the heated wafer indicates no evidence of oxidation after thermal cycling to 400°C for 20 min.

Figure 4.6 is an example of two, 200-mm blanket Cu-coated wafers that were bonded after formic acid passivation. The acoustic image is defect-free except for manual handling issues, and subsequent trials with 3D via-bonded wafers have yielded excellent electrical properties. Clearly, Cu diffusion bonds benefit greatly from surface preparation methods.

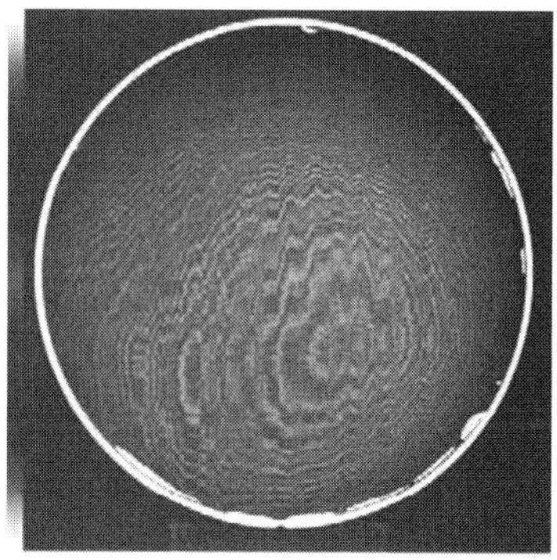

Fig. 4.6 Acoustic image of 200-mm blanket layer Cu-to-Cu diffusion bonding with vapor pre-cleaning. Edge quality is compromised by manual mechanical handling. Image was taken with a Sonoscan D9000 C-SAM

4.4 Bond Aligner – Equipment Working Principle

Bond aligners used in wafer-to-wafer bonding are very similar to the mask aligners used for advanced packaging of 300-mm wafers. Figure 4.7 shows the major components and where these components are located in the actual equipment.

To perform the alignment process the first wafer is placed face-down on the bond-aligner chuck and introduced to the aligner. The wafer is then moved up the *z*-axis of the aligner until it is held by the transport fixture using vacuum. This first wafer being held by the transport fixture becomes the foundation for the alignment process. It will define the origin of all motions.

In standard processing each wafer will have an alignment key on the left- and right-hand side of the wafer. The keys should be located as far apart as possible so that the accuracy of the alignment is enhanced with respect to sensitivity to rotational errors as well as *x*-axis and *y*-axis positioning and run-out. Run-out is the error due to stepper patterning of the CMOS device layers. In general run-out will cause a center to edge-shift in the die-to-die pitch that may be as large a 0.1 μm. Run-out is identified when the right- and left-aligned wafer keys are both shifted inward or both shift outward along the *x*-axis.

The microscopes are used to find the alignment keys on the first wafer. The microscopes can move in *x* and *y* directions as well as in *z* direction to focus on the alignment keys. The range of motion of microscopes is confined to an area at the left and right periphery of the chuck and is ±20 mm from the centerline along the *y*-axis and a radial distance of 35–95 mm each side of center on the *x*-axis (note that these numbers are based on a 200-mm wafer/chuck). Custom opening size and location can be accommodated. The openings in the load slide chuck are visible in Fig. 4.7 and determine where the alignment keys need to be positioned. In automated systems the location of the keys are stored in the recipe so that when wafers with

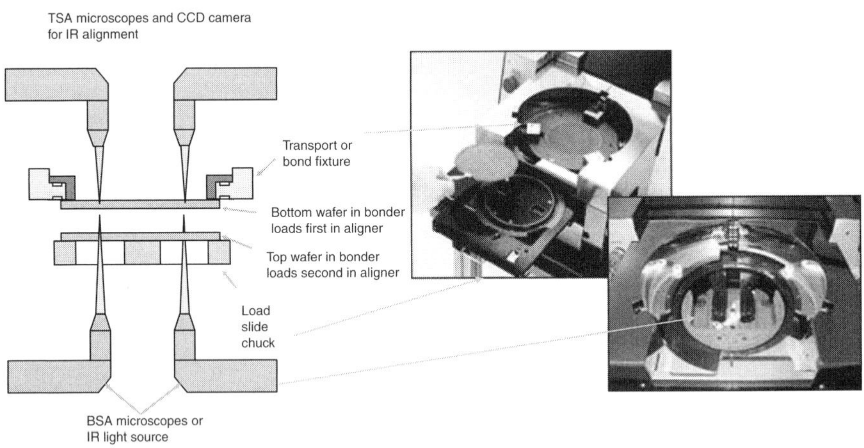

Fig. 4.7 Major components in a bond aligner. Upper cameras and objectives removed for photographs

the same alignment pattern are loaded, the search algorithm will begin very near to the known location of the fiducials.

The second (upper) substrate is loaded last, in a face-up position so that the wafers are face-to-face. Separation flags, if used, are inserted at this time and should not rub or scratch the surface of the substrate. The second wafer is held by vacuum to the chuck on a movable alignment stage. The alignment stage is then moved in x, y, or θ (rotation) directions to find the keys on the second wafer and ensure that the keys on both wafers are properly aligned or overlaid. The stage x and y drives have 30 mm of travel while the θ drive allows the stage to move through a total of $\pm 3^{\circ}$ of rotation to align the wafers. Depending on the alignment type, the second wafer keys are either aligned to the actual keys on the first wafer or overlaid to the stored image of the keys from first wafer.

Once alignment is complete the bottom wafer is raised into "contact" position and the wafers are clamped into the fixture from the outside of the stack. This is when the spacers and clamps first make contact with the surfaces and this should happen only in the exclusion zone at the edge of the wafer. Proper fixture adjustments are used to ensure there is no movement of the wafers during the clamping operation.

Proper adjustment of the aligner entails verification that the optics are aligned and that the motion of the slides is smooth and parallel. During the operation of the alignment a wedge error compensation (WEC) step will occur. During this procedure the second wafer is held on the chuck by vacuum and is moved up and brought in contact with the first wafer held by the fixture. The WEC mechanism allows the motion of the bottom wafer to achieve parallelism by applying $\sim$1000 g spring force against the top wafer. This can be done either in contact or in proximity when using the three spacers. Once both the surfaces are parallel, the chuck is locked in position. The WEC ensures that the fixture holding the bottom substrate and the chuck holding the top substrate are parallel. Manufacturers supply detailed instructions on how to properly adjust the WEC for their equipment.

Calibration wafers are glass and silicon wafers with alignment keys with additional graduated scales with verniers. These wafers can be aligned over and over again, clamped and removed from the aligner, and checked for overlay accuracy by simply looking at the verniers in an optical microscope through the glass. If there is a systematic problem, offsets can be used to compensate for final alignment. This is commonly used if the incoming wafers have a known amount of run-out. Then the aligner will initially align the left and right keys as close as possible and then shift the bottom wafers "out of alignment" by the amount needed for the run-out compensation. This will assist in the overlay accuracy of both the inner- and outer-die locations.

4.5 Alignment Strategies

Alignment challenges define the fundamental difference between using wafer-to-wafer bonding for 3D IC integration and employing the techniques for MEMS manufacturing. In MEMS fabrication the feature sizes are large and the alignment requirements are only now evolving toward 2–5 μm in accuracy requirements for most products. This is due in part to the fact that there are theoretical limits to the scaling of mechanical systems. In 3D IC integration the scaling is significantly reduced and the goals set 5 years ago at ±2 to 3 μm are today at ±1 to 1.5 μm. Roadmaps for both the end-users and the equipment suppliers call for submicron alignment accuracies in production equipment for 300-mm application by late 2008 [14, 15].

Alignment techniques are separated into those that align to live images and those that align to stored images. The generic process of wafer-to-wafer alignment involves placing the first substrate into the aligner. A left and right objective lens is used to find alignment marks at the periphery of the wafer. Next, the second wafer is inserted and the procedure is repeated and the second wafer is moved until overlay of the alignment keys is achieved. During live image alignment the operator can simultaneously view the alignment keys on the upper and lower substrates. This means that the first wafer is loaded and the keys are identified. But while the second wafer is loaded and positioned into alignment using the keys, both sets of alignment marks are in view, real time. Any mechanical shifting is viewed in real time. This facilitates closed-loop feedback systems that allow for repeated alignment attempts and correct for shifting as the wafers are secured in the bond fixture for transfer to the bonder.

There are several types of bond-alignment methods. They are topside alignment and bottomside alignment (TSA/BSA), IR, and intersubstrate alignment (ISA) or face-to-face alignment. Topside alignment and bottomside alignment are used extensively in MEMS fabrication and use only one set of objective lenses in the aligner. The method requires that one of the substrates is transparent so that the image of the upper and lower keys can be viewed simultaneously from one side of the stack. Figure 4.8 depicts an overview of lens and wafer pair locations for each method, discussed in detail below.

Infrared transmission imaging is used extensively in 3D wafer-level bonding. A key advantage of IR imaging is that it aligns to live images and metal traces in

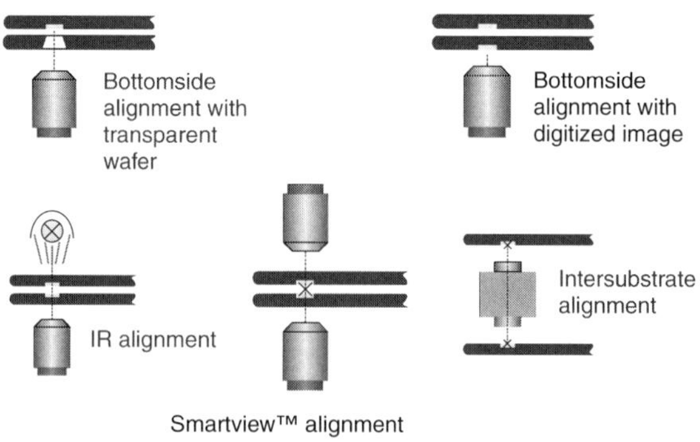

Fig. 4.8 Substrate alignment methods. All techniques can be used to image upper and lower wafers simultaneously while SmartView™ and bottomside alignment with digitized image techniques use stored images. Courtesy of EV Group

silicon, typically giving very sharp, high-contrast, images that are easily recognized by human and computer pattern-recognition software. However, there are a few practical limitations to IR alignment methods. The first has to do with the adsorption of light in silicon. If the photon energy of the light, in electron volts, is greater than the semiconductor band gap, there will be strong or total adsorption [16]. Infrared (with photon energy less than the band gap of silicon) images are generally possible in silicon as long as the doping levels result in a resistivity greater than 0.01 ohm-cm (thus minimizing free carrier absorption) and no solid metal layers are overlaid with the alignment key.

The second limit for using IR imaging for alignment has to do with the thickness of the wafer. In Figure 4.9 thick and thin wafer-aligned IR images are compared. Attenuation is exponential with material thickness so the image from a pair of 200-mm wafers is brighter than that from a pair of 300-mm wafers with exactly the same optical system. In 3D ICs most CMOS wafers are not heavily doped and the IR optical absorption associated with up to 3-mm thickness is not a general problem. However, repeated layer stacking of bulk Si wafers (i.e., not SOI) may limit the viability of IR alignment due to additional scattering from device layers unless the alignment key regions are open fields.

The third limit is the smoothness of the exit surface of the wafer stack. In Fig. 4.8, the light source is positioned on one side of the wafer stack and the charge-coupled device (CCD) camera and objectives on the opposite side. If there is significant backside roughness the image will be blurred due to light scattering at the exit surface toward the CCD camera. This intensity loss and image degradation is not a consideration for 300-mm 3D IC integration since all 300-mm wafers are double-side polished.

In the meantime, however, a solution is needed for 200-mm products and development efforts. Figure 4.10 shows the improvement that can be gained by switching to LED light sources. The LED light source is a brighter light source which mitigates the attenuation loses. In addition, the wavelength distribution of the light output is less adsorbed than for the halogen source.

Another alignment mode that is necessary for 3D wafer bonding when the other method requirements are not satisfied is face-to-face alignment. There are presently two options available. The ISA camera offered by SUSS MicroTec inserts the optics between the wafers during the alignment step. The SmartView^{TM} alignment method from EV Group uses image capture of the first wafer to a digital storage file and then brings in the second wafer aligning it to the stored image [17, 18].

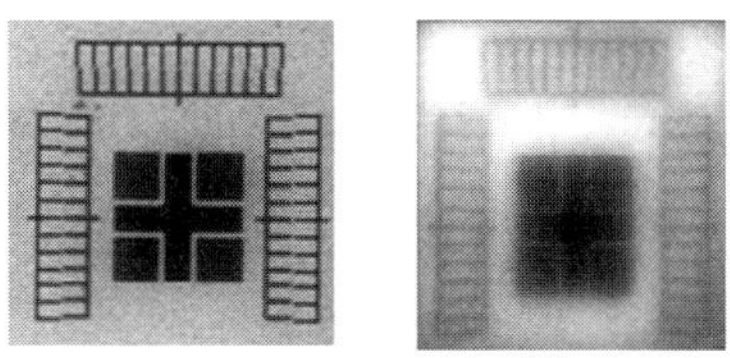

Fig. 4.9 Infrared image quality from a thin 200-mm aligned wafer pair (*left*) compared to a standard-thickness aligned pair (*right*). Both wafers had standard roughness on the backside of the wafer

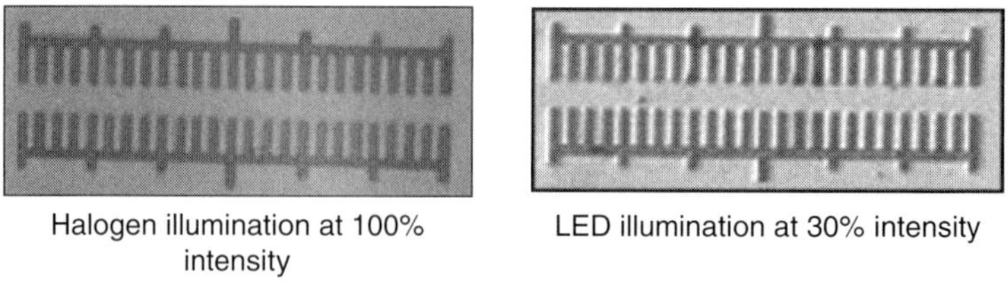

Fig. 4.10 Improved image quality achieved with light-emitting diode (LED) illumination compared to halogen illumination in a bond-aligner system

The objectives are outside the wafer stack and the lower lenses are used to image the upper wafer first and then that wafer is removed and the upper optics image the lower wafer. In both of these methods the bond aligner is equipped with two pairs of objective lenses. The advantage of the SmartView™ system, in principle, is that the objectives are located "outside" of the wafer stack. So there is limited movement of the objectives in the *z*-axis and the wafer separation distance is also minimal. Figure 4.11 schematically illustrates the three-step process for this alignment method [18]. This method typically achieves ± 1 μm or slightly better in alignment accuracy [18, 19].

ISA is a SUSS MicroTec offering which evolved from high-precision die-to-wafer stacking technology. The main drawback of the ISA alignment is that the separation distance between wafers is very large. If there is mechanical precession in the *z*-axis then errors will occur as the wafers are brought into proximity and

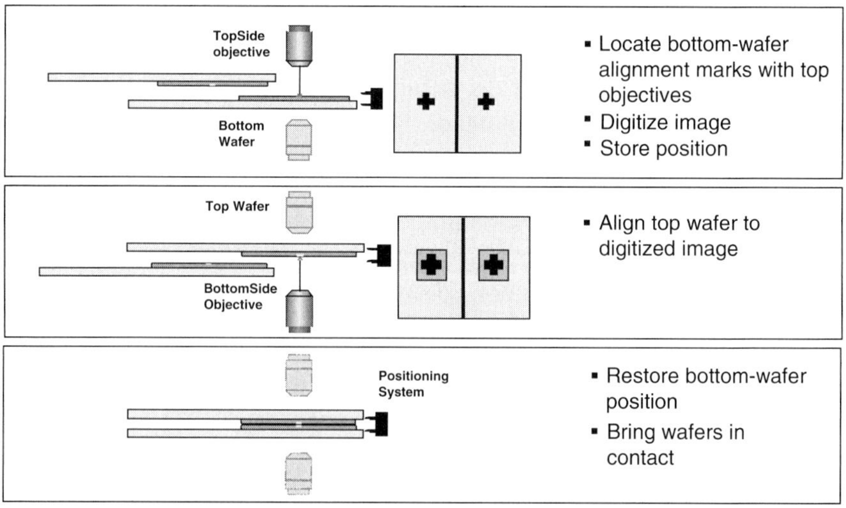

Fig. 4.11 SmartView™ aligner working principle. © 2001 by IMAPS – International Microelectronics And Packaging Society. Reprinted with permission from [18]

clamped to the transport fixture after the camera is extracted. The ISA camera has demonstrated micron-level alignment but is presently considered a ±1.25- to 2-μm tool depending upon sample type.

It is not possible to determine presently the repeatability of the existing systems because the R&D volumes have been low and interrupted process flows do not lend themselves to statistically significant numbers although both vendors will claim 3-sigma repeatability. The roadmaps of both manufacturers include programs to reach deep submicron-level alignment by the end of 2008. This is consistent with the ITRS roadmap goals for 3D IC integration and both equipment suppliers will introduce deep submicron alignment tools for 200-and 300-mm wafers.

Roadmaps also indicate that the optical-based alignment schemes presently used in wafer-to-wafer bond alignment will reach theoretical limits at approximately the 0.25-μm level. At that time it may be that bond-alignment techniques will migrate to higher-precision motion controllers, lasers, or holographic alignment keys. Electrostatic chucks may also be considered. Certainly the large potential market and anticipated growth will drive novel and revolutionary changes in wafer-to-wafer alignment

The ability to achieve good alignment is more than just the optics. Other factors such as the symmetry of the keys, contrast adjustments, and geometry all play key roles in the cumulative error analysis of bond alignment.

Any shape can be used for alignment keys but several options will enable higher accuracy and automation. Alignment marks always come in male and female pairs. The most common alignment keys are derived from crosses and boxes, with two

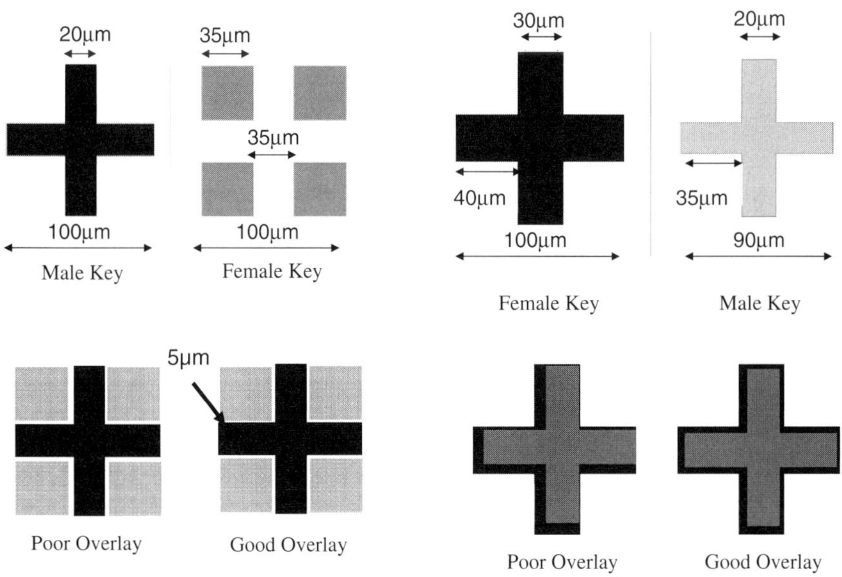

Fig. 4.12 Standard alignment keys: (*left*) cross-and-box style and (*right*) cross-on-cross

standard examples shown in Fig. 4.12. In the case on the left, the cross will be positioned between the boxes as shown during alignment. For the dual-cross scenario, the smaller male key is overlaid onto the female image during alignment. Misalignment is measured by comparison of the gap or open space between the two targets using a calibrated microscope. However for cross-on-cross alignment keys, the two overlaid opaque layers are indistinguishable after bonding; therefore, both the examples shown would yield the same quality control value if post-bond alignment accuracy measurements were required. In addition, the relative positioning method used to determine post-bond alignment accuracy with the cross-in-box method is not truly quantitative.

The ideal alignment key should have geometric steps and verniers that enable overlay offset determination. Examples of verniers combined with geometric keys are shown in Fig. 4.13. In this example, each division of the vernier is 0.2 μm and can be used after bonding as a quantifiable post-bond metrology metric. Whenever there is a possibility to measure both the x-axis and y-axis movements throughout the align-and-bond process, systematic shift can be followed and analyzed for root cause.

The principle of the vernier is straightforward as shown in Fig. 4.14. A graduated scale composed of 5-μm wide lines with a pitch of perhaps 18μm is used for the level 1 scale. The second scale has the same width of lines but the pitch is increased by whatever amount is desired for the precision of vernier, typically 0.5 or 0.2μm. The scales are placed on both the x-axis and y-axis which then allows for shift and rotational data extraction. Figure 4.15 gives an example with full dimensions indicated.

The method of summed projection has been used in high-resolution IR alignment process flows for the production of aligned fusion-bonded wafers. In this method, overlay of the upper and lower alignment marks is not used. Figure 16 shows the IR images of the light-colored female keys and the dark-colored male keys after alignment. During the alignment process the pixel intensity from the CCD camera-

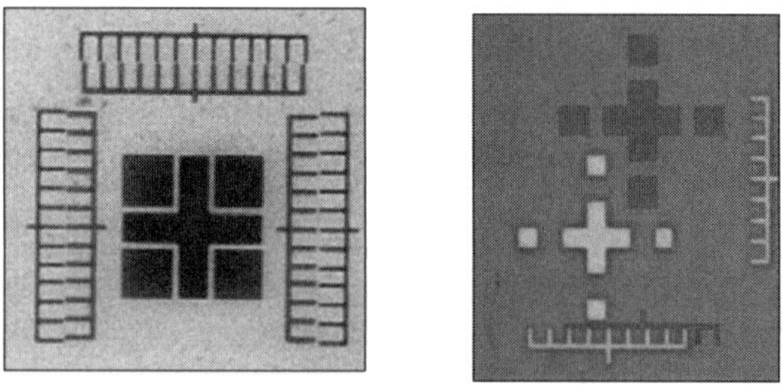

(a) (b)

Fig. 4.13 (**a**) The overlay of upper (cross) and lower box is used by the operator or pattern-recognition system to align the wafers. The verniers on the x- and y-axis provide quantitative analysis. (**b**) A similar process can be done with the overlay of male and female cross but this method only works with stored images

Level 1 vernier is a set of equidistant 5-μm lines with a pitch on 18 μm

Level 2 verniers is a set of equidistant 5-μm lines with a pitch on 18.2 μm if the accuracy is to be 0.2 μm or 18.5 μm if the accuracy is to be 0.5 μm

To determine the amount of misalignment you find the location where the level 1 and level 2 verniers line-up and knowing the interval (0.2 μm) you can determine shift. This example is 1.2 μm shift to the left.

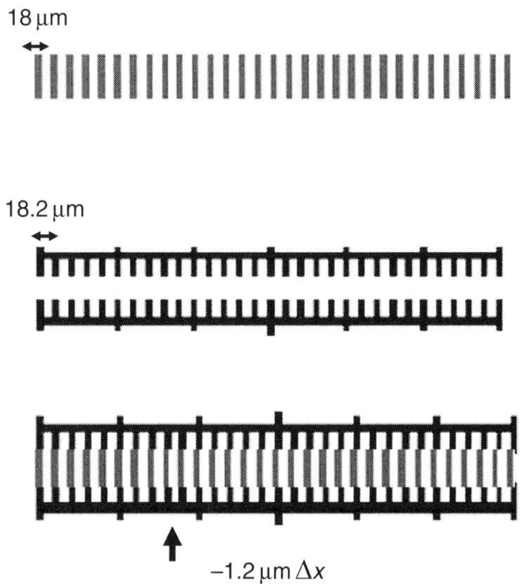

Fig. 4.14 Verniers are constructed in pairs, level 1 verniers on one of the wafers and level 2 on the other wafer. The resolution of the vernier is determined by the scaling of the two graduated scales

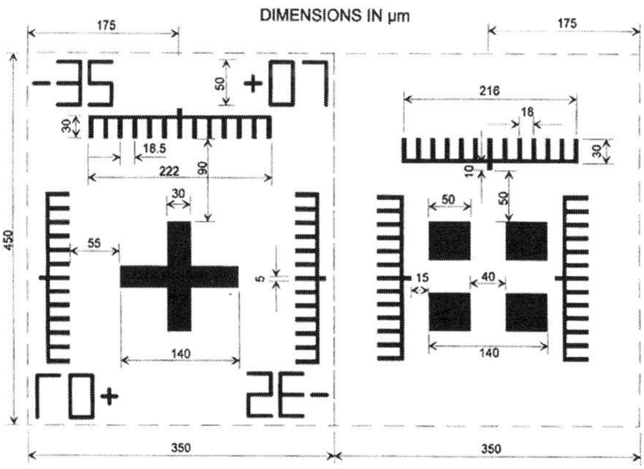

Fig. 4.15 Dimensions for a 0.5-μm per division vernier using cross-in-box keys

captured image for the upper wafer with the open female keys is mapped in the x- and y-axis by summing pixel intensities in a given row or column. This yields the intensity plots shown in Fig. 4.16. The process is repeated for the male key. By averaging the maxima positions of the intensity plots, the overall image center is determined for both the male pair of tic-tac-toe patterns and for the female keys. Due

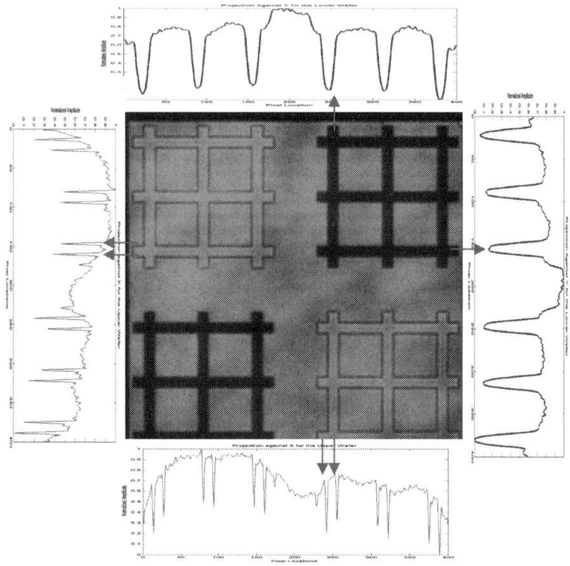

Fig. 4.16 Summed projection uses offset male and female tic-tac-toe grids. The intensity profile in the x- and y-plane is used to mathematically calculate the center of the female keys (upper wafer) and the male keys (lower wafer)

to the presence of multiple lines and good separation distances the calculated center position is precisely known and deep submicron alignment is achievable. These keys are not used for manual alignment since there is no overlapping of features that allow for human differentiation.

4.6 Wafer Transfer Fixtures

Mechanical clamping fixtures, which hold the wafers in alignment on the aligner, are used to transfer the aligned stack to the thermal compression bonder. This is a critical process to overall alignment accuracy, because the transfer fixture must support both the upper and lower wafers during the transport from the aligner to the bonder. Fig. 4.17 shows several types of fixtures offered by leading suppliers of 3D bonding equipment. It is very important that the mechanical clamps hold the wafer in three positions, thereby defining a plane of reference, and not bend the wafer stack about a diametrically opposed set of clamps. The clamp motion must lower the clamp onto the wafer stack exactly perpendicular to the surface and not induce a shear force to the wafer that would automatically shift the substrates out of alignment. In some designs the contact to the wafer stack is made with a roller cylinder so that if there is a minute change in the position height the roller will adsorb the shear momentum rather than transfer those forces to the wafer stack.

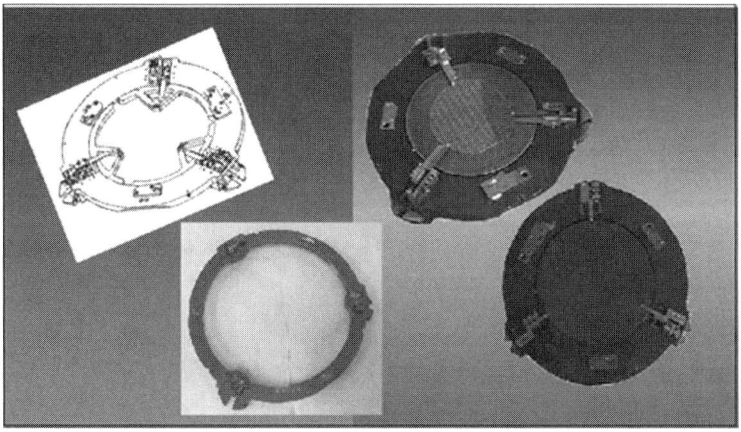

Fig. 4.17 Open and closed fixture design used for wafer-level bonding. Courtesy of SUSS MicroTec

The force adjustments, height adjustments, and motion adjustments are all provided to effectively clamp the wafers into alignment. Using either conformance test wafers from the supplier or IR alignment test pairs the fixture performance can be tested and measured both at the aligner and in the bonder. Equally important are the synchronization of the actuators that mechanically operate the clamping operation in the aligner and the unclamping operation in the bonder. The motion must be smooth and continuous and when all spacers are released simultaneously as is the case in some process flows, the release should be simultaneous.

In addition to clamping mechanisms all fixtures come with the option of using separation flags or spacers between the two substrates. The spacer is inserted after the alignment marks are found on the first substrate and it is vacuum-chucked into position. The spacer, which varies in thickness from 30 μm to 100 μm are made from a variety of materials and are adjusted to penetrate the interface only in the exclusion area of the wafers. After the spacer is inserted the second wafer is aligned and during clamping the second wafer will rest against the spacer leaving a fixed separation distance between upper and lower wafers. The separation distance is used if the interface will need to be out-gassed during the heat-up cycle for bonding as is the case for many adhesive bonds and usually results in higher yields on some benzocyclobutene (BCB) processes. The space may also be needed if a hydrogen-forming gas treatment is used during heat-up to remove oxide layers.

Figure 4.18 shows a close-up view of the clamp and spacer mechanisms presently being used in 3D bonding application using SUSS MicroTec fixtures. The retraction of spacers must be adjusted so that the movement is linear. The spacer should slide out radially from the interface and not drag or scratch the surface.

It has been shown that the fixture and spacers can dramatically affect the final bond alignment. One study recently concluded that a sevenfold reduction in shifting can be accomplished by using what is termed sequential spacer removal [15].

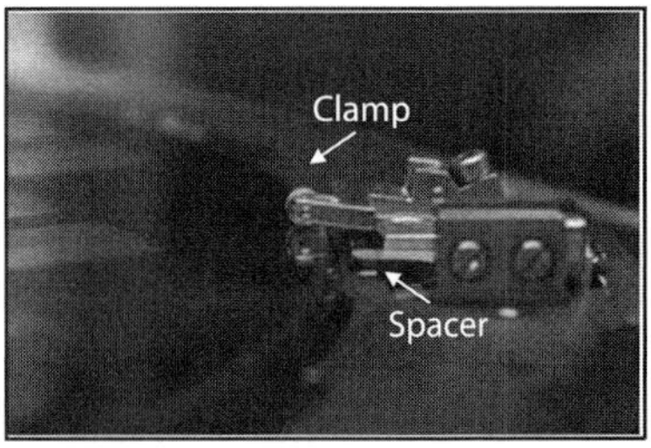

Fig. 4.18 Close-up view of the clamp and spacer mechanism currently used in 3D bonding. In this view the spacer is fully retracted

In many older process flows a center pin was used in the bonder to press the center of the upper wafer against the lower wafer while the clamps were released. The principle was to pin the wafers together in the center, then release the clamps and retract the spacers. Unfortunately center pins only secure the wafer stack in at one location and as the spacers are pulled shifting and rotation is known to occur.

Sequential spacer removal is possible when using three-point clamping fixtures and modified software control of the clamp and spacer movements. While the stack is held down at two clamp locations the third clamp is raised and the spacer is retracted. Then the first clamp is put back in place and the next clamp is raised and the second spacer removed. The process is repeated at the third location and at any given moment the wafer is always held in alignment by two contact points.

Because of the importance the fixture plays in the alignment processes and the ever-increasing alignment specifications, conformance standards as well as setup and testing beds for fixtures are used by equipment vendors [15]. The testing not only guarantees the flag and spacer adjustments are accurate but also establishes that the flatness of the fixture is met and that surfaces are smooth and contamination-free.

Thermal expansion is a considerable source of concern during 3D bonding. Therefore, there is an advantage of manufacturing the fixtures from materials closely matched to the expansion coefficient of silicon. In addition, if the wafers are in contact directly with a thermally matched heater surface that is thermally matched, there is less potential for differential expansion between the fixture and the wafer or the heater and the wafer to induce shear forces. Many experts recommend SiC fixtures over stainless steel for this reason. There have also been suggestions to manufacture fixtures from silicon, but thus far no one has tested this possibility.

Electrostatic chucks, common in deposition chambers and implanters, are being explored as possible alternatives to bond fixtures that embody clamps and spacers. These developmental fixtures need to be qualified for cleanliness and also for compatibility with sensitive device structure but may hold significant opportunities in the future for higher precision alignment.

4.7 Wafer-Bonding Technology

Wafer-bonding technology for vertical integration has explored every type of thermal compression bond and silicon direct bonding technology as the overview shows in Fig. 4.19 [20]. Needless to say the glass frit and anodic bonding are not compatible with CMOS processing due to the contamination from mobile ions in the glasses or from the voltages used in field-assisted diffusion. All other methods are still in serious contention for market share and have been detailed in subsequent chapters of this book. In this section we will focus directly on the bonder-related process flow and not the device needs or challenges associated with 3D wafer-to-wafer bonding.

The wafer bonder is a mechanical system which uses an applied force to press two substrates into direct contact while heating the materials in controlled atmospheres to achieve a desired metallurgical or chemical reaction. The challenge then becomes creating the most uniform thermal and mechanical system possible.

4.7.1 Bond Chamber Design and Components

Figure 4.20 is a schematic of the major components of two commercial wafer-bonding systems. These systems use pressure plates that can be easily removed for maintenance. The pressure plates come in contact with the fixtures or wafers placed inside the bond chamber and are used to press the surfaces together. The heaters are located behind the pressure plates and are embedded with thermal couples at several locations to provide closed-loop feedback to the recipe and controller circuits. The force is applied to the wafer stack either by pneumatic or hydraulic actuation of the upper bond head assembly (EVGroup EVG500 series bonders) or by a full-area air

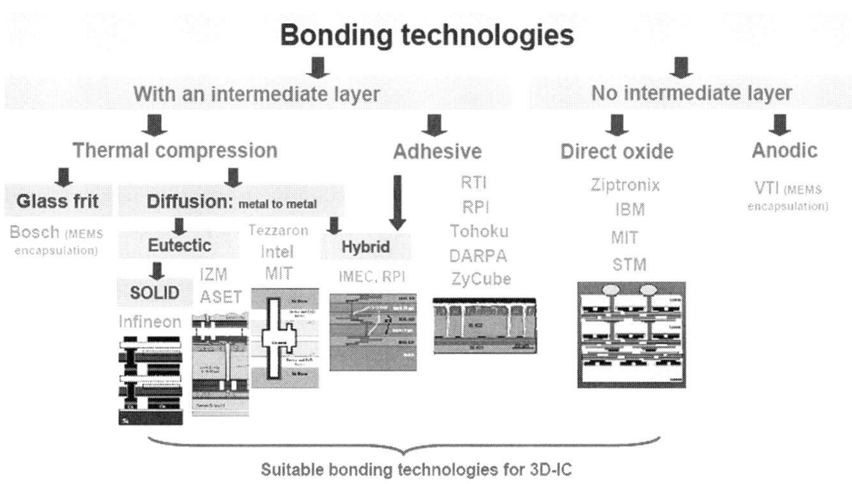

Fig. 4.19 Overview of bonding technologies used in 3D vertical integration. From [20] © 2007 Yole Development

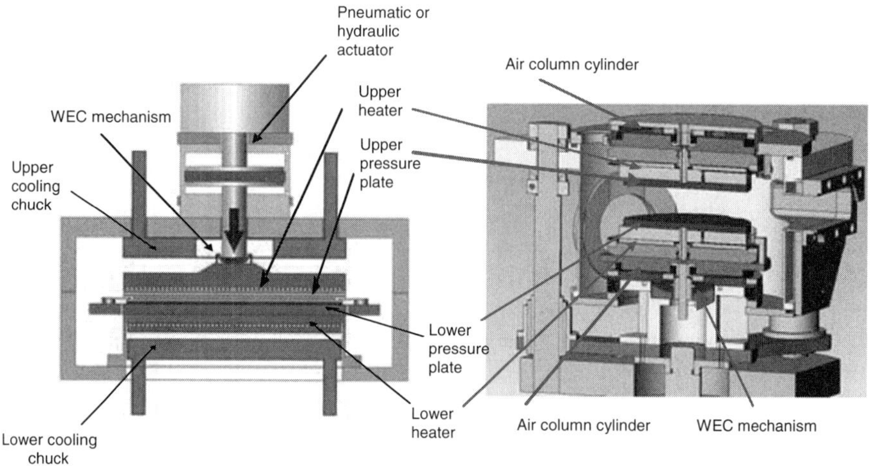

Fig. 4.20 Major mechanical components in a wafer bonder offered by EV Group (*left*) and SUSS MicroTec (*right*). Courtesy of EV Group and SUSS MicroTec

column approach (SUSS MicroTec CB series). The force is calibrated to the incoming air or hydraulic pressure in the cylinders and compared to the recipe demands. The air column approach includes three pressure transducers in the force column that provide direct feedback of the applied force against the pressure plates.

These mechanical components are inside of the bond chamber which is a stainless steel chamber with atmospheric controls. These systems can be differentially pumped by mechanical and turbo molecular vacuum pumps. Base pressures are typical of turbo molecular pumped systems and reach 1e-5 to 1e-6 mbar in less than 10 min depending upon configuration. The chambers can be overpressurized to up to 3 bar by forming gas mixtures (H_2/N_2, H_2/He, H_2/Ar), inert gases such as argon, or standard process gases such as nitrogen and oxygen. Downstream control of the vacuum system is also available on these systems.

The EVG 500 series tools are loaded through top-opening lids. The robot hands off the fixture to a set of lift pins that lower the stack onto the lower pressure plate and heater assembly. In the SUSS CB series of equipment the side-opening VAT door opens and the robot hands off the stack to similar support pins that guide the wafers and fixture onto the lower assembly. Both systems maintain a positive pressure during lid- or door-open operations to assist in keeping the chambers clean.

All process and handling steps are computer controlled through the bond recipe. Once the wafers are inside the chamber the recipe will usually run through a series of pumping and purging steps as needed to remove residual moisture and air. The pressure plates are guided onto the wafer stack and the force applied. The multizone heaters ramp to the process set point and hold the desired temperature from both sides. Control of the within-wafer thermal variations as well as the matching of upper- and lower-heater performance is absolutely critical in 3D applications.

Once the heating cycles are complete, the samples can be cooled from both upper and lower sides with controlled cooling rates. The applied force is released and the wafers removed.

The application of heat and force under controlled environments is the primary step in the wafer bonder. In 3D applications it is critical that these steps should be completed in a way that will minimize additional misalignment provided by the precision bond aligner. These detail steps vary with bonding method and are discussed in the ensuing sections.

Bonder setup requires periodic testing of the vacuum system to ensure that the chamber is not leaking. Metal diffusion processes are very sensitive to oxygen, and thin metal films will rapidly oxidize during the thermal ramp if the system is not vacuum-tight. The force uniformity is checked using pressure-sensitive paper at room temperature. While testing at process temperature is desirable, no metrology methods have been developed that are compatible with temperatures above ambient.

Figure 4.21 shows the results obtained from taking pressuregram images of the force uniformity. The manufacturers of the pressure-sensitive paper also supply software that scans the imprints and correlates the pixel density and intensity to a pseudo quantified result as shown at the far right. This test is excellent for daily qualifying procedures and quickly identifies pressure plates that have distorted due to thermal cycling or have been scratched or damaged.

Temperature uniformity should only require testing periodically and is done using a special commercial wafer called a thermal couple (TC) wafer. The TC wafer has several thin-film TCs on the surface with signal leads brought out through a vacuum port or other chamber openings and connected to the readout electronics. A standard TC wafer will have 13 sensors located at the position shown in Fig. 4.22. Also shown in Fig. 4.22 is a temperature profile taken on a bond recipe that cycled to 320°C. The performance of the bonder must achieve less than 1% deviation to be considered for use in a 3D IC application. This data demonstrates <0.5% deviation including transients.

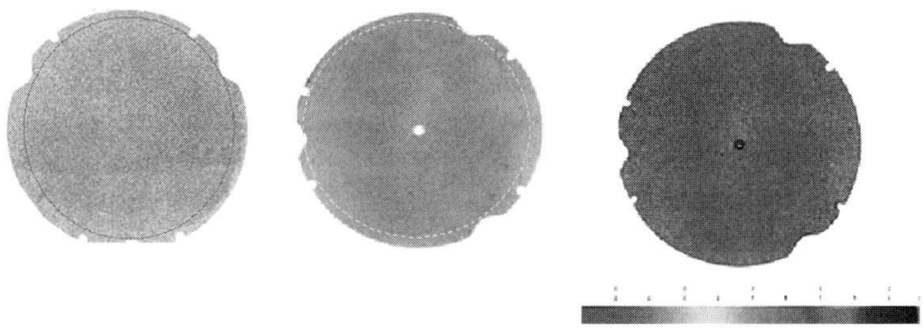

Fig. 4.21 Pressure uniformity data. From left to right, pressuregrams from bonder without center pin, pressuregram with center pin, and quantized data from pressuregram with center pin. Wafer contact area is indicated by highlighted dashed circles

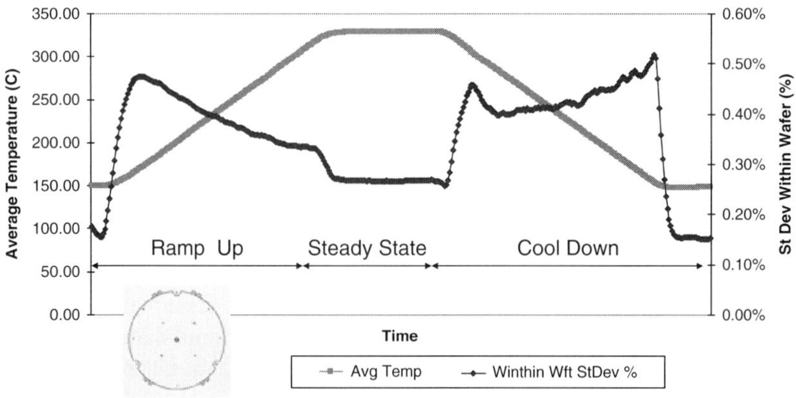

Fig. 4.22 Average temperature within wafer and the standard deviation based on 13 point measurements taken with commercial thermal couple wafer placed between wafers in the fixture to mimic bond interface conditions

4.8 Wafer-Bonding Techniques for 3D

Many of the wafer-bonding techniques for 3D integration are covered in detail in other chapters. This section presents the author's perspective from the equipment sector and involvement with alternative technology platforms.

4.8.1 Silicon Direct Bonding

The thermal expansion coefficient of silicon is 3.2 ppm/°C, which translates to about 50 μm of radial growth for every 100°C increase in temperature on a 300-mm wafer. To ensure that thermal misalignment does not affect the final result (1) both the top and bottom wafers must be thermally matched substrates with equal coefficients of thermal expansion (CTE), (2) the wafers must be heated at equal rates to the same set point, and (3) there can be no radial variations in the temperature uniformity for either substrate. Under these conditions expansion will be isotropic and the wafers will remain aligned.

Therefore, bonding methods that complete at lower temperature offer advantages compared to higher-temperature processes, assuming adequate bond strength is reached during the bonding step or during subsequent processing steps. The silicon direct or fusion bonding method is the lowest temperature possibility. As described in the section on surface treatments, either plasma activation or wet chemicals can be used to hydrate the wafer surfaces. When this technique is used, bonding in the aligner itself is usually possible, using IR or face-to-face alignment methods. After the alignment is completed, the wafers are not clamped into the transfer fixture. Instead a gentle force, ~1 N, is used to create a point of contact and the bond wave is initiated and prebonds the wafers with the van der Waals attractive forces pulling the surfaces together and expelling the air. Once the wafers are prebonded, the wafers are actually annealed in a batch process and the bonder is not needed.

In this process flow several advantages are immediately recognized: no fixture-induced shifting from clamps, no differential thermal expansion since the room temperature bonds establish the overlay accuracy, and batch processing for the anneal to improve throughput dramatically. Overall equipment costs are potentially reduced because the throughput of the aligner is about 4–10 times faster than a thermal compression bond process.

The drawback of the silicon direct bonding method is that the surface must be flat and atomically smooth. It is generally accepted that the surface roughness must be approximately 1 nm or less, the total indicated run out (TIR), which is a measure of the surface waviness, must be $<2\ \mu m$ and the wafer bow needs to be $<60\ \mu m$ [21]. The process is sensitive to particles which will result in voids one to two orders of magnitude larger than the particle itself. Surface adsorption of hydrocarbons will lead to post-bond anneal voiding, so that silicon direct bonding should never be located physically near to lithography processes or other areas that pose a hydrocarbon or aerosol risk to the surfaces.

Table 4.1 shows a production run of 24 IR-aligned silicon direct bonded wafers. This product used the summed projection alignment keys described earlier and the prebonded wafers are annealed in batch process ovens. Similar alignment results have been obtained in other 3D applications using silicon direct bonding methods [1].

4.8.2 BCB Bonding

The next lowest thermal processes are those using adhesives of which the most work has been focused on benzocyclobutene (BCB) bonding [22, 23]. BCB is a polymer developed by Dow Chemical and comes in a photosensitive formulation as well as a nonphotosensitive product. Most often in 3D IC stacking the wafers have been planarized and the BCB is applied as a blanket layer. The BCB needs to be precured to drive off solvents and also to strengthen the polymer. This prevents the distortion of the polymer layer during compression and improves the overall alignment results. A typical precure is done at 190°C for 1.5–3 min and is done in the coating step upstream to the bonding processes. The bonding can be done at atmosphere or under vacuum; however, vacuum processing eliminates trapped air and if there are residual solvents released during the thermal ramp to bond temperature the gases will be pumped out. The bonding is usually done at 320–220°C or higher for less than 30 min. An early example recipe is shown in Fig. 4.23. Several variations on the basic recipe can be employed to tailor the bonding process to the application and throughput requirements.

For example time and temperature are variables that can be exchanged. Higher temperatures will translate into shorter anneal times as the cross-linking occurs more quickly. The spacers can be left in between the wafers until the temperature rises over 150°C and surface moisture is fully desorbed. Benzocyclobutene bonding is

Table 4.1 Post-bond alignment accuracy using IR alignment and direct bonding in the aligner. No fixture or bonder is required

Wafer #	dXI in Pixels	dYI in Pixels	dXr in Pixels	dYr in Pixels	Accuracy in Pixels	dX in Pixels	dY in Pixels	dR in Pixels	dT in Pixels	Alignment in μm
1	0.5117	−0.0939	−0.3483	−0.1078	0.1353	0.0817	−0.1008	0.4300	0.0070	0.1701
2	0.3503	−0.0546	−0.3620	−0.0582	0.0585	−0.0058	−0.0564	0.3562	0.0018	0.0736
3	−0.6013	0.1309	0.0515	−0.0471	0.1310	−0.0049	0.0419	−0.0564	0.0890	0.1648
4	0.1474	0.0613	0.0544	−0.0533	0.0769	−0.0465	0.0040	−0.1009	0.0573	0.0967
5	0.0200	0.0034	−0.0268	−0.1261	0.1261	−0.0034	−0.0614	0.0234	0.0647	0.1586
6	−0.0007	−0.0032	0.0450	−0.0172	0.0280	0.0222	−0.0102	−0.0228	0.0070	0.0353
7	−0.0073	−0.0827	−0.0666	−0.0332	0.0906	−0.0370	−0.0579	0.0296	−0.0248	0.1139
8	−0.0061	0.0573	0.0253	−0.0501	0.0581	0.0096	0.0036	−0.0157	0.0537	0.0730
9	−0.0495	0.0091	0.0120	−0.0724	0.0748	−0.0188	−0.0317	−0.0308	0.0407	0.0941
10	−0.0154	0.0713	0.0252	0.0230	0.0715	0.0049	0.0472	−0.0203	0.0241	0.0899
11	0.0746	−0.0556	−0.1892	−0.0825	0.1005	−0.0573	−0.0691	0.1319	0.0135	0.1264
12	0.0804	0.0395	−0.1468	−0.0057	0.0516	−0.0332	0.0169	0.1136	0.0226	0.0649
13	0.5979	−0.0369	−0.6194	−0.0348	0.0385	−0.0107	−0.0359	0.6086	−0.0011	0.0484
14	0.4887	−0.0782	−0.4537	−0.0586	0.0801	0.0175	−0.0684	0.4712	−0.0098	0.1008
15	0.4940	−0.0965	−0.4090	−0.0892	0.1054	0.0425	−0.0928	0.4515	−0.0036	0.1326
16	0.3946	−0.1509	−0.3916	−0.1429	0.1509	0.0015	−0.1469	0.3931	−0.0040	0.1898
17	0.4385	0.0019	−0.4179	−0.0622	0.0630	0.0103	−0.0302	0.4282	0.0320	0.0793
18	0.3970	−0.0788	−0.4291	−0.0936	0.0950	−0.0161	−0.0862	0.4130	0.0074	0.1195
19	0.1094	−0.0827	−0.1604	−0.0701	0.0866	−0.0255	−0.0764	0.1349	−0.0063	0.1088
20	0.0958	0.0075	−0.1854	−0.0363	0.0577	−0.0448	−0.0144	0.1406	0.0219	0.0725
21	0.0960	−0.0946	−0.1713	−0.0672	0.1018	−0.0377	−0.0809	0.1336	−0.0137	0.1280
22	0.0190	0.0358	−0.0258	−0.0756	0.0756	−0.0034	−0.0199	0.0224	0.0557	0.0951
23	0.0383	0.0758	−0.0206	−0.0997	0.1001	0.0089	−0.0878	0.0295	0.0119	0.1258
24	0.0852	−0.0894	−0.0366	−0.0627	0.0926	0.0243	−0.0760	0.0609	−0.0134	0.1165

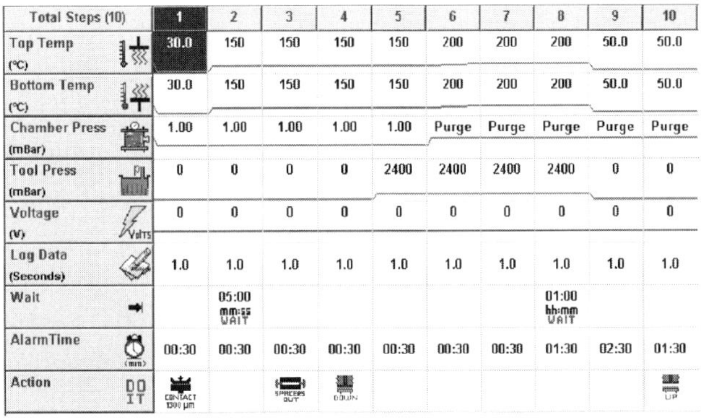

Total Steps (10)	1	2	3	4	5	6	7	8	9	10
Top Temp (°C)	30.0	150	150	150	150	200	200	200	50.0	50.0
Bottom Temp (°C)	30.0	150	150	150	150	200	200	200	50.0	50.0
Chamber Press (mBar)	1.00	1.00	1.00	1.00	1.00	Purge	Purge	Purge	Purge	Purge
Tool Press (mBar)	0	0	0	0	2400	2400	2400	2400	0	0
Voltage (V)	0	0	0	0	0	0	0	0	0	0
Log Data (Seconds)	1.0	1.0	1.0	1.0	1.0	1.0	1.0	1.0	1.0	1.0
Wait		05:00 mm:ss WAIT						01:00 hh:mm WAIT		
AlarmTime (min)	00:30	00:30	00:30	00:30	00:30	00:30	00:30	01:30	02:30	01:30
Action DO IT	CONTACT 1300 µm		SPACERS OUT	DOWN						UP

Fig. 4.23 Early example of BCB bonding recipe

known to be sensitive to moisture on the surface and if wafers are coated and put in a queue with a long wait period then moisture on the surface should be removed before bonding. In addition, it is always wise to bring the force down gently and in a controlled ramp to prevent abrupt contact with the surface and introduction of shear movements, although recent results with partially cured BCB (i.e., partially cured prior to bonding) reduce shear movements significantly [24].

Table 4.2 lists other adhesives that are used in 3D bonding. Many are more popular with image sensor bonding to CMOS controller wafers than for the IC-to-IC stacking. Nearly all of the available choices are produced in photosensitive or non-photosensitive formulations. The lower-modulus materials and the thermal plastic products are useful when the wafer shape is less than perfect and some viscous flow is desired to achieve a conformal interface. Figure 4.24 is the photograph of a CMOS wafer bonding to transparent wafer using the HD3003 polymer in preparation for thinning and bonding to the next wafer in a back-to-front integration scheme.

4.8.3 Coppert-to-Copper Diffusion Bonding

Permanent bonding with Cu-to-Cu diffusion bonds has benefited greatly from the CMOS industry development and infrastructure for Cu damascene processing. Without the process know-how for deposited dielectric passivation and CMP planarization of Cu embedded in the various dielectric layers, it would not be possible to achieve surfaces with the morphological conditions conducive to wafer-level bonding. In order for diffusion reactions to proceed, the Cu surfaces must be in intimate contact.

Atomic motion (diffusion) proceeds from one crystal lattice location to another and cannot "leap" free space. Hence, high applied forces in the bonder can be used to press out bow and warp and plastically deform to some degree the surface asperities in the Cu films. However, fundamentally the CMP processing must create flat films with a microroughness on the order of 2–5 nm RMS.

Table 4.2 Comparative listing of materials properties used for permanent adhesive bonding

Company	Dow	Toray	Sumitomo	Sumitomo	Dow corning	HD-micro	HD-micro
Trade Name	Cyclotene	PWDC-1000	CRC-8000	CRX 2580P	WL-5000	HD-2771	HD-3003XP
Material	BCB	PI	PBO		Silicone	PI	PI
PhotoPatternable	Both Negative	Positive	Positive	Yes	Yes Negative	Yes Negative	Both
Residual Stress (MPa)	28	28	60		<6.4		
Moisture Uptake (%)	0.23	0.6	0.3–0.9	0.06	~0.2	>1.0	0.08%
Coefficient of Thermal Expansion (ppm/°C)	52	36	51		<236	42	124
Glass Transition Temperature (°C)	>350			295		294	188
Cure Temperature (°C)	210–250	250+	320	200	<250	>350	220
Dielectric Constant	2.65	2.9		2.65	<3.3	3	3.4
Modulus (GPa)	2.9	2.9	2.9	1.6	1.15–0.335	2.7	2.4
Thermal Stability (%loss at 350C/1 hr)	2		<1		<6	<1	<1
Shrinkage During Cure(%)					<2	40–50	<0.04%
Minimum Thickness (μm)	1	3	3		2	4	1
Storage Temperature (°C)	−15	4			−15	r.t. or −18	
Shelf Life (mos.)	6	6			6	12 @ −18°C	

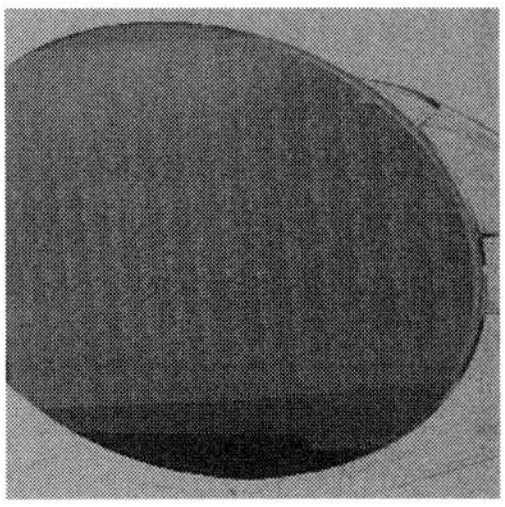

Fig. 4.24 CMOS wafer bonded to glass in preparation for a back-to-front integration process after thinning of the backside of the CMOS wafer. Permanent adhesive material was HD3003

Copper-to-copper diffusion bonding recipes run at 400–450°C and will stay at this temperature for 20–40 min. The forces required to maintain intimate contact are 20–80 kN depending upon wafer diameter and the density of the Cu features. The best results are obtained by annealing in a forming gas ambient. Either H_2/N_2, H_2/Ar, or H_2/He at a 4% hydrogen mixture is used to prevent oxidation during thermal cycling. The partial pressure atmosphere also aids in thermal heat conduction and helps to prevent thermal gradients vertically across the stack.

Special care must be taken to ensure the ramp to temperature of both heaters is matched and center-to-edge uniform so that both wafers expand synchronously. In many cases it has been discovered that after the formic acid vapor cleaning the use of spacers between the wafers is not necessary. The wafers are then in direct contact and heat transfer between the wafers at the interface is improved resulting in better post-bond alignment accuracy.

Another very important variable in Cu–Cu bonding is the applied force and force uniformity. Up to 60 kN force can be required for a 300-mm wafer to ensure intimate contact across the entire wafer surface. However, if there is as much as a 30% variation in the force uniformity the total force required maybe 90 kN in order to reach the minimum requirements at all locations. Therefore, a high degree of force uniformity can not only diminish the total force needed but can alleviate stress gradients and sensitivity to fragility in thinned devices.

There have been several efforts to reduce the temperatures needed for Cu diffusion bonding which would minimize the thermal misalignment potential. Although diffusion will occur at lower temperatures the time to completion of the reaction is longer. Therefore, surface modification methods have been explored.

One method that holds some promise is described by Dr. Suga and his team of researchers in Japan as well as Dr. Schmidt at MIT [2, 3, 25]. This process uses plasma surface pretreatments to create very clean surfaces. The microroughening of the Cu surface may also provide accelerated diffusion rate due to increased surface-to-surface contact. It is well know that surface diffusion is the most rapid mobility path for atomic diffusion (followed by grain boundary diffusion and lastly bulk diffusion).

There have also been studies conducted using nanorod Cu layers although it is not clear that these films are readily integrated into mainstream high-volume manufacturing lines [26, 27].

4.8.4 Eutectic Bonding

Eutectic bonds are a special class of metal bonding in which the metal melts when it reaches the eutectic temperature. The alloy composition corresponds to a triple point on the phase diagram in which the solid-to-liquid transition is spontaneous and does not involve the coexistence of both liquid and solid phases simultaneously. The most common eutectic used in wafer bonding is the Au–Sn system and most 3D demonstrations use this alloy. The eutectic temperature for this 80/20 alloy is 280°C. Figure 4.25 illustrates the reaction pathway taken in two embodiments of Au–Sn bonding [15, 28].

Path 1 is the direct approach in which the bonding metal is deposited as an alloy in one layer by electroplating, composite alloy sputtering, or dual-source evaporation. The wafers are aligned and put in the bonder and heated. The entire alloy layer melts and resolidifies precisely at 280°C. Bonding recipes will use a bond

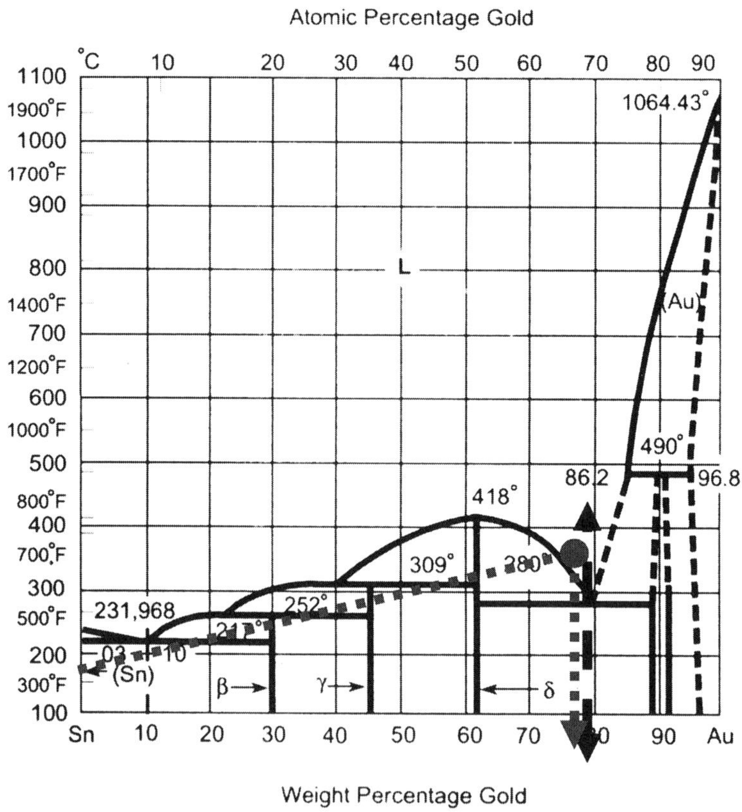

Fig. 4.25 The Au–Sn phase diagram. Processing path 1 (dashed) is a direct melting of the eutectic alloy while path 2 (dotted) is the diffusion path taken through several two-phase regions until the melt reaches the eutectic composition and is solidified [31]. Reprinted with permission of ASM International®. http://www.asminternational.org

temperature of 290°C so that the melt becomes less viscous and readily flows to areas of higher roughness that might not be in physical contact. A low bond force is used to prevent pushing of the melt into unwanted areas [29].

Alternatively pure gold and pure tin layers can be deposited on opposite wafers. When the wafers are brought together the tin will melt and the gold will diffuse into the liquid layer. This will move the interface composition along the pathway (shown in Fig. 4.25 as blue). As the diffusion proceeds the temperature is gradually raised so that each successive two-phase liquid and solid region is observed. The liquid phase increases the diffusion reaction but is not essential. Eventually the melt reaches the eutectic composition and the bonder temperature is dropped to resolidify the interface.

While gold is not a front-end friendly material and rarely welcome in IC fabrication foundries, the Cu–Sn eutectic is gaining in acceptance. The phase diagram for Cu–Sn is shown in Fig. 4.26. The eutectic point at 99.3% Sn is barely visible, but the

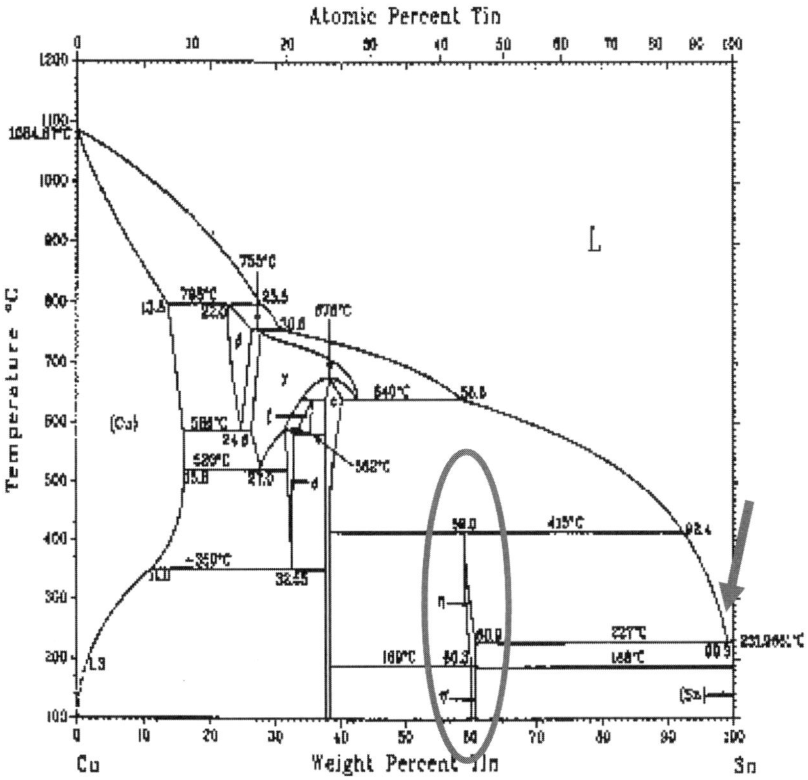

Fig. 4.26 The Cu–Sn phase diagram. The arrow identifies the eutectic reaction at 99.3%. The extended two-phase region extending to the right of the eutectic point is used to continue the diffusion of Cu into the melt until the intermetallic phases highlighted in the center of the phase diagram are formed [31]. Reprinted with permission of ASM International®. All rights reserved. http://www.asminternational.org

important feature of this reaction is that the two-phase liquid-plus-solid equilibrium region extends from 99.3 wt.% Sn to the intermetallic phase boundary at 60.9 wt.%. In practice the eutectic phase is melted and facilitates rapid diffusion of additional Cu into the melt until the intermetallics form.

Eutectic bonds hold potential to reduce the overall temperatures compared to Cu methods and compete with BCB methods because of less sensitivity to surface topography and roughness. In addition the eutectic alloys provide for electrical conduction and interfacing with redistribution layers. The most important requirement for eutectic bonds is the surface preparation to remove oxide layers and extremely good temperature and pressure uniformity.

4.9 Bond Quality Testing

The bond quality is the verification of the alignment accuracy, the bond strength, and the lack of interface voids. Quality assurance should begin in the production cluster with inline metrology so that as many problems can be identified as early as possible before any waste of resources and materials occurs. Inline metrology is a software flexibility and data-logging system on the cluster bonding tool. Figure 4.27 illustrates how 3D cluster tools attempt to provide safety nets for misprocessing both at the alignment stage and at the bonding stage. The aligner recipe will have tolerances ranges for the limits on acceptable misalignment. If the image quality is poor or the wafer is flawed and cannot be aligned to specifications, the wafers are returned to the FOUPs and the next pair is selected. The engineer can attempt to bond the wafers later with relaxed tolerances or determine if only one wafer is bad and use the good wafer in the next lot.

Similarly if there is a problem with the bonder, it is not wise to continue to process 50 pairs without checking the results. Therefore, the second fail-safe loops allow the bonded wafers to be inspected on the aligner or metrology station and

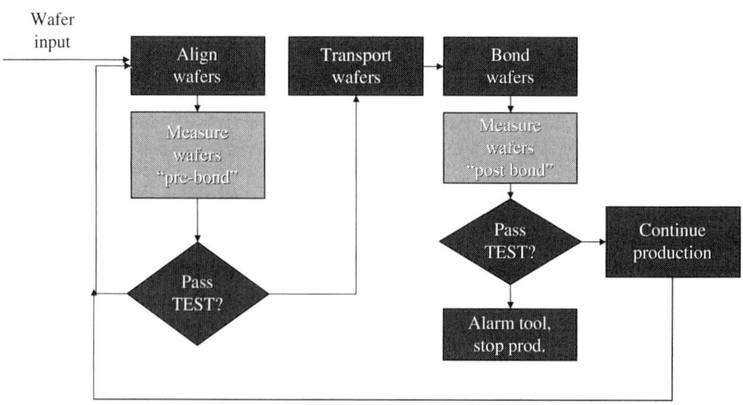

Fig. 4.27 Inline metrology process flow in which one checkpoint is in the aligner loop and the other node is at between bonded pairs coming out of the annealing step

if the alignment does not conform to recipe tolerances the bonding stops until the problem is resolved. The recipe determines the frequency of the metrology checks and the acceptance range. Lot, FOUP, and wafer ID logging are also provided to support statistical process control in the factory.

Alignment accuracy can be verified using the IR alignment microscope or the BSA alignment microscopes if one of the wafers is transparent. Otherwise the alignment accuracy cannot be verified until wafers are thinned to transparent thicknesses or destructively verified by cleaving the interface and examining the interface in cross-section.

Bond strength testing is discussed extensively by Tong and Gösele in [23]. The methods are destructive and involve pull-testing or crack-opening measurements. All the bond technologies described in this chapter using the methods described will reach full strength in the described processing time. This means the wafers will have achieved an interfacial energy of 1.5 J m^{-2} or more and will successfully survive back-grinding and other backside processing steps.

Void imaging can be done with IR whenever the materials are IR transparent and a rapid qualitative examination is desired. This simple low-cost method will identify any void that is laterally resolved by the CCD imaging camera ($\sim$1 mm or larger) and must have a void opening (height) of 1 μm since both a light and dark Fresnel fringe are required to see the void. If this level of resolution is not adequate for the needs of the process control then acoustic microscopy is the next best option.

Acoustic microcopy uses high-frequency sound waves to image bonded interfaces. Acoustic images can resolve voids with 4-μm lateral resolution and 10-μm vertical interface gaps. The sample must be immersed in a liquid (usually DI water) or have a liquid flow between the acoustic transducer and the wafer to couple the sound waves into and out of the sample. When the interface is not continuous, capillary forces will pull the water into the interface starting at the wafer edges and penetrate between each die. This will yield very unusual images that are not easily interpreted. The water will also prevent additional post-processing since the moisture at the interface is difficult to expel and leads to detrimental problems later.

Other methods of void identification include X-ray tomography and destructive analysis using cleaved wafers and interface etching to decorate the voids.

4.10 Summary

Wafer-level 3D bonding holds great potential for lowering the cost of vertically integrated devices. All of the alignment methods and bonding techniques described in this chapter have achieved excellent results as reported in numerous conference proceedings. Table 4.3 summarizes the methods discussed in this chapter. The best-case alignment results are from conferences and papers references within this book and have been achieved with commercial bonders and aligners. The alignment methods that result in the very best accuracy are the IR and face-to-face alignment methods. Naturally, fixtureless silicon direct bonds completed at the alignment stage have given the best results as demonstrated also in Table 4.2.

Table 4.3 Overview of the bonding methods, processing parameters, and alignment results

Method	Temperature	Applied force (8″ & 12″ Wafers)	<2μm Alignment capable*	Demonstrated best case
LT Silicon Direct Bonding	<350C	DNA	Yes	0.1μm
Plasma Bonding	RT–300C	DNA	Yes	0.2–0.5μm
Metal Diffusion Bonding	400–450C	20-60 KN	Yes	0.35–0.7μm
Metal Eutectic Bonding	250–325C	1–5 KN	Maybe	1μm
BCB & Polymer Bonding	200–250C	<3 KN	Yes	0.5–1μm

*Alignment capability depends on substrates, alignments marks, alignment method and

The joint efforts between equipment suppliers, device manufacturers, materials suppliers, and research institutions throughout the world are pushing to establish wafer-to-wafer bonding methods as true 3D fabrication methods for volume manufacturing. Although many challenges remain, the goals of the roadmap appear within reach.

References

1. Enquist P (2006) Direct Bond Interconnect (DBI™) – technology for scalable 3D SoCs. In: RTI conference proceedings on semiconductor integration and packaging, Burlingame, California.
2. Suga T (2006) Cu–Cu room temperature bonding – current status of surface activated bonding (SAB). ECS Trans 6:155–163.
3. Takagi H, Takagi H, Maeda R, Ando Y, Suga T (1997) Room temperature silicon wafer direct bonding in vacuum by atomic beam irradiation. In: Technical Digest of MEMS Workshop, p 191.
4. Stengl R, Ahn KY, Gösele U (1989). Jpn J Appl Phys 27:2364.
5. www.suss.com (see for example CL8, ABC200, and CBC models).
6. www.evgroup.com (see for example EVG301, EVG Gemini).
7. Gabriel M et al (2005) Capabilities of an ambient pressure plasma for activation in LT wafer bonding process. In: Hobart KD, Hunt CE, Baumgart H, Suga T, Bengtsson S (eds) Semiconductor wafer bonding VIII: science, technology and applications, 2005, ECS PV2005-02, p 50.
8. Farrens S (1995) Chemical free room temperature wafer to wafer bonding. J Electrochem Soc 142:3949.
9. Farrens S, Dekker J (1995) Van der Waals model of wafer bonded substrates. In: ECS proceedings, PV95-7, p 184.
10. Swalin RA (1961) Thermodynamics of Solids. John Wiley and Sons, New York, p 116.
11. Darken LS, Gurry RW (1953) Physical chemistry of metals. McGraw Hill, New York.
12. Laine E et al (2007) C4NP – reliability and yield data for lead free wafer bumping. In: IMAPS device packaging conference, Scottsdale, March 17.
13. Ruhmer K et al. (2006) C4NP: lead-free and low cost solder bumping technology for flip chip and WLCSP. In: IWLPC pan pacific conference, San Jose, CA.
14. Matthias T (2007) 3D integration schemes using temporary and permanent wafer bonding. In: 3D architectures for semiconductor integration and packaging, Oct 22–24, 2007, Burlingame, CA.
15. Farrens S (2007) Processing solutions for reproducible submicron 3-D integration. In: 3D Aarchitectures for Ssemiconductor Iintegration and Ppackaging, Oct 22-–24, 2007, Burlingame, CA.

16. Kittel C (1995) Introduction to Solid State Physics, 7th edn. Wiley, New York.
17. Lindner P et al (2003) Advances in aligned wafer bonding for 3D interconnects and wafer level packaging. In: SEMICON Europa, advanced packaging conference, Munich Germany.
18. Brubaker C et al (2001) Novel alignment technologies for wafer level packaging. In: IMAPS 2001 – the 34th international symposium on microelectronics, Oct 9–11, Baltimore, MD.
19. Matthias T (2004) Advances in aligned wafer bonding for 3D interconnect technologies. In: CREMSI micropackaging days, ST University, Fuveae France, Sept 15–16.
20. Yole Development (2007) 3D IC report. Lyon, France, p 79. www.yole.fr
21. Tong QY, Gösele U (1999) Semiconductor Wafer Bonding. Wiley, New York.
22. Gutmann RJ, McMahon J, Lu J (2006) Damascene patterned metal/adhesive redistribution layers. In: Bower CA, Garrou PE, Ramm P, Takahashi K (eds) MRS proceedings, vol 970.
23. Kwon Y et al (2001) Wafer bonding using low-K dielectrics as bonding glue in three-dimensional integrations. In: ECS Joint Meeting, San Francisco, CA, Sept 2–7, 2001 and Chip and WLCSP Pan Pacific Conference Oct. 2006, San Jose, CA.
24. Kumar RJ, Niklaus F, Yu J, McMahon JJ, Lu JQ, Persans PD, Gutmann RJ (2005) Wafer bonding with partially-cured benzocyclobutene (BCB) for 3D integration applications. In: Proceeding of the EMC 863, Santa Barbara, USA.
25. Schmidt M (1998) Wafer-to-wafer bonding for microstructure formation. Proc IEEE 86(8):1575.
26. Wang P, Karabacak T, Yu J, Li H, Pethuraja GG, Lu J, Lu T (2006) Low temperature copper – nanorod bonding for 3D integration. In: MRS fall meeting, Boston, MA.
27. Karabacak T et al (2006) Low temperature melting of copper nanorod arrays. J Appl Phys 99:064304–1.
28. Farrens S (2007) High performance bonding simplifies high tech 3D options. In: IWLPC conference, Sept 17–19, San Jose, CA.
29. Farrens SN (1992) Mechanical testing of bonded silicon on insulator wafers. Mat Res Soc Proc 239B:287.
30. Farrens S (2007) Stop the Squish! How to control eutectic flow in hermetically sealed devices. Chip Scale Rev:44.
31. Massalski T (1990), Binary alloy phase diagram, 2 nd edn. ASM International, Materials Park, OH. Sylvester D, Hu C (2001) Analytical modeling and characterization of deep-submicrometer interconnect. Proc IEEE 89(5):634–664

Chapter 5
Through-Silicon Via Fabrication, Backgrind, and Handle Wafer Technologies

Sharath Hosali, Greg Smith, Larry Smith, Susan Vitkavage, and Sitaram Arkalgud

5.1 Introduction

The important method of bonding wafers to wafers or die to wafers has been discussed in an earlier chapter. In this chapter, we will examine the formation and filling of through-silicon vias (TSVs) and the post-bond process of thinning wafer-to-wafer pairs to further process TSVs and build metallization on the final exposed surface. In the section on wafer thinning, the hydrogen-induced splitting of thin silicon layers developed for silicon-on-insulator (SOI) will also be described. In a sense, this is a deviation from the basic processes to stack wafers or dies, but it describes a method to create a type of 3D, which was first envisioned in the 1980s, namely "stacked complementary metal-oxide semiconductor (CMOS)". Finally, we will comment on processing with handle wafers and mention some of the current methods and areas of research to make this a more cost-effective way of transferring device layers.

5.2 Through-Silicon Vias Compared with Wirebonds

Through-silicon vias are an area array technology. Compared to wirebonds, they have many of the same advantages and disadvantages of flip chip technologies because they connect directly from the "surface" of one die to another. Wirebonds have to cross the "edge" of a chip; this can limit the number of connections and increases their length. Wirebond pads are typically placed at pitches of 30–50 μm or greater, and typically in no more than 2–3 rows. Through-silicon via pitches for wafer-to-wafer applications are projected to be 10 times smaller and can be implemented in a full-area array. Thus, the potential number of TSVs on a given die size is many orders of magnitude greater than the potential number of wirebonds.

Wirebonds, however, have the advantage of programmability and flexibility. Die revisions and die shrinks can be easily accommodated. For applications using

S. Hosali
SEMATECH, Austin, TX, USA
e-mail: Sharath.Hosali@SEMATECH.Org

C.S. Tan et al. (eds.), *Wafer-Level 3D ICs Process Technology*,
DOI: 10.1007/978-0-387-76534-1_5, © Springer Science+Business Media, LLC 2008

standard memory parts, wirebonding can accommodate die from different suppliers that have similar but not identical IOs. Using TSVs, in contrast, the two die must line up exactly. Reusing an existing die may require creating a new design by modifying one or more layers. Wirebonds also offer an additional degree of flexibility, in that they can connect a chip to either another chip in a stack or to the substrate, whereas TSVs connect directly only from one chip to the next.

Through-silicon vias offer better electrical characteristics than wirebonds. Typical TSV depths are ~10–100 μm, compared with typical wirebond lengths of ~1–5 mm. Since wire resistance, capacitance, and inductance all scale with length, TSVs generally have smaller values for each of these. For example, interconnect inductance on power leads is important because of switching noise, which scales as $L\times \mathrm{d}I/\mathrm{d}t$, where L is the effective inductance of the power/ground leads and $\mathrm{d}I/\mathrm{d}t$ is the current slew rate. Wire inductances are proportional to length but depend weakly (logarithmically) on the wire diameter and pitch. Typical values are ~0.5–1 nH mm^{-1}. Through-silicon vias thus offer 100 times smaller inductances per connection, as well as more connections. This may substantially improve switching noise, depending on the overall design of the power network,

Through-silicon via capacitance depends on the k value of the silicon (~11) and the liner (~4) as well as the liner thickness, via diameter, and pitch. Typical values are ~0.0005 pF $\mu\mathrm{m}^{-1}$. For a TSV ~20-μm long, capacitance would then be ~0.01 pF. For wirebonds, the wire-to-wire capacitance is typically ~0.05 pF mm^{-1}, with the capacitance of a 2-mm long wire ~0.1 pF, which is approximately 10 times larger than the TSV. Also, a pad capacitance must be added to this.

Through-silicon via resistance is proportional to resistivity. Typical resistivities of commonly used metallurgies are as follows: gold, 2.5 μΩ-cm; copper, 1.8 μΩ-cm; aluminum, 2.7 μΩ-cm; tungsten, 5.6 μΩ-cm; and polysilicon, ~190 μΩ-cm. For copper and tungsten TSVs, resistance is not usually a serious issue. However, for polysilicon, a TSV 3 μm in diameter and 50 μm in length would have a resistance ~10 Ω. While this is acceptable for some memory applications, it may be too high for other applications. Bond wires are typically gold or aluminum.

5.3 Through-Silicon Via Hole Formation

When stacking die to improve packaging densities, each die can be separately wirebonded to the substrate from bond pads in the periphery, if the die-stack exposes the pads on all the dies. This kind of approach works well for memory, which has a small number of pads per unit area of circuitry. Realizing the performance advantage of short wiring distances, however, requires building high-density vertical interconnection between the dies using TSVs. Early implementations of TSVs for 3D had a via density of a few hundred per die in a die-to-wafer integration. With a via density this low, the vias tend to be large (a few tens of microns across) and deep (a hundred microns or more). This can be scaled with reductions in the thickness requirement for die handling.

Higher-density vias can be realized with wafer-to-wafer integration. Here the donor wafer is thinned to 10 μm or less and aligned to within a micron or so. Vias 1 μm in diameter can be used to form up to hundreds of thousands of vertical interconnections per die.

5.3.1 TSVs for Die-to-Wafer Stacking: Laser Drilling

Die-to-wafer stacking requires pick-and-place technology and fast die-by-die alignment and bonding. Die tend to be at least 70–100-μm thick, which means that the TSVs need to be this deep. This limits the TSV diameter to more than about 10 μm because of limitations in the aspect ratio of the vias to fill them with metal. If the number of vias per die is sufficiently small, laser drilling can be used to fabricate the holes. Laser drilling provides a significant cost advantage over patterning and etching by eliminating lithography and resist removal steps, as long as the number of vias is small enough. In one embodiment of 3D die-to-wafer stacking, TSVs ranging in diameter from 50 μm to 200 μm were fabricated through 155-μm-thick silicon using a 30 kHz Q-switched 355-nm laser with a 15-μm spot size. After the vias were drilled, they were filled with polymer dielectric, and the polymer was ablated using the laser and aligned to the original hole in the silicon, forming dielectric-lined vias [1].

Vias larger than the collimated beam diameter can be drilled using a technique called trepanning. The beam is displaced optically from its normal axis and rotated about this axis to "machine" the inside of the via [2]. This technique can also be used to minimize damage and produce a cleaner top to the via.

A rule of thumb for usefulness of laser drilling is that if 10 000 vias are to be made or their diameter is less than 10 μm, it is more favorable to use patterning and etching to form them because all the vias are formed at once. Vias smaller than 10 μm are difficult to form by laser.

5.3.2 TSVs for Die-to-Wafer Stacking: Etching

Vias can be etched using photoresist or a silicon dioxide hard mask. For vias 70–100-μm deep with an aspect ratio of 5–10, a high etch rate and sidewall passivation are key. The deep silicon etching generally uses SF_6 with additives to control the profile, such as O_2 and HBr. The etch equipment used for TSV etching usually involves a high-density plasma (HDP) source such as inductively coupled plasma (ICP), electron cyclotron resonance (ECR), or magnetic enhancement of the parallel plate configuration to achieve a high throughput. The etching is a combination of sputtering, chemical etching, and simultaneous sidewall passivation with ion bombardment removing the passivation on the bottom [3]. The fluorine etches the silicon, resulting in volatile SiF_4 products. When O_2 is added for sidewall passivation, volatile $SiOF_4$ is also generated, with Si_xF_yO deposited on the via sidewalls, inhibiting lateral etching [4, 5]. If HBr is added to the SF_6/O_2 mixture, the fluorine/oxygen ratio is reduced, presumably by reaction of fluorine with hydrogen-forming HF, which improves the passivation of the sidewalls. According to Gomez et al. [7], the etch rate of the bottom of the trench is increased by adding HBr, due to fluorine neutral focusing by the more efficiently oxide-passivated trench sidewalls. Over the range of HBr flows studied in our experiments, this component decreased the vertical etch rate and had little effect on the profile (see Fig. 5.1).

Because oxygen and fluorine attack photoresist, a silicon dioxide hard mask has been used to define the vias in the silicon. Selectivity of an SF_6/O_2 etch to silicon dioxide can be around 50. Thus, a 2- or 3-μm-thick hard mask can allow a 100-μm deep silicon via to be etched. Generally, it is favorable to have a high plasma density and low bias voltage to preserve the mask. The typical etch rate for SF_6 is from 3 μm to 6 μm per minute; consequently, 100-μm-deep TSVs are etched at a rate of 2–4 wafers per hour, not impressive by wafer fab throughput standards.

The profile of such a via is determined by forward sputtering of the mask, ion bombardment of the bottom of the trench, isotropic etching of the free fluorine, and protection of the trench sidewall from fluorine etching by passivating gases. Forward sputtering of the mask results in a bowed profile at the top. The profile can be controlled by adjusting the gas flows so as to achieve sidewall passivation. However, poor mask selectivity, undercut, and general profile control limitations plague one-step etching.

The Bosch etch process [6] was developed in the early 1990s for deep silicon etching. It is capable of very high selectivity to either resist or an oxide hard mask. The etch profile is controlled by alternately depositing polymer on the silicon using C_4F_8 and removing it on the surfaces orthogonal to ion bombardment using an SF_6 etch. The fluorine from this step undercuts the sidewall polymer from the C_4F_8 deposition, creating a "scalloped" sidewall, as shown in Fig. 5.2. The number of cycles required depends on the depth of the via and the desired period of the

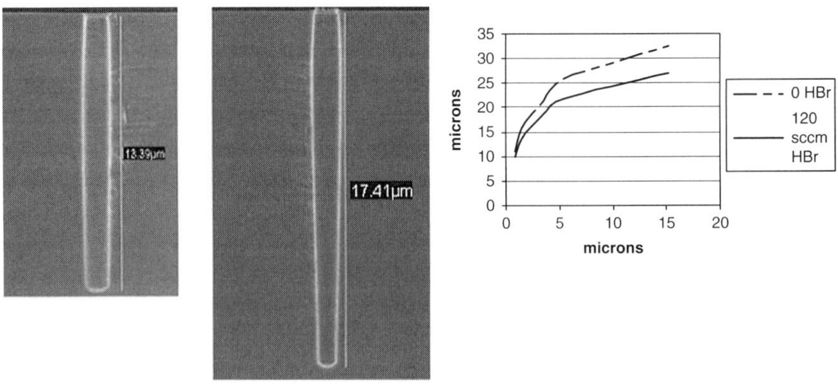

a. b. c.

Fig. 5.1 (**a**) 1.5-μm-diameter via etched with $SF_6/HBr/O_2$. (**b**) 1.5-μm-diameter via etched with SF_6/O_2 only. (**c**) reactive ion etching (RIE) lag of these etches

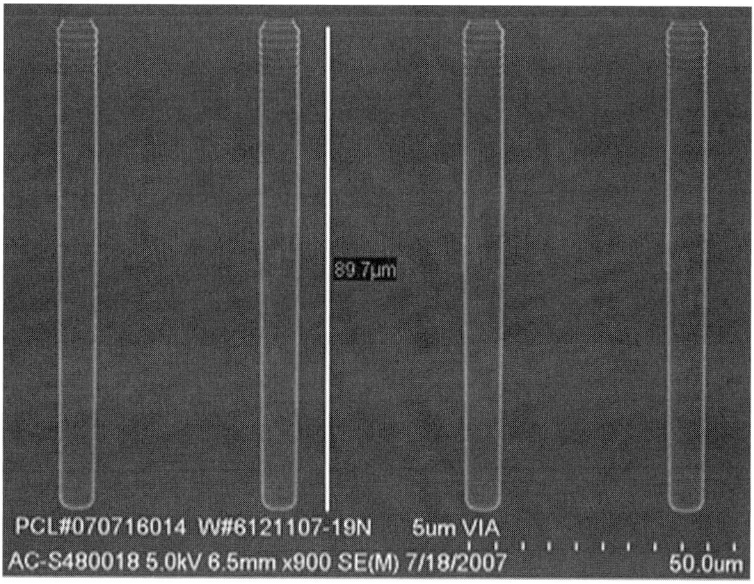

Fig. 5.2 90-nm-deep, 5-μm-diameter vias using a nonoptimized Bosch etch process
Note the scalloping at the top and the mask undercut. The SiO_2 mask loss was negligible.

scalloping on the walls. The overall profile can be effectively controlled by varying the ratio of the C_4F_8 polymer deposition and SF_6 etching times.

The amplitude of the scallops can be controlled by regulating the total time in each cycle and adjusting the number of cycles to hit the required depth.

The SF_6 step can remove the polymer at the bottom of each cusp with relatively low substrate bias. This means that Bosch etch processes with low ion bombardment are possible. Since oxide etching requires ion assistance, high selectivity to an SiO_2 hard mask (more than 200:1) can be achieved. This means that 100-μm-deep vias, which would be from 10 to 20 μm across to lend themselves to filling with plated or CVD metal, can be easily fabricated with a 1-μm-thick hard mask. The plated fill metal can then be polished using the remaining mask as a stop.

5.3.3 TSVs for Wafer-to-Wafer Stacking

In wafer-to-wafer bonding, the stacked wafer can be thinned until it has no structural integrity, since it is supported by the substrate wafer (in the case of face-to-face bonding) or by a handle wafer. This allows the vias to be much shallower than those in die-to-wafer processing, in some cases less than 10 μm. The limits to depth of the via are the total thickness variation (TTV) of the substrate wafer and bond layer, the accuracy of the grinding tool, and the depth of any trench capacitors, and any silicon depth buffer that might be required to separate damaged silicon from the devices. Bond layer thickness is usually less than a few microns with a TTV of a few tenths of microns. Since the allowable aspect ratio of the via as limited by chemical vapor deposition (CVD) or plated filling capability is probably about 10:1, the minimum usable via size is submicron for devices other than dynamic random access memory (DRAMs), which have deep trench capacitors.

If the vias are fabricated first, then revealed from the back by grinding and polishing, techniques similar to DRAM trench etching can be used for fabrication. Again, $SF_6/O_2/HBr$ chemistry is useful for fabricating vertical trench walls and limiting mask undercut. The shallower vias mean higher etch throughput than with die-to-wafer structures, but the etch rate tends to be suppressed in smaller vias due to reduced ion flux with increasing aspect ratio. Depending on the via size in a particular wafer-to-wafer application, etching a 10 μm deep submicron diameter via can require the same time as etching a 25 μm deep 20 μm diameter via.

Since the requirement for trench depth is relaxed, extreme mask selectivity is not required; therefore, the Bosch etch need not be used. The advantage here is the absence of sidewall scalloping, which in small vias can account for a large fractional variation in the via diameter and potential problems with sputtered barrier or seed layers.

5.4 Dielectric Liner for TSVs

Isolation is crucial for functional high-performance 3D interconnection. The silicon surrounding the via may be very lightly doped if the devices are built in bulk silicon or very heavily doped if they are made in epi. The vertical interconnect might only be through dielectric, as is the case with SOI, in which the substrate is entirely removed after face to face bonding. If the via goes through substrate silicon a dielectric liner is required to isolate the vias from the silicon and limit parasitic capacitance. It is noted that the poor scaling of this capacitance in copper and low-*k* interconnect is one of the attractions of 3D interconnect. It is important that the liner deposition process is conformal so that the entire sidewall of the high aspect ratio (HAR) via is covered. Typical plasma-enhanced CVD (PECVD) processes result in films that are not sufficiently conformal for lining HAR vias. One film that can successfully line 10:1-aspect-ratio vias is produced by ozone activation of tetraethylorthosilicate (TEOS) and is deposited with the substrate held at 400°C. The reaction used is [8].

$$Si(OC_2H_5)4 + 8O_3 \rightarrow SiO_2 + 10H_2O + 8CO \qquad (5.1)$$

A scanning electron microscopy (SEM) image of the coverage of this film in the bottom of a 1.2-μm-diameter, 13-μm deep blind via is shown in Figure 5.3. A comparison of the coverage of this film as a function of depth in the via with that of PECVD TEOS and silicon nitride (formed from SiH_4 and NH_3) is shown in Fig. 5.4. Step coverage of the film is adequate at the 400°C deposition temperature,

Fig. 5.3 Step coverage of a TEOS/03 liner layer at the bottom of a 1.2 μm diameter 13 μm deep hole

but it can be significantly improved at higher temperature (500°C). This, however, may be at the expense of its usability as a liner for a TSV produced after metal layers are on the wafer.

Atomic layer deposition (ALD) techniques have been used to produce extremely conformal alumina thin films with a dielectric constant of about 8. These may be of limited interest as dielectric liner films for 3D since ALD throughputs are extremely

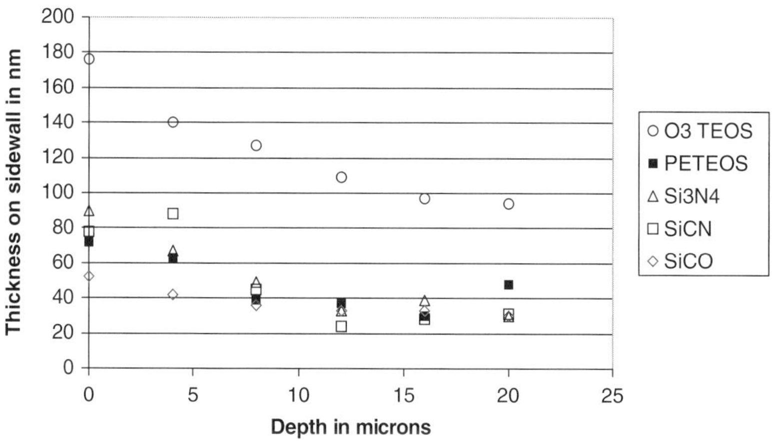

Fig. 5.4 Sidewall coverage of three inorganic CVD liner layers in a 3-μm silicon via hole, 20-μm deep

low, favoring very thin films. These liners would then either have a very high capacitance or very low throughput.

5.5 Tungsten Via Fill and Polish

5.5.1 *Contact/Seed/CVD Tungsten Depositions*

Tungsten is generally used to fill the small HAR vias used in wafer-to-wafer stacks. The process is well known in the wafer fabrication community. The first applications of the technology for 3D are for 1–2-μm vias, 10–15-μm deep for use in face-to-face wafer-to-wafer flows. Tungsten was first introduced to wafer fab processing in the late 1980s to fill 1-μm diameter contacts and vias up to 2-μm deep. The process has evolved to fill small high aspect ratio self-aligned contacts in DRAMs. For this reason conformality is not a limitation for 3D applications. The deposition requires a contact and liner (barrier/seed) layer, which is normally sputtered titanium and reactively sputtered or CVD titanium nitride. Atomic layer deposited titanium nitride from $TiCl_4$ and NH_3 allows the liner process to be scaled to even higher aspect ratios if necessary. The deposition of the tungsten is conformal, meaning that somewhat more than half the via diameter must be deposited so that the via will be filled with tungsten. The two steps of the deposition are silane reduction of tungsten hexafluoride (WF_6) to cover the titanium nitride, followed by hydrogen reduction of WF_6 to deposit the conformal film. This is a standard wafer fab process with a relatively high throughput.

The tungsten can be polished using a colloidal silica slurry, with a standard urethane closed cell hard pad for minimum dishing and plug recess.

5.6 Copper Via Fill

5.6.1 *Barrier and Seed Layers for Copper Plating*

Metallic barrier layers for copper plating include titanium, titanium nitride, titanium–tungsten, tantalum, and tantalum nitride among others. High aspect ratio TSVs demand special attention to the step coverage of the barrier layer to prevent copper from contaminating the silicon surrounding the vias. The copper seed layer must be continuous to allow subsequent electroplating to fill the vias.

The prevalent deposition method for barrier and seed layers is physical vapor deposition (PVD), specifically sputtering. The step coverage of films deposited by this method has been enhanced dramatically over the last two decades by using long throw system designs, ion metal plasma (IMP) [9], hollow cathode magnetron (HCM) [10], and argon ion bombardment during or sequenced with the processing. Difficulties can still be encountered using PVD to cover a via with retrograde sidewalls, but it is possible to cover the walls of a 10:1-aspect-ratio via with a continuous sputtered barrier film.

Atomic layer deposition makes possible nearly perfect step coverage of the barrier layer, TiN, TaN, or WN. This technique can use metal organics [11] or metal halides and is characterized by a low deposition rate since the chamber must be pumped and purged after each half cycle (one cycle deposits only one molecular layer). However, very thin barrier layers, less than 10 nm, can be used since the sides and bottom of the via are of very uniform thickness. The metal organic barrier layers tend to have from 1 to 10% carbon incorporation due to incomplete scission of the precursor molecules. The resistivity of these films tends to be far higher than bulk. Barrier properties are not degraded by the carbon incorporation.

The seed layer must support copper plating. This means that it must be sufficiently conductive to pass the plating current into the via and to reduce the copper ions. Copper seed layers have been fabricated by the aforementioned PVD methods for 3D TSV applications. These can be successful even at an aspect ratio of 10:1 if the via walls can be assured not to be retrograde. For the larger vias typical of die-to-wafer applications, seed layers much thicker than are currently applied in standard dual-damascene metallization can be used. The thick seed layers as deposited on the top surface are more likely to be continuous as they thin on the sidewalls of the HAR vias. Copper seed layers in dual-damascene high-density interconnects are less than 10 nm, but in 3D applications, particularly in die-to-wafer methods, they can be up to five times thicker, limited only by their tendency to cause a void to form during subsequent copper plating. If the vias are small (about 1 μm or less), it may be impossible to sufficiently cover the seed layer by PVD to support electroplating. However, if the seed layer is not properly nucleated, the plating will be unsuccessful, as in Fig. 5.5.

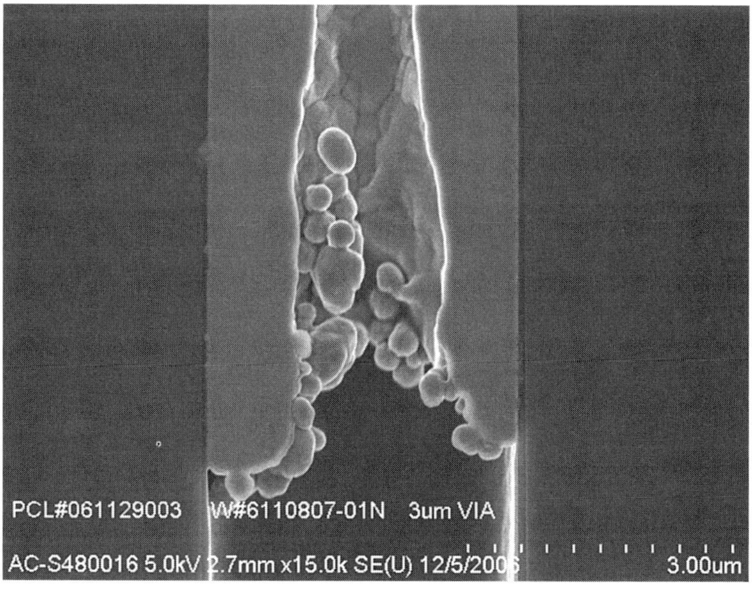

Fig. 5.5 Effect on copper plating of a subnucleated copper seed layer

Chemical vapor deposition of copper has been used to deposit seed layers for small HAR vias in wafer-to-wafer applications. The organic precursor copper hexafluoroacetylacetonate hydrate can be used to deposit copper oxide, which was subsequently reduced to copper by exposure to ethanol [12]. Copper CVD tends to result in highly stressed films; surface preparation of the underlying layers is critical to developing adhesion, which is critical for reliability.

Other methods of seed formation include sputter deposition and "repair" of discontinuous seed layers by electroless plating [13]. The electroless deposition allows isolated islands of copper to grow together to form a continuous layer for subsequent electroplating steps. It involves another plating bath, with generally more demanding stability issues than electroplating baths. However, only very thin electroless films must be deposited to ready the sputtered films for successful electroplating.

Another seeding technique that has been developed for high-density 2D interconnection technology is direct copper plating on an ALD ruthenium layer deposited using organic precursors such as bis(ethylcyclopenta-dienyl) ruthenium [Ru(EtCp)2] [14]. The ruthenium is generally deposited on another barrier layer such as tantalum nitride, but serves itself as a barrier and a seed layer. Copper can be successfully plated on ruthenium [15].

The more exotic (and expensive) methods that are required for high-density metallization are applicable to 3D in only particular integrations with very high aspect ratio small vias, because the features employed in 3D are one to two orders of magnitude larger than in planar metallization. In common usage, the barrier and seed deposition will be done using PVD, with thicker layers than are used in planar interconnect systems until wafer-to-wafer applications evolve to submicron vias.

5.6.2 Electroplating Copper

Copper electroplating presents challenges in 3D technology because of the aspect ratio of the vias. In planar dual-damascene structures, the aspect ratio and size of the vias is generally less than 2. The trenches above the vias also have an aspect ratio of up to 2, but the lack of gas flow restriction along the trench means that the effective aspect ratio of the via is less than 4. The vias for 3D, however, can have an aspect ratio from 6 to more than 10; this presents a different set of challenges than those with planar metallization. In die-to-wafer stacking, which involves 5–50-μm vias that are 50–150-μm deep, conventional filling used in planar interconnect applications would create an overburden 50–150-μm thick. This poses serious problems for polishing throughput and dishing. Plating solutions with copper sulfate in sulfuric acid are used in both planar interconnect and 3D applications. Plating processes optimized for 3D using strong suppressor and accelerator additives can plate copper at the bottom of the via much faster than at the top. Even with these enhancements to the chemistry, the process must be run at very low throughput to assure bottom-up filling. Plating 50-μm wide, 100-μm-deep vias can take 2 h; 10-μm wide, 100-μm-deep vias can take 1 h [16]. Pulsed plating with short reverse current pulses can help to improve the filling of the plated film, but reverse current plating adversely affects throughput [17], giving about the same overall rate. Thus, plating TSVs demands batch plating equipment to meet throughput requirements without having to purchase many new tools. The batch electroplating tools are generally designed for the semiconductor packaging industry rather than wafer fabs.

5.6.3 Chemical–Mechanical Polish (CMP) of Copper

Even if efficient bottom-up plating is used, the overburden due to filling the deep vias is many microns thick. This requires a two-phase polish. The first phase uses a high removal rate slurry with good planarization and low nonuniformity. Low dishing is not a concern at this stage. The second phase uses a slurry or process with a lower removal rate and acceptable dishing, usually selective to the barrier. The second step is to clear the barrier and compensate for any topography from the earlier steps. The topography requirements are set forth by the integration scheme. If the TSVs are etched in the earlier metal levels, topography control is critical since metal layers will be routed above these vias and high topography will lead to metal shorts at upper levels. If the TSVs are etched at the final metal level, dishing specifications can be relaxed.

5.7 Wafer Thinning

Wafers have been thinned for many decades, first to deposit metal films on the back to promote backside contact to the device, allow illumination through the back of silicon optical sensors, increase packaging density, and now decrease the TSV depth. Silicon substrates can be thinned after being bonded together to allow the interconnect to be formed on the back of the bonded wafer or to reveal existing interconnect for bonding to another substrate. If vias have been filled with metal in the wafer to be bonded, the thinning method must be capable of revealing the metal without removing it. Often this is optimally done in a multistep process that may include grinding, polishing or wet etching for stress relief, and/or dry etching for revealing the bottom of vias without removing the metal.

5.7.1 Backgrind

A generic process for wafer thinning via backgrinding is shown in Fig. 5.6.

In preparation for grinding, to protect the front (device) side of the device wafer, a "backgrind tape" is applied to the front. The backgrind tape is about 100–300-μm thick; it comes in many heat-resistant versions depending on the need for subsequent processing. The wafer is then loaded face-down onto the backgrind table. If a rigid handle wafer is used, the backgrind tape is not necessary and the wafer stack is

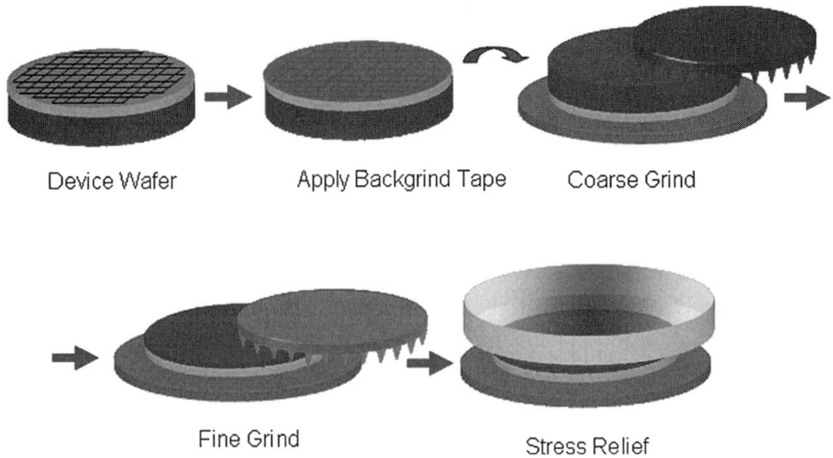

Fig. 5.6 Generic process flow for wafer thinning

placed with the handle wafer on the holding chuck such that the backside of the device wafer is facing up.

A grinding wheel/disk with embedded diamonds performs the grinding itself. The diamonds are of a size needed to cut the silicon at a rate that is determined by the pressure applied and the rotational speed of the chuck/grinding wheel. The grinding assures good across-wafer planarity since the head is held relatively rigidly with respect to the chuck table. A profiler constantly measures the height difference from the chuck table to the top of the silicon wafer being ground and provides the thickness of the removed or remaining silicon.

The grinding is done in two steps: a "coarse" grind to remove the bulk of the silicon, followed by a "fine" grind. Since the grinding is a mechanical abrasion process, it introduces surface and subsurface damage. The depth of damage corresponds to the grit size. For a grit size of 50 μm, which is used in the coarse grind, the subsurface damage layer is 10–20-μm deep. The subsurface damage layer is reduced to about 2-μm deep by the fine grind. With this level of damage, especially for wafers less than 100-μm thick, any crack can propagate through the bulk silicon with even a little bit of external stress. Consequently, even handling thinned wafers can propagate cracks. Hence a stress relief step is necessary to remove this damaged silicon layer and increase die strength.

During the coarse grind step, a disk containing coarse diamonds (320 grit or sized about 40 μm) is used. The silicon removal rate at this stage starts out very high at 4–6 μm s^{-1}, then slows down to 2–4 μm s^{-1} when target removal is approached. During this entire process, water is used to both cool the wafer during grind as well as to flush away the silicon dust that is produced. During the fine grind step, a disk containing fine diamonds (2000 grit or about 11 μm) is used. The silicon removal rate at this stage starts about 0.5–11 μm s^{-1} and ramps down to about 0.1–0.3 μm s^{-1} to minimize the surface damage. New innovations, such as Poligrind™, with a very fine abrasive particle size to further minimize surface damage have been found to improve die strength [18].

To achieve greater die strength, stress can be relieved by a number of processes: dry polish, wet polish or CMP, dry etch, or wet etch. Currently, advanced grinding tools incorporate wet or dry polishing stations to perform this function. With a dry polish, a pad is used to buff off the damaged layer without any liquids. Chemical–mechanical polish uses silica abrasives under alkaline conditions to remove the damaged layer. A plasma dry etch process can also be used to blunt the deeper crack tips in addition to etching the damage layer. About 2–3 μm of material must be removed to improve the die strength. An alternative is wet etch, which is more independent of topography. It requires about 20 μm of removal to obtain the best mechanical strength (details are provided in the following sections). It has been observed that all of these stress relief techniques show improved die strength [18].

The strength of the wafer is typically measured by the die strength, which should be an estimate of the resistance of the die to cracking during packaging. Strength of the die is measured using the ball-and-ring test or the three-point bend test. Schematics of these tests are shown in Fig. 5.7a and 5.7b. In the ball-and-ring test, the die is mounted on a ring; a ball is lowered on the center of the die, and a force applied until the die cracks. In the three-point bend test, the die is supported by two cylindrical rods; a third rod, with axis parallel to the other two but above the die, pushes on the die from above, midway between the support rods. Force is ramped until the die cracks [19]. The distribution of forces is approximated by the Weibull distribution:

$$P(\sigma) = 1-\exp\{-(\sigma/\sigma_0)^m\}, \tag{5.2}$$

where σ is the die strength and σ_0 is the scale parameter. The value of$m$, the Weibull parameter, is a measure of the breadth of the distribution. When m is high, the distribution is tight. The failures in the three-point bend test stem from edge defects and surface defects, whereas the ball-and-ring test failures are primarily from surface defects [20].

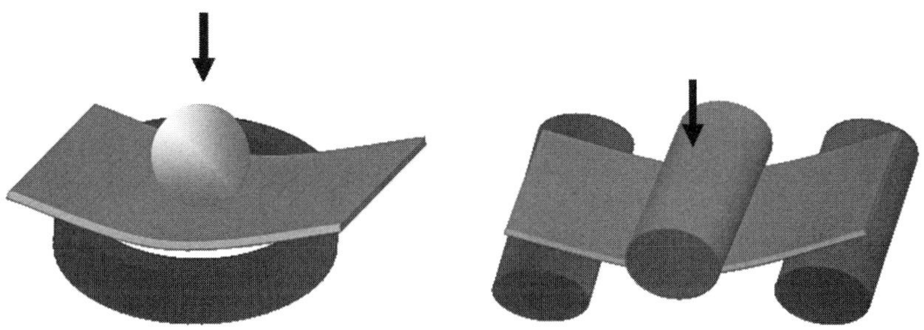

Fig. 5.7 (**a**) Ball and ring. (**b**) The three-point bend test

5.7.2 Wet Etching Silicon to Thin the Wafer and Relieve Stress from Grinding

<100> silicon etches in various bases including potassium, ammonium and tetramethyl ammonium hydroxides (i.e., KOH, NH_4OH, and TMAH), and acid mixtures, such as mixtures of nitric acid or hydrogen peroxide with concentrated hydrofluoric acid, in which the silicon is oxidized and simultaneously removed as oxide. Some of the acidic solutions etch quite quickly and have been used in single-wafer spin etch equipment. The bases are somewhat slower and require long batch etches in a heated tank. For example, the silicon etch rate in potassium hydroxide decreases with increasing concentration between 20% and 60% and increases with temperature. The dependence of the etch rate on the KOH and water concentration, respectively, is $R \sim [H_2O]^4 \times [KOH]^{1/4}$ and the Arrhenius activation energy is about 0.86 eV, giving an etch rate of about 78 $\mu m\ h^{-1}$ at 80°C and 30% concentration [21]. Masking materials for the back of the substrate wafer (which is not thinned) include thermally grown silicon dioxide, to which 250:1 selectivity can be achieved, furnace-deposited silicon nitride, which does not etch, even in hot potassium hydroxide, and special polymer films such as BCB (benzocyclobutene). The wafer must be rinsed well after this etch since the potassium is mobile in silicon dioxide.

Tetramethyl ammonium hydroxide etches silicon somewhat more slowly than KOH, but with more than an order of magnitude higher selectivity to oxide. At 80°C and 30% concentration, the silicon etch rate is only 21 $\mu m\ h^{-1}$, with a selectivity to thermally grown silicon dioxide of 4400:1 [22]. This may be important for etching the back of wafers that are dielectric fusion-bonded to prevent the bond between the wafers from being attacked. Tetramethyl ammonium hydroxide is widely used in the MEMS industry.

5.7.3 Fracturing Wafers at a Hydrogen Implant-Damaged Plane

Three-dimensional integration involves stacking semiconductor device layers, as opposed to simply adding layers of wiring, which has been practiced since the 1970s. Early attempts at 3D integration in the 1980s took the form of building device quality layers of silicon over the first layer of transistors, but before the first layer of aluminum interconnect wiring was applied. Many methods to form these layers were tried, including growing single-crystal silicon layers from amorphous and polysilicon layers seeded from contacts through the pre-metal dielectric over the first layer of transistors [23]. This is a form of SOI. These layers tended to be highly defective, insufficient for building high-performance CMOS devices. Efforts began to transfer intact single-crystal layers of silicon from one substrate to another to allow for dielectric isolation of the silicon.

Various ways now exist to form a thin high-quality single-crystal silicon layer on SiO_2. This can help solve electrical isolation problems with scaling MOSFETs, such as substrate leakage, short channel effects, and junction capacitance. The SOI industry has embraced a technique of bonding and fracturing wafers called Smart

Cut^{TM}, pioneered by M. Bruel at LETI, who realized that hydrogen or noble gas atoms implanted into the silicon would form microvoids, to which interstitial gas atoms would migrate, generating small pressurized "platelets" in the silicon, eventually generating sufficient pressure to cleave the crystal [24]. This ultrahigh-dose implantation had been a problem in some applications, causing the silicon surface to bubble and locally delaminate. Bruel's patent describes bonding the silicon surface to a stiffener or performing the implantation through a stiffening layer such as silicon dioxide to allow the voids to coalesce and sufficient pressure to be developed to uniformly remove the whole surface layer of silicon. The commercial realization of his invention is Smart Cut^{TM}, a trademark of SOITEC. It involves oxidizing an undoped (seed) wafer and a substrate wafer, implanting the seed with hydrogen and helium at a total fluence of about 5×10^{16} cm^{-2} and sufficient energy to penetrate the oxide and the silicon to be transferred, treating this (seed) wafer and the substrate wafer in a hydrogen peroxide and ammonia clean to leave the surfaces hydrophilic, bonding the wafers at room temperature (oxide fusion bond), and cleaving the silicon seed by heating at high temperatures. The resulting gas pressure in the microvoids (or platelets) separates the thin silicon layer from the seed wafer. The thin silicon layer is bonded very well to the original oxide on the seed layer, which is firmly attached by dielectric fusion bonding to the substrate wafer. The surface roughness of the cleaved surface is a few nanometers, which can be reduced to bulk silicon roughness levels with a polish, resulting in device quality layers of silicon that can be as thin as 10 nm [25]. Implications of this technology for 3D include the fabrication of high-quality PMOS layers over NMOS layers, such as stacked static random access memory.

5.8 Singulation

After wafer-level processing, the wafers are ready to be diced into individual die to be either packaged or assembled onto another die. There are several ways to singulate the die. For thin wafers, however, the wafer is supported on a tape frame and techniques such as scribe and cleave are not applicable. Dicing is therefore accomplished by a mechanical saw, laser cut, or dry etch.

Mechanical methods involve a rotating saw blade that is used to cut a calibrated depth into the wafer and a few microns into the tape. The blade has micro-diamonds that actually do the cutting, and the debris from the cut silicon is flushed away with a high-pressure jet while the wafer is being cut. The width of the diamond blade determines the width of the cut as well as the cut rate and damage to the sidewalls of the die. A significant issue with dicing by mechanical saw, especially with thin wafers, is chipping on the backside. An example is shown in Fig. 5.8. The thin wafers, supported by a tape frame on a die attach film (DAF), do not have much mechanical rigidity and consequently break (or chip) before being cut cleanly with a saw. This can significantly damage the die and reduce the die strength.

A typical dicing process uses the already thinned (ground) wafer on the DAF mounted on a tape frame. Hence the wafer is subjected to "grind before dice." One

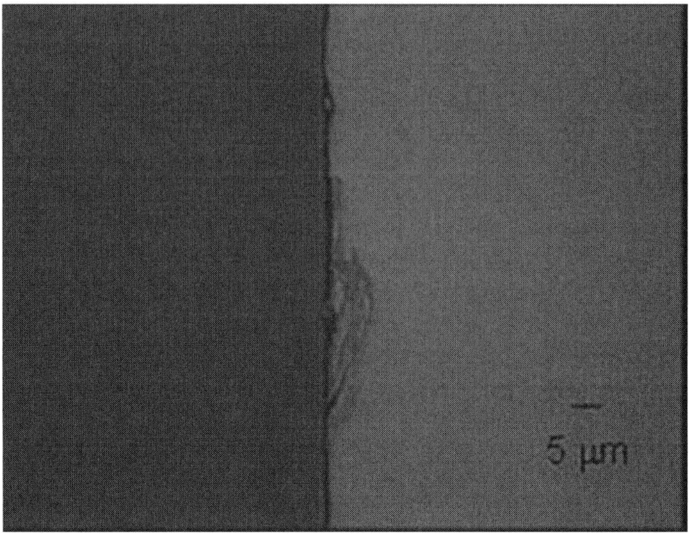

Fig. 5.8 Backside chipping using a mechanical saw for dicing a 50-μm-thick wafer

method that has been identified to minimize backside chipping is a process called "dice before grind" [26]. In this process, the wafer is partially diced (to a depth of the final thickness plus a margin of safety, approximately about 10 μm) while still in wafer form, before any thinning operation. Once this partial cut is performed, the wafer is then subjected to the backgrind as shown in Fig. 5.6. When the wafer is thinned down to the final thickness, the dies are revealed. This process has been shown to reduce chipping on the back of the die. Figures 5.9 and 5.10 illustrates the two approaches.

Techniques such as laser dicing have also been developed to address these issues. A number of sources for the laser, such as CO_2, excimer, solid state, etc., are

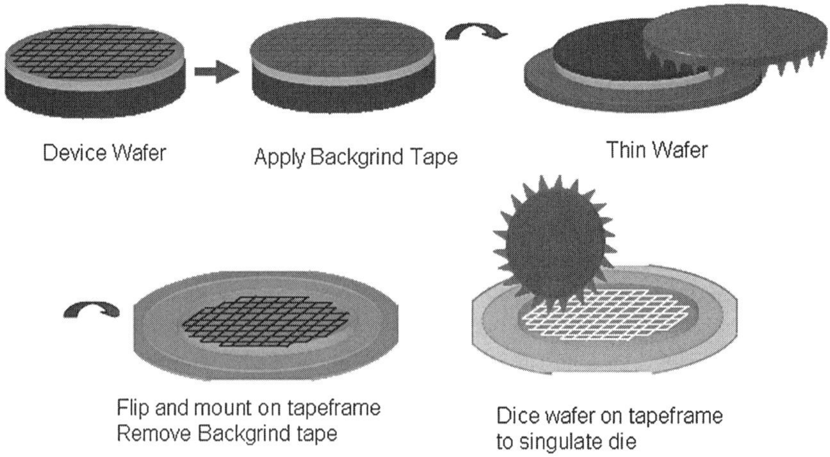

Fig. 5.9 Schematic of grind before dice

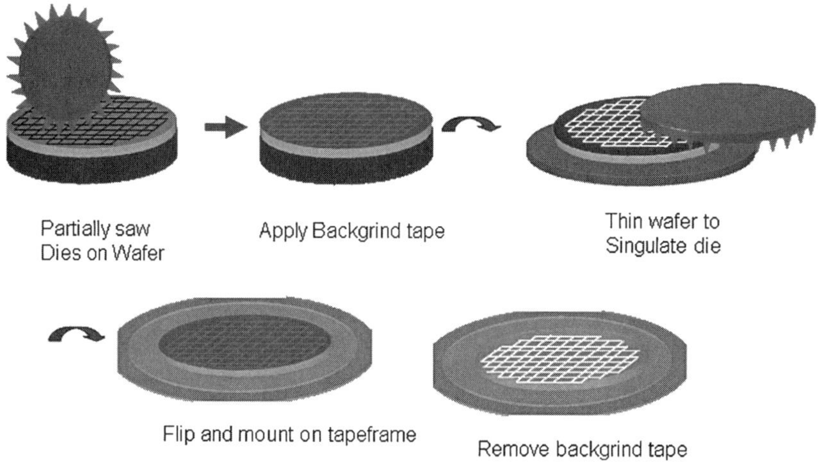

Fig. 5.10 Schematic of dice before grind process

currently in use [27–32]. This technique essentially uses a high-power laser to melt or vaporize the silicon, thus separating the die. Since a great amount of energy is to be injected into the wafer, there is still a "damage zone" around the edge of the die. This zone increases with thicker wafers. Tools are available to minimize the damage by splitting the beam into multiple sequential beams of lower power [33]. Multiple passes of the laser are needed to dice thick wafers, which reduces the throughput. For these reasons, lasers are well suited to dice thin wafers. Since some lasers vaporize or otherwise eject a molten stream of silicon, these substances can condense back on other parts of the wafer being diced. Therefore, a protective layer, which can be easily removed later, is usually applied to the wafer before dicing and any particles on the wafer can be washed away after the dicing operation. Some technologies also use a water jet to both guide the laser and remove debris.

5.9 Handle Wafer Technologies

One of the key requirements for 3D integration is the need to thin the wafer to less than 100 μm. At these thicknesses the silicon wafers lose their rigidity and tend to bend and buckle under their own weight. Therefore, handling of this wafer (robotically or otherwise) as well as subsequent processing becomes impossible without the help of a supporting or a "handle wafer".

Handle wafers have been successfully used in fabrication of microelectro mechanical Systems (MEMSs) devices [34]. This technology is also at the core of fabricating SOI wafers. In their application to 3D TSV technology, handle wafers play a critical role in both die-to-wafer as well as some wafer-to-wafer schemes.

Table 5.1 Comparison of process flows for die-to-wafer and wafer-to-wafer bonding

Die-to-Wafer	Wafer-to-Wafer
Starting device wafer	Starting device wafer
Frontside processing	Frontside processing
Attach frontside to handle wafer	Attach frontside to handle wafer
Thin wafer to reveal TSV	Thin wafer
Additional backside processing (CVD, Litho, Etch, CMP)	Additional backside processing (if needed) (CVD, Litho, Etch, CMP)
Release handle wafer to tape frame	Permanent bonding to device wafer
Dice wafer	Release handle wafer
Attach dice to wafer	Additional frontside processing or dicing

The typical process flows for a device wafer used for a die-to-wafer and wafer-to-wafer are shown in Table 5.1:

As can be seen from above, the handle wafer system needs to be chosen to perform a number of processes including backgrinding, CMP, lithography, etch and metallization, in addition to the final permanent bonding step.

The primary components of a handle wafer system are:

- a rigid wafer support substrate
- a bonding layer
- optionally, a release layer

5.9.1 Substrate

The substrate needs to be made out of a suitably rigid material. Common materials used are silicon, glass, pyrex, borosilicates, sapphire, etc. Depending on the scheme used, the substrate wafers may need to be aligned to the other wafer during the bonding processes. For these applications, handle wafers transparent to visible light (so that the frontside device patterns can be recognized for alignment) are needed. If infrared alignment schemes are used, then undoped silicon wafers can possibly be used. Additionally, the handle wafer material should be extremely chemically resistant to all the common chemistries that are used in the interconnect processes. For example, a chemical etchant such as TMAH used to thin the Silicon on an SOI device will place restrictions on the use of a silicon handle wafer unless it has a coating that resists the etch. Mechanical specifications of the handle wafer such as planarity and TTV, etc. should also be considered depending on the final requirements.

Some processing tools require a notch to help align wafers before processing. For such systems, the substrate wafer needs to have a notch of similar dimensions, so that the device wafers can be notch-aligned to the substrate wafer during bonding. The use of silicon wafers as substrates will also address this issue.

A final consideration is reusability. Sapphire wafers are more resistant to mechanical damage than the softer glass wafers, but are more expensive. However, they may be reused more times. Consequently, the ease with which the substrate wafer can be cleaned and the number of times it can be recycled needs to be considered in the choice of the substrate.

5.9.2 Bonding Layer

The bonding layer is the medium that holds the device wafer and the substrate wafer together. The adhesive is critical in establishing the usability of the handle wafers for any specific integration or unit process (e.g., CMP, lithography, wet etching, vacuum deposition, and plasma processing). The bonding layer can be either a liquid that can be hardened by thermal [35] or UV treatment [36] or it can be a solid film applied by a lamination technique [37] or by other techniques such as electroplating. In any case, several properties need to be considered, as described below, when choosing this material.

5.9.2.1 Bond Strength

Backgrinding and CMP provide the highest shear force during the handle wafer operation and the handle wafer system must be able to resist delamination under these conditions. Although backgrind is typically done only once, multiple CMP operations (dielectric as well as metal) may be needed and the bond strength should hold during all these operations. The material will undergo multiple heating and cooling cycles and hence bonding robustness during these cycles is important. While no specific metrology has been developed to determine the bond strength of temporary bonding materials, traditional three- and four-point bend as well as shear tests should serve as a good starting point. The bond strength of materials has so far been functionally tested; it is estimated that a bond strength of 1 J m^{-2} is required to survive the backgrind process.

5.9.2.2 Maximum-Use Temperature

The maximum-use temperature and time of use at this high temperature often determines the appropriate bond material for each process. Dielectric deposition, the degasification step before PVD of the liner deposition, anneal after electroplating, and tungsten CVD are the potential high-temperature steps in the interconnect processes. Achieving a permanent bond between the device wafer and the 3D stack before the handle wafer is released is also a temperature-and-time-sensitive operation. In general, a higher temperature when using bond material is preferred; currently available bonding materials can retain good bond strength to 220°C (organics) [38] and even upwards of 400°C (inorganics) [39].

5.9.2.3 Chemical Stability

The chemical resistance of the adhesive and the bond layer to normal processing chemicals used in wafer-level interconnects is important so that the bond layer does not dissolve or disintegrate during these operations. Chemicals are used in photoresist develop and strip, wet etching, and CMP and post-CMP cleans. Even chemical degradation of small areas at the edges of the wafers can lead to significant delamination in downstream processes. Lastly, the bond material should not absorb moisture easily as its release in vacuum systems can cause process issues.

5.9.2.4 Vacuum Compatibility

The handle wafer will encounter vacuum chambers during deposition (CVD, PVD, sputter) as well as dry etch and dry stripping processes. It is imperative that there be no voids or air pockets in the bonding layer that can cause blistering or bubbling that can lead to subsequent delamination and other issues.

5.9.2.5 Final Thickness and Planarity

The TTV of the bonding material and the TTV of the substrate together determine the final TTV and planarity of the device layer during backgrinding. These thickness variations need to be smaller than the desired final device wafer. Bond layer thickness is controlled by spin speed for liquids and by control of the tape thickness for laminates. The bond layer should be thick enough to cover the topography so that the device wafer and the substrate wafers do not come into direct contact as this may damage the device wafer. Typically, die-on-wafer schemes have high frontside topography, requiring the use of thick bonding layers, on the order of tens of microns. For wafer-to-wafer schemes, the frontside topography is typically submicron and hence the bond layers need to be thin (in the few microns to submicron range) so that the thickness tolerances are tighter. Within-wafer thickness variation and avoidance of an edge bead at the periphery of the handle wafer are important considerations. Waxes and spin-on liquids provide a variation of about 1%, while laminates are usually higher at about 2–5%.

5.9.2.6 Releasibility

There are many different methods to release the handle wafer from the post-bonded wafer – chemical, mechanical, thermal, ultraviolet (UV), and laser treatment among others. The general mechanical requirements are that the release be a gentle, shear-free process with minimal stress or scratching on the thin wafer. Chemical release mechanisms should remove the bonding material cleanly on both the handle substrate as well as the device wafer, and the chemical used to dissolve the glue layer should not affect the materials in the devices. Thermal and UV irradiation weaken the bonding interfaces so that the handle wafer can be easily separated from the device wafer with minimal mechanical force. The residual material on either side of the bonding material needs to be removed completely without causing any surface defects on the support or device wafers.

5.9.3 Optional Release Layer

The bonding and the ease of release quite often need opposing properties. For bonding, a high-strength interface needs to be created, while for release a low-adhesion interface is necessary. An approach taken in some systems is to use two different materials: one to facilitate bonding and another to facilitate release. With this method, the bonding and debonding mechanisms can be separated. The release mechanisms of the optional release layer currently are laser treatment and use of a highly etchable chemical. If this method of release is chosen, the additional benefits and ease of application of these processes must be weighed against the added complexity and cost.

5.10 Wafer-Bonding Methods

The wafer-bonding/debonding process intimately depends on the choice of the bonding material. Several systems have been developed in the industry. The process below shows a few generic applications that have been commonly practiced.

5.10.1 Thermoplastic Materials

Thermoplastic materials are organic polymers that are deformable; they melt into a liquid when heated and solidify when cooled back to room temperature. Natural and synthetic waxes as well as proprietary wafer-bonding formulations liquify at temperatures between 120°C and 250°C. The exact shape and other thermal characteristics of these bonding materials vary depending on the material, but Fig. 5.11 shows a generic viscosity dependence on temperature. At low temperatures, the viscosity is

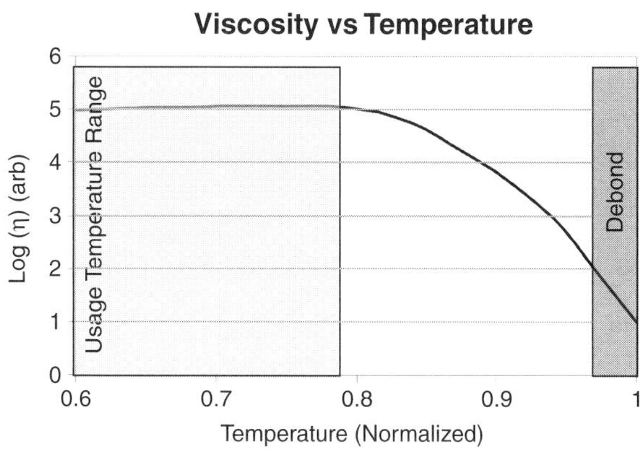

Fig. 5.11 A generic plot of viscosity (or adhesion strength) as a function of temperature for thermoplastic materials

high enough to be considered a solid although the material could be in a glassy state. As temperature is increased, the viscosity decreases. The temperature regime for use of the material, in practice, is dependent upon this curve – it needs to be substantially lower than this melting or softening temperature.

Materials that are spun on, are of two types:

1. Waxes (natural or synthetic) are spun onto heated substrate wafers at temperatures beyond the melting point. The device wafer (with a protective coating, if necessary) is brought in contact with the substrate wafer ensuring notch-to-notch alignment as well as parallelism with the substrate. A few psi of pressure is applied and held at temperature for a short amount of time. The system is then slowly cooled so that the wax solidifies, thus bonding the substrate to the device wafer. The sequence is shown in Fig. 5.12.
2. Solvent-based proprietary formulations [38] are spun on to substrate at room temperature, followed by a hot plate bake to evaporate the solvent. Control of the

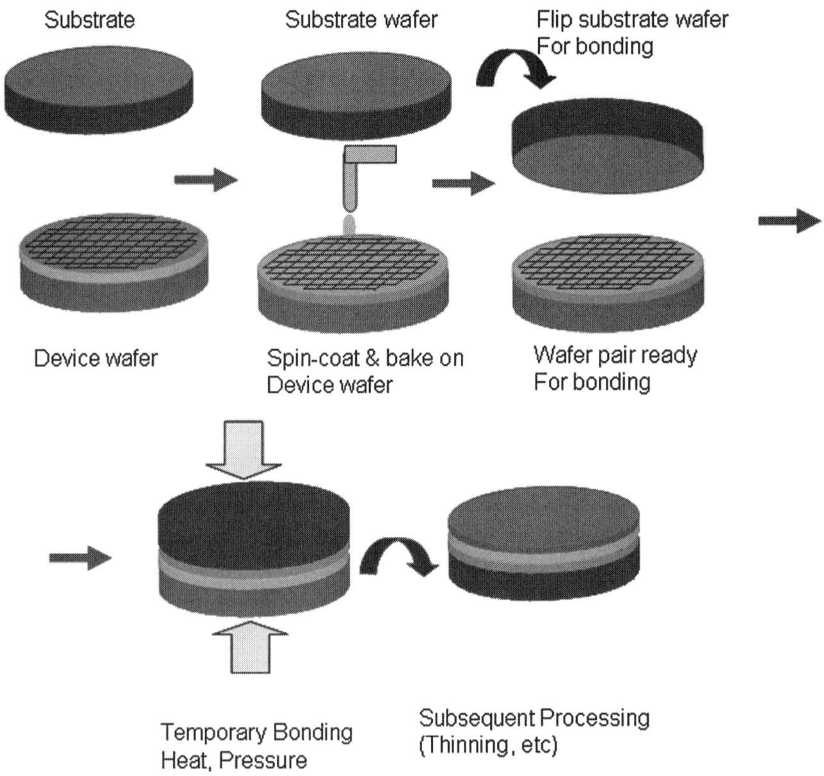

Fig. 5.12 Schematic of process flow for wax bonding

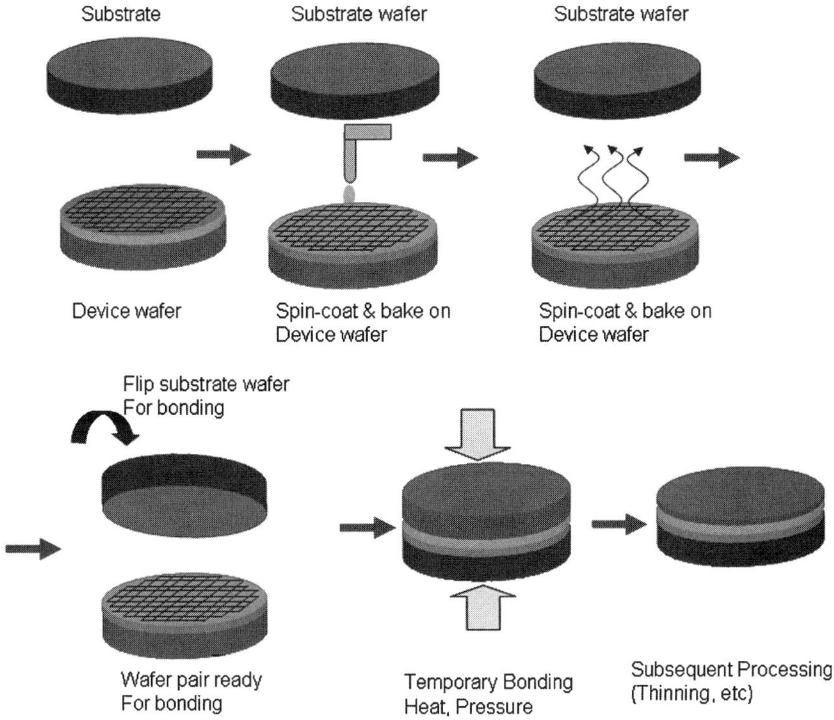

Fig. 5.13 Schematic of process flow for solvent-based thermoplastic bonding materials

temperature and time is critical at this juncture since the material should be in a bondable stage and not densify and stiffen up excessively. The device wafer (with a protective layer, if necessary) is then brought in contact with this substrate, aligned to the notch, and subjected to a reasonably high pressure of a few tens of psi of pressure is applied at reasonably high temperatures (120°C–180°C) but well below the liquification point, for a few tens of minutes. Figure 5.13 is a schematic illustration of this process.

5.10.2 UV-Curable Materials

Since the UV-cured material is quite stable toward thermal softening and chemical dissolution, a release layer that is releasable by a different (non-UV) mechanism is usually used on substrate wafers prior to spin-on on UV-curable materials. The UV-curable bonding materials are specially formulated liquids wherein the polymers crosslink upon exposure to UV radiation [36]. The device wafers are brought into contact with the substrate wafers and the bonding material is irradiated with UV light through the substrate. It is a critical requirement that the substrate be

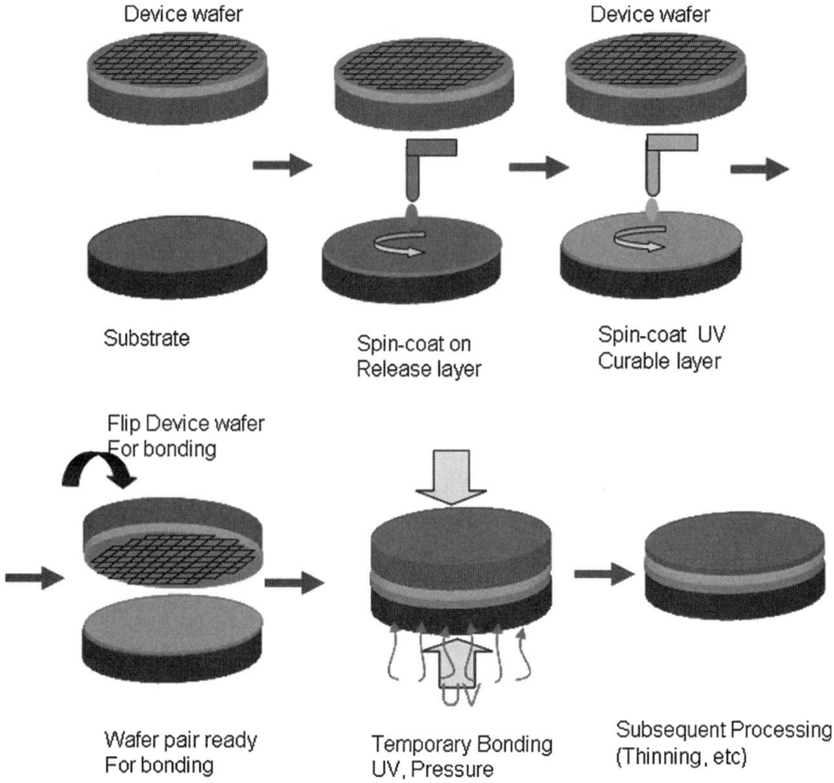

Fig. 5.14 Schematic of process flow UV-curable bonding materials

transparent to UV light of the right wavelength. The UV light crosslinks the bonding material and densifies forming a fairly rigid bond. A schematic of this process is shown in Fig. 5.14.

5.10.3 Laminates

This technique uses processes similar to what is used in other applications, such as tape frame, backgrind tapes, etc., and is characterized by its ease of use. Dry film adhesive tapes for these applications would typically be a laminate consisting of two layers of adhesive sandwiching a base film and liner layers on the outside protecting the adhesives. One adhesive layer would be a pressure-sensitive adhesive while the other would be a release adhesive. For the particular application of bonding onto handle wafers, the process sequence is as shown in Fig. 5.15. The liner layer is removed and the laminate is applied onto the substrate wafer. The laminate can optionally be pre-cut to the size of the substrate wafer or can be cut after application

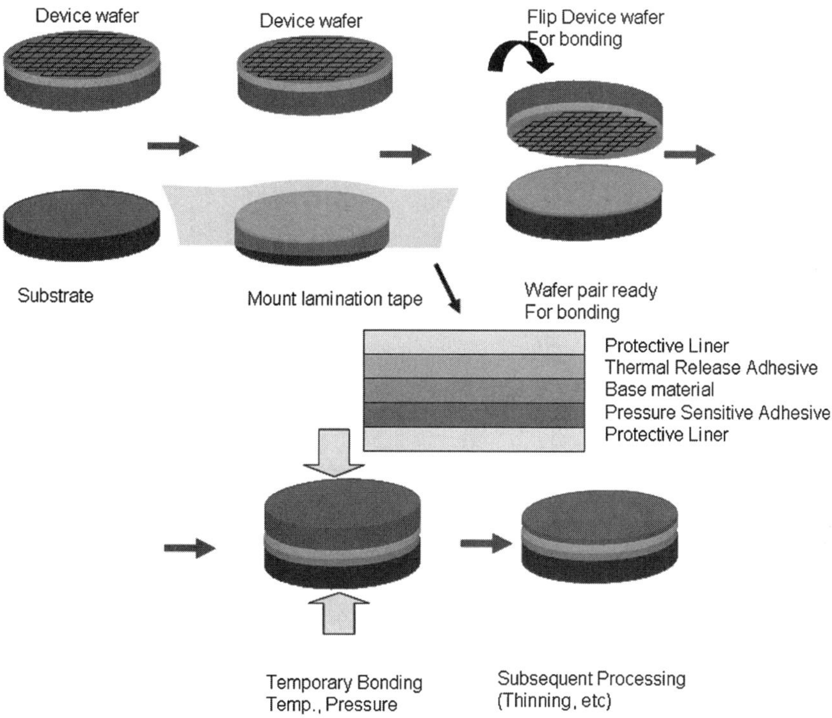

Fig. 5.15 Schematic of process flow for laminated bonding materials

by either a blade or a laser. The application process itself is a pressure process via a roll or a plate, after the other liner layer is removed.

5.10.4 Metal Based

Metal thermocompression bonding has also been used for handle wafers [39]. The use of Cu–Cu bonding facilitates processes that are resistant to high temperatures, pressures, and (to a limited extent) high pH solutions. A sandwiched Zr layer has been used as a release layer (as explained later). This enables direct bonding at higher pressures and temperatures than that can be used for thermoplastics or laminates. The process is schematically illustrated in Fig. 5.16.

5.11 Handle Wafer Use

Subsequent to wafer bonding, the stacked wafer is available for all processes. The first step is usually thinning of the silicon on the backside of the device wafer to thickness in the range of 150–25 μm. For wafers thinner than about 75 μm, the

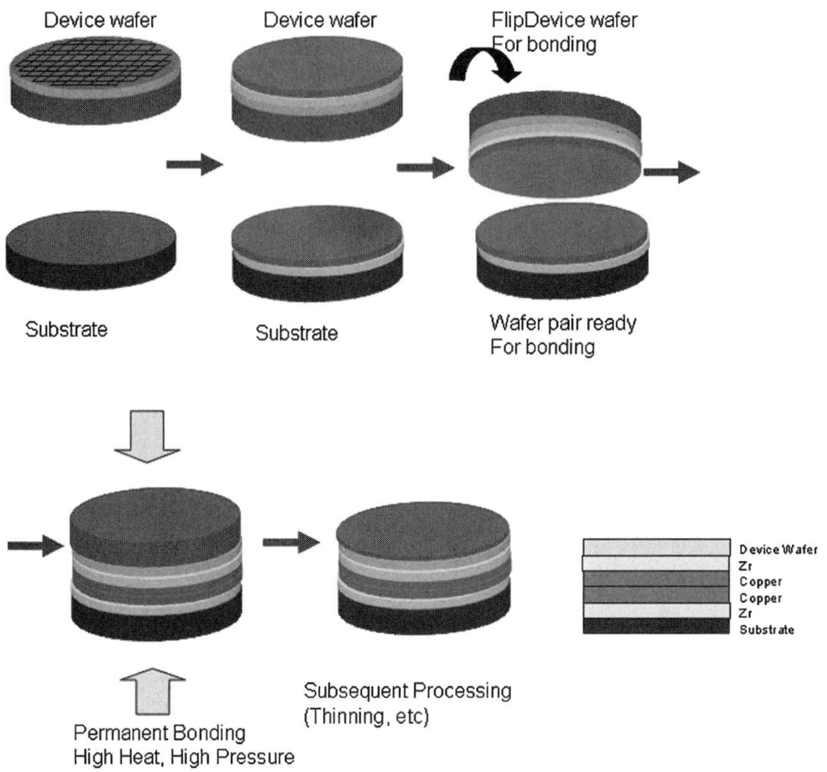

Fig. 5.16 Schematic of process flow for metallic bonding materials

backgrind is followed by a wet/dry or a plasma stress relief. The bonding layer can be subjected to the same forces and chemicals as the silicon.

The choice of bonding layer affects the thinning capability as showed in Fig. 5.17. Since the silicon wafer has a bevel, without the use of a supporting layer, the edge is a cantilever and tends to chip-off leading to cracking. One advantage of the liquid glue layer processes is its ability to support the edge bevel of the wafer when the wafer is thinned into the bevel and this prevents major chipping at the edge [40].

In some cases, as in the use of SOI device wafers, the backgrind is followed by a wet etch in hot alkaline solution and the interface between the bonding material

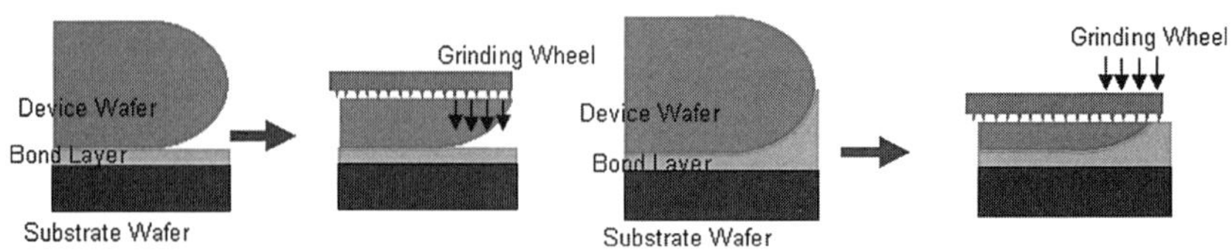

Fig. 5.17 Edge-cracking issues with unsupported wafer edge

as well as the bond material itself should be resistant to attack from this caustic medium for a few hours.

Following thinning, the handle wafer/device wafer system can be subjected to the normal interconnect processes, as allowed by bond layer properties.

5.12 Wafer Debonding Methods

Once all the required processes have been completed, the handle wafer must be removed from the device wafer. The thinned device wafer, then, is supported by either another device wafer (with a permanent bond, as in wafer-to-wafer bonding) or by tape (with an adhesive bond, as in a tape frame for dicing, for die-to-wafer bonding). Several release mechanisms have been developed, depending on the characteristics of the bonding layer or the release layer.

5.12.1 Chemical Methods

Some bond materials can be removed with very specific chemicals that are not used during the normal handle wafer/device wafer application. In such cases, simply dissolving the bond material in a chemical tank can separate the device wafer from the handle wafer. Temperature can be increased to speed up the chemical dissolution. Two examples of this kind of release are as follows:

- Dissolution of waxes or thermoplastics in organic solvents under warm conditions [35, 41];
- Rapid dissolution of the Zr layer (in the metal bonding) with dilute HF, while minimizing attack on the device materials [39].

In either method, dissolution starts from the edge of the substrate/device wafer pair and proceeds into the center. Some agitation is required so that a boundary layer or chemical gradient does not accumulate and slow down the dissolution rate. Additionally, the handling equipment should be such that the delicate wafers are held in place after release (particularly thin, unsupported wafers).

The major challenge with the chemical release is the process time, which can be many hours. Although the chemicals can potentially be recycled or used for multiple wafers, the material cost as well as chemical disposal cost need to be taken into consideration.

5.12.2 Thermal Methods

The device wafer can be easily separated from the substrate wafer when the glue layer is liquified. This is especially useful for thermoplastic bonding materials, because the viscosity decreases upon heating. As shown in Fig. 5.11, when the glue layer viscosity is reduced to about 1–2 poise, the wafer stack is ready for release. Either one of the following two mechanisms can be used [37, 38]:

- slide-off mechanism wherein the two wafers are pulled apart horizontally to separate.
- Wedge lift-off mechanism wherein the two wafers are pulled apart vertically to separate.

These two mechanisms are shown in Fig. 5.18.

Most laminated adhesives are also released by the wedge lift-off method [37]. Some versions of the thermal release material used in laminates contain proprietary formulations that expand when heated into a foam-like material. This reduces the contact area between the device wafer and the liner. The material is also quite weak due to its porosity. In other embodiments, the adhesive mechanical integrity breaks down upon exposure to UV radiation and the bond becomes weaker. The separation between the wafers therefore needs minimal mechanical effort.

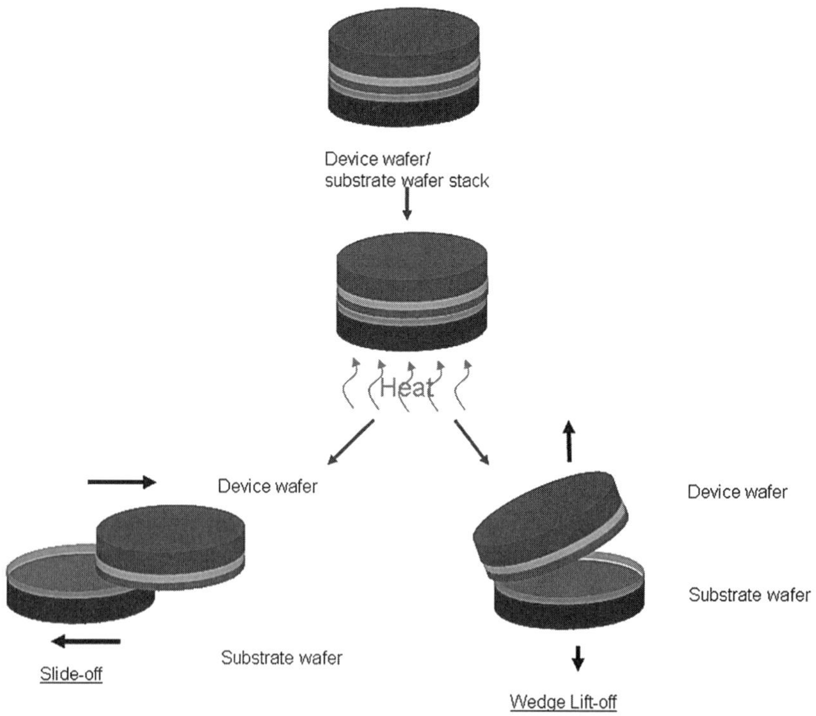

Fig. 5.18 Schematic illustration of slide-off and wedge lift-off methods for removing handle wafers

5.12.3 Laser Treatment

For the case of bond layers that cannot be etched back or heat-treated to a releasable state (e.g., UV-cured bond layers), release layers are employed. An innovation in this field is the use of a light-to-heat conversion layer (LTHC) as a release layer. This layer, when exposed to a laser of the appropriate frequency, absorbs the laser energy to produce microvoids. The weakened layer can then be split up and separated by wedge lift-off method [36].

5.13 Post-release Treatments

When the device wafer is released from the substrate, the substrate wafer undergoes a cleaning cycle to remove any residual adhesives and particles. The reusability of the substrate wafer depends on a number of substrate as well as process requirements: defect levels, bowing/warpage, surface finish, etc. The tool, consumable, and yield cost of this recycle must be considered in calculating the overall cost of this process.

5.14 Summary, Conclusions, and Future Projections

Fabrication of TSVs involves extension of typical wafer fab processes. However, the larger feature sizes and HARs dictate processes that are well outside the mainstream CMOS requirements. The key technical challenges with this module are

- *deep silicon etching*: Profile control, undercut minimization, selectivity to hard-mask;
- *barrier/seed liner*: Continuous barrier and seed coverage in the HAR vias;
- copper plating: Void-free copper filling with minimum overburden in larger-sized vias;
- CMP: Low dishing performance, especially for some via-first integrations.

Along with these technical issues, the throughput offered by some of these processes is very poor compared to traditional BEOL standards. Throughput of less than 10 wph is common for etch, plate, and some CMP steps, which results in greatly increased cost of ownership of this module. Much higher etch rates are currently being attempted through radically different chamber and process designs. Multicell plating, common in the packaging industry, along with more advanced electroplating chemistries, will be needed to productively fill deeper, larger vias with thinner overburden to decrease impact on CMP.

Backgrind and handle wafer technologies have been borrowed and adapted from packaging and MEMS industry processes. The currently available backgrind tools have been proven to be adequate for back-end packaging-related thinning needs. For the ultrathin requirements of 3D wafer stacks, more attention needs to be focused on edge-chipping and defectivity concerns. The current trend toward integrated inline stress relief in backgrind tools by dry and wet polish will be continued along with the introduction of new techniques such as plasma etch to improve the mechanical properties of thinned wafers. Defectivity of post-ground surface has so far not been concentrated on, and this has to be addressed especially if multiwafer stacks need to be built. The throughput of the backgrind module is acceptable for fairly thick silicon (30–50 μm), but needs to be enhanced for achieving thinner silicon.

This chapter also reviews many different handle wafer materials and methodologies to bond and debond device wafers with the use of substrate. A significant challenge for these handle wafer technologies is the maximum-use temperature of the materials, which is currently limited to about 220°C. The dielectric deposition and most permanent bonding processes are most affected by this. Materials are currently in development which allow for a use temperature of 300°C–350°C. The consumable materials and equipment will add costs to the flow, which need to be incorporated along with throughput, and the recyclablity of the substrate wafers to obtain cost-effective integration flows.

References

1. Lo WC, Chen YH, Ko JD, Kuo TY, Chien CW, Chen YC, Chen WY, Leu FJ, Hu HT (2005) Development and characterization of low cost ultrathin 3D interconnect. ERSO/ITRI, Taiwan. Proceedings of the 2005 Electronic Components and Technology Conference, IEEE, pp 337–342
2. Miller WD, Gassman RA, Keicher DM (1993) Laser drilling of vertical vias in silicon. Sandia Report SAND93-1772, UC-704 for Unlimited Release
3. Flamm DL (1990) Mechanisms of silicon etching in fluorine and chlorine-containing plasmas. University of California, Berkeley, Pure Appl Chem 62 (9):1709–1720
4. Rosli SA, Aziz AA, Hamid HA (2006) Characteristics of RIE $SF_6/O_2/Ar$ plasmas on n-silicon etching. Nano Optoelectronics Research and Technology Laboratory, Universiti Sains Malaysia, Proceedings of the International Conference on Semiconductor Electronics 2006, IEEE, Kuala Lumpur, Malaysia, pp 851–855
5. Figueroa RF, Speisshoefer S, Burkett SL, Schaper L (2005) Control of sidewall slope in silicon vias using SF_6/O_2 plasma etching in a conventional reactive ion etching tool. University of Arkansas, Fayetteville, AR, J Vac Sci Technol B 23(5):2226–2231
6. Laermer F, Schilp A (1996) Method of anisotropically etching silicon. US Patent 5,501,893, assigned to Robert Bosch GmbH, issued March, 1996
7. Gomez S, Belen RJ, Kiehlbauch M, Aydi ES (2005) Etching of high aspect ratio features in Si using $SF_6/O_2/HBr$ and $SF_6/O_2/Cl2$ plasma. University of California Santa Barbara, Lam Research Corporation, and University of Minnesota, J Vac Sci Technol A (23) 6: 1592–1597
8. Shareef IA, Rubloff GW, Anderle M, Gill WN, Cottke J, Kim DH (1995) Subatmospheric chemical vapor deposition ozone/TEOS process for SiO_2 trench filling. IBM, NC State University, CMBM, RPI, J Vac Sci Technol B 13(4):1888–1892
9. Dickson M, Hopwood J (1997) Axially-resolved study of highly ionized physical vapor deposition. Northeastern University, J Vac Sci Tech A 15(4):2307–2312
10. Wang Z, Cohen SA (1999) Hollow cathode magnetron. Princeton University, J Vac Sci Tech A 17(1):77–82

11. Van Der Straten O, Yu Z, Dunn K, Eisenbraun E, Kaloyeros AE (2004) Atomic layer deposition of tantalum nitride for ultrathin liner applications in advanced copper metallization schemes. J Mater Res 19(2):447–453
12. Lee C, Lee HH (2005) A potential novel two-step MOCVD of copper seed layers. National Taiwan University of Science and Technology, Electrochem Solid-State Lett 8(1):G5–G7
13. Sukamto JH, Webb E, Andryushchenko T, Reid J (2004) An evaluation of electrolytic repair of discontinuous PVD copper seed layers in damascene vias. Novellus Systems, J Appl Electrochem 34(3):283–290
14. Kwon OK, Kim JH, Park HS, Sang-Woo (2004) Atomic layer deposition of ruthenium thin films for copper glue layers. J Electrochem Soc 151(2):G109–G112
15. Josell D, Wheeler D, Witt C, Moffat TP (2003) Seedless superfill: copper electrodeposition in trenches with ruthenium barriers. NIST and International SEMATECH, Electrochem Solid-State Lett 6(10):C143–C14
16. Keigler A, Wu B, Liu Z (2006) Copper deposition for pillars and vias. Nexx Systems, Semiconductor Manufacturing Magazine
17. Spiesshoefer S, Schaper L, Burkett S, Vangara G, Rahman Z, Arunasalam P (2004) Z-axis interconnects using fine pitch, nanoscale through-silicon vias: process development. Proceedings of the 2004 Electronic Components and Technology Conference, pp 466–471
18. Sandireddy S, Jiang T (2005) Advanced wafer thinning technologies to enable multichip package. 2005 IEEE Workshop on Microelectronics and Electron Devices, pp 24–27
19. Kroninger W, Schneider L, Wagner G (2004) Creating stable and flexible chips for thin packages. Infineon Technologies and SEZ AG, Semiconductor International
20. Bohm C, Hauck T, Juritza A, Muller WH (2004) Weibull statistics of silicon die fracture. Freescale Dutschland, GmbH, Proceedings of the 2004 Electronics Packaging Technology Conference, pp 782–786
21. Seidel H, Csepregi L, Heuberger A, Baumgarte H (1990) Anisotropic etching of crystalline silicon in alkaline solutions I. Orientation dependence and behavior of passivation layers. J Electrochem Soc 137(11):3612–3626
22. Sonphao W, Chairikul S (2001) Silicon Anisotropic etching of TMAH solution. KMITL, Proceedings of the 2001 IEE International Symposium on Industrial Electronics, Pusan, Korea, pp 2049–2052
23. Bradbury D, Tsao CW, Kamins TI (1985) CVD lateral epitaxial growth of silicon over insulators. US Patent 4522662
24. Bruel M (assigned to Commissariat a l'Energie Atomique, Grenoble, France) (1994) Process for the production of thin semiconductor material films. US patent 5,374,564
25. Celler GK, Cristoloveanu S (2003) Frontiers of silicon-on-insulator. SOITEC USA and Institute of Microelectronics, UMR CNRS, INPG & UJF, Grenoble, France, J Appl Phys 93(9):4955–4978
26. Roesner M, Wallis D (2001) Grinding and dicing 300 mm wafers. Future Fab Int 11 http://www.future-fab.com
27. http://www.discousa.com
28. http://www.alsi-international.com
29. http://www.xsil.com
30. http://www.synova.ch
31. http://www.accretechusa.com
32. http://www.oxfordlasers.com
33. Müller MC, Hendriks R, Chall HP (2006) Significant step in wafer yield optimization and operation cost reduction due to dicing innovation. Proceedings of CS MANTECH Conference, p 141
34. Madou MJ (2001) The MEMS handbook: Chapter 16 – MEMS fabrication. CRC Press, Boca Ratan
35. Moore J, Smith A, Nguyen D, Kulkarni S (2004) High temperature resistant adhesive for wafer thinning and backside processing. MANTECH

36. http://www.3 m.com
37. Brubaker C, Wimplinger M, Malzer A, Lindner P (2005) Advances In Processing of Compound Semiconductor Substrates. MANTECH Proceedings, p 261
38. Puligadda R, Pillalamarri S, Hong W, Brubaker C, Wimplinger M, Pargfrieder S (2007) High-performance temporary adhesives for wafer bonding applications. Mater Res Soc Symp Proc 970:239–249
39. Reif R, Fan A, Chen KN, Das S (2002) Fabrication technologies for three-dimensional integrated circuits. Proceedings of the International Symposium on Quality Electronic Design, p 33
40. Kazemi H, Tran L, Xin H, Deakin D, Greer J, Hacker J (2004) Novel via planarization scheme for high resolution backside wafer processing. MANTECH, p 97
41. Munck K, Vaes J, Bogaerts L, Moor P, Van Hoof C, Swinnen B (2006) Grinding and mixed silicon copper CMP of stacked patterned wafers for 3D integration. Mater Res Soc Symp Proc 970:275–280

Chapter 6
Cu Wafer Bonding for 3D IC Applications

Kuan-Neng Chen, Chuan Seng Tan, Andy Fan, and L. Rafael Reif

6.1 Introduction

This chapter describes a fabrication method for three-dimensional (3D) integrated circuits (ICs) initially proposed and developed at the Microsystems Technology Laboratories (MTL) at MIT. This work was further refined at IBM T. J. Watson Research Center. Currently, IBM, NTU, MIT, and other institutes are working on this method for use in real 3D applications. This approach to 3D integration is based on metallic wafer bonding, that is, low-temperature direct copper-to-copper (Cu-to-Cu) thermocompression bonding. In subsequent sections of this chapter, we describe low-temperature wafer-to-wafer bonding using Cu as the bonding medium. While most research on the blanket Cu wafer bonding were carried out at MIT, the patterned Cu-bonding studies were developed at IBM. Fundamentals of the bonding mechanism and development of a reliable bonding process are presented. Building on work with Cu wafer bonding, process flow of multilayer stacking of silicon thin films is presented. The salient features of this approach are also included.

6.2 Classification of Cu-Bonding Techniques

There are two approaches to achieving a successful Cu bond. The first method, surface-activated bonding method, enables direct Cu-to-Cu bonding at room temperature [1]. The second method is thermal compression Cu bonding, which is a more popular technique in 3D integration application considering its simplicity and less cost. The following sections describe the concepts behind these two techniques.

6.2.1 Surface-Activated Cu Bonding

The bonding energy originates from adhesive force between two atomically clean solid surfaces under contact [2, 3]. The bonding process is completed by contact

K.-N. Chen et al.
IBM T. J. Watson Research Center, Yorktown Heights, NY, USA
e-mail: chenku@us.ibm.com

C.S. Tan et al. (eds.), *Wafer-Level 3D ICs Process Technology*,
DOI: 10.1007/978-0-387-76534-1_6,

between two surface-activated wafers (chips) under ultrahigh vacuum conditions ($\sim 10^{-8}$ torr). The activated surfaces can be obtained by dry etching process such as ion-beam bombardment or radical irradiation in a clean atmosphere, for example, in ultrahigh vacuum. Since the surface contamination and native oxides are removed during the surface-activation process, the bonding process can be completed under ultrahigh vacuum conditions at room temperature.

The bonding results show strong bond strength with void-free interface. This bonding technique has one advantage compared to others in that it can be applied to all materials. This is because that the origin of bonding is from the surface adhesive force. In addition, the room temperature bonding can overcome the thermal mismatch problem on the bonding of dissimilar materials. However, the process complexities such as integration of the cleaning equipment, such as Ar ion beam, and ultrahigh vacuum conditions in the bond chamber are key factors affecting mass production consideration.

6.2.2 Thermal Compression Cu Bonding

In addition to the surface-activated bonding method, the second option is thermal compression Cu-to-Cu bonding (also known as diffusion Cu-to-Cu bonding), which uses high pressure to hold two Cu surfaces of the wafers/chips at elevated temperatures. The heat and pressure permit the deformation of microscopic contact between two Cu regions; thus, the contact area can increase. Copper structures with enough thermal energy from the two surfaces can diffuse into each other to finish the bonding process under certain bonding temperatures. With the use of higher bonding temperatures, this method does not have strict requirements on surface cleanliness and chamber vacuum conditions.

Thermal compression Cu-to-Cu bonding method seems more attractive and has received more development in industry and academia because of its simple and less cost requirements. The most important parameter for this method is temperature. Since Cu-to-Cu bonding is used for electronic devices and applications, the bonding temperature has to be compatible with back-end-of-the-line (BEOL) process temperatures in order not to affect the device performance.

Because the mainstream Cu bonding in 3D integration is thermal compression, in the following description we will focus on developments and achievements in this technique.

6.3 Fundamental Properties of Cu Bonding

This section describes the fundamental properties of Cu-bonded layer, including the morphology, oxide contents, microstructure evolution, and orientation evolution. Understanding these properties is important because it is the key to achieve a good Cu bonding for 3D integration. Most of the research was completed at MIT.

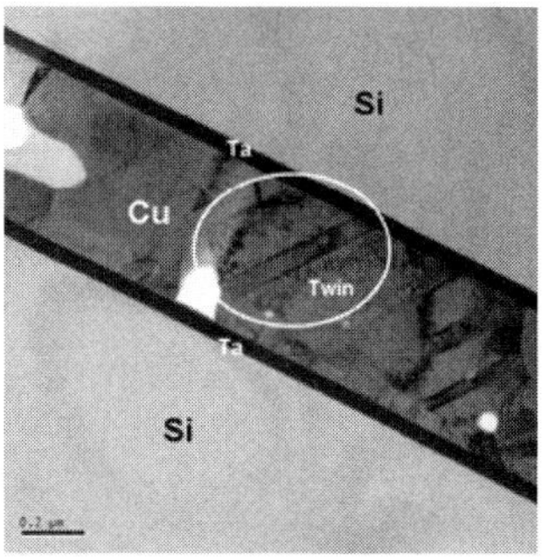

Fig. 6.1 Transmission electron microscopy morphology of Cu-bonded layers [4]. Used with permission from TMS/Springer

6.3.1 Morphology of Cu-Bonded Layer

A well-bonded Cu-to-–Cu layer is shown in a transmission electron microscopy (TEM) image of Fig. 6.1 [4]. The bonding process was performed in an EVGroup AB1-PV bonder with a pressure of 4000 mbar, a vacuum of 10^{-3} torr, and bonded at 400°C for 30 min, followed by nitrogen anneal at 400°C for 30 min. Before bonding, wafers were soaked in HCl solution for 30 s followed by a deionized (DI) water rinse and spin dry to remove the surface oxide. From this image, we cannot observe the original bond interface but grain structures are shown, as well as twins across the Cu-bonded layer. Because of the disappearance of the bonded interface, the two Cu layers may have merged into a single layer after bonding with strong bond strength. In addition, continuous twins which penetrate the bonded layer were observed. These twins could not have been formed and would have stopped at the interface instead if there were still a distinct interface or defect areas between the two original copper layers. Especially, since there is no evidence of truncated twins in this area, it implies that twins were formed after the uniform bonded layer was formed [5].

6.3.2 Oxide Examination of Cu-Bonded Layer

With energy dispersive spectroscopy (EDS) examination, a uniform oxide distribution across the bonded layer is shown in Table 6.1 [5]. These results were detected from various beam sizes of X-ray as large as 500 nm or as small as 5 nm. The quantities

Table 6.1 Energy dispersive spectroscopy examination for oxygen in different areas of the bonded layer [5]. Used with permission from TMS/Springer

Beam size (nm)	Oxygen(wt.%)	Location Description
500	2.67	Whole bonded layer
25	2.13	Near tantalum layer
5	2.22	Near tantalum layer
25	2.53	Near bonded interface
5	2.78	Near bonded interface
25	2.98	Near tantalum layer
5	2.89	Near tantalum layer

of oxygen in the layer were always lower than 3 wt.%, which is below the detection threshold of EDS. Therefore, the original oxides on the surface of copper layer might have been removed during HCl treatment, and the remaining of the oxides may be considered negligible during bonding. It may be also possible that the surface oxides on copper layer were only partially removed during HCl treatment (or new oxides formed between the HCl treatment and bonding). However, during bonding, under the high temperature and pressure the oxygen atoms diffused into the layer easily and quickly and then distributed everywhere in the bonded layer. Both scenario and the results show that under this condition and HCl treatment, the quantity of oxide in the bonded layer could be effectively controlled for bonding to proceed [5].

6.3.3 Microstructure Evolution During Cu Bonding

In order to understand the bonding mechanism, the evolution of the bonded layer, including the microstructure morphology and grain orientation, was investigated. As shown in Fig. 6.2a–c, the bonded Cu layer reaches a steady state after the anneal process [6]. Figure 6.2a shows that before bonding the Cu grains have an average size of 300 nm, and a major (111) orientation. In Fig. 6.2b, the bonded interface is clearly observed but it is not straight. Different kinds of morphologies such as twins are present in the bonded Cu film. The grain size distribution ranges from 300 nm to 700 nm. In addition, no major grain orientation but (111), (200), and (222) grain orientations are observed. However, Fig. 6.2c shows a well-developed grain texture with an 800-nm grain size profile and a major (220) orientation after further anneal [6].

During bonding and annealing, there is strong grain growth, based on TEM image results. The two original Cu layers can still be distinguished. The jagged Cu–Cu interface shows that interdiffusion between two Cu layers takes place during bonding. However, the grain growth is incomplete and the bonded layer still contains defects such as twins in the individual layer, as shown in Fig. 6.2b. Enough energy during the post-bond annealing was provided to the bonded layer to complete the grain growth and finally achieve a stable microstructure. As a result, the two Cu layers can no longer be distinguished in Fig. 6.2c. From TEM images under different bonding conditions, the distribution of average grain size is estimated in Fig. 6.3. A large distribution of grain sizes after 30-min bonding is observed due to the incom-

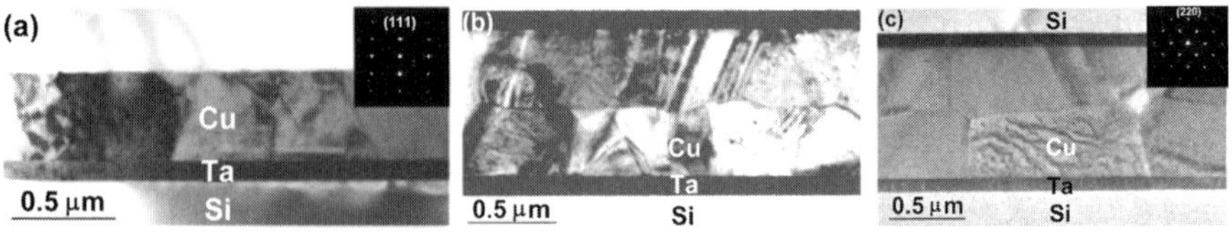

Fig. 6.2 Transmission electron microscopy images of Cu-bonded layers (**a**) before bonding, (**b**) after 400°C bonding for 30 min, and (**c**) after 400°C bonding for 30 min followed by nitrogen anneal at 400°C for 30 min. Reprinted with permission from [6] © 2002, American Institute of Physics

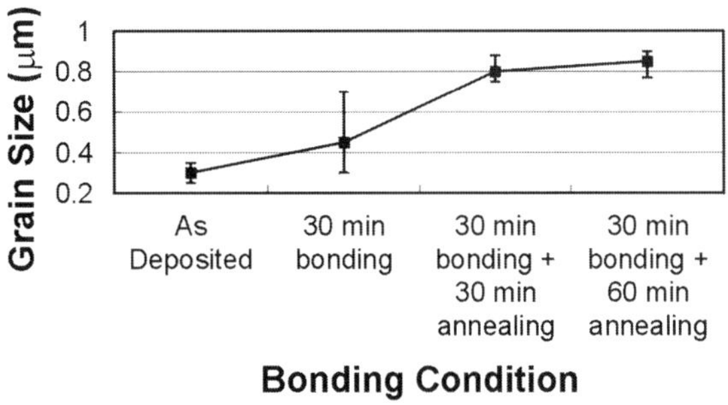

Fig. 6.3 Average grain size as a function of different bonding conditions. Reprinted with permission from [6] © 2002, American Institute of Physics

plete grain growth. The grain growth then accelerated during the first 30 min of post-bond annealing. However, the grain size did not increase significantly beyond 30 min of annealing [6].

6.3.4 Orientation Evolution During Cu Bonding

An identical result of preferred orientation changes in Cu film from (1 1 1) to (2 2 0) can be also observed in X-ray diffraction (XRD) analysis, as shown in Fig. 6.4. This abnormal grain growth can be explained with yield and energy minimization of the whole system [6]. Because both Cu layers before bonding are attached to the much thicker Si substrate, Cu-bonded layer is in a state of biaxial strain. Since the in-plane stress in a grain is the product of the biaxial strain and biaxial modulus, which is a function of grain orientation factor C_{ijk}, the yield stress of the grain also varies depending on its orientation. It is reported that the orientation factor C_{ijk} of (111) has the largest value of 3.46 while that of (220) has the smallest one of 1.42 [7]. This means that the yield stress of (220) is much smaller than that of (111). For different grains with equal initial sizes, (220) grains will yield before (111) grains, thus the (220) grains have an energetic advantage for further growth [7]. At the same time, the yielding process leads to strain energy minimization [7, 8]. During

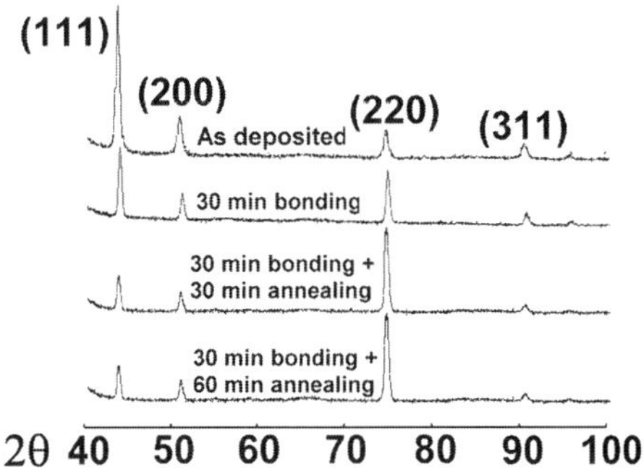

Fig. 6.4 X-ray diffraction patterns for different bonding conditions. Reprinted with permission from [6] © 2002, American Institute of Physics

the further annealing process, the bonded wafer is heated again from room temperature to high temperature. Therefore, the in-plane strain increases again. Then, the secondary grain growth occurs at high temperature. At the same time, (220) grains yield first and take this energetic advantage to minimize the strain energy in the bonded layer. This explains the abnormal (220) grains growth and becomes the final preferred orientation [6].

6.4 Cu-Bonding Development

To achieve the best bond quality, the structural design of Cu bond pads, Cu pad fabrication and preparation, and bonding parameters all play important roles. Each aspect also includes several variables. The following is the summary of Cu-bonding quality development on the wafer-level scale. While most research on the blanket Cu film bonding was carried out at MIT, the patterned Cu-bonding studies were developed at IBM. Most results and guidelines are also applicable for die-level Cu-bonding applications.

6.4.1 Structural Design

The structural design of Cu bond pads determines not only the circuit placement but also bond quality since it is related to the available area that can be bonded in a local region or across the entire wafer/chip. Cu bond pad size (interconnect size) and pad pattern density (total bond area) are two factors that need to be considered:

1. *Copper bond pad size:* Cu bond pad sizes larger than 2 μm have been investigated for the corresponding bonding strength under the same bonding density (total bonding area). It has been suggested that the bonding quality does not have a strong dependence on the bond pad size [9].
2. *Pattern density:* Several of pattern densities from <1% to 35% have been investigated for their bond qualities. Higher bond density results in better bond quality. Based on existing results, the bonded area rarely fails in the dicing test when pattern density is larger than 13% [9]. Figure 6.5 shows the bond quality for different bond pad sizes and pattern densities [9].

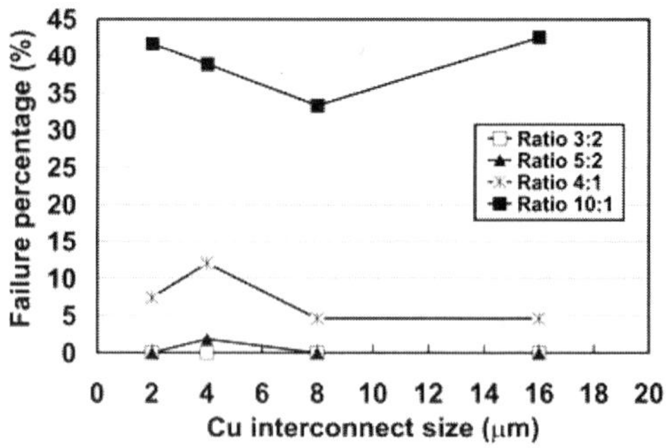

Fig. 6.5 Failure percentage results of dicing test for various Cu interconnect pattern geometry. Ratio is taken as pitch-to-diameter. From [9] © 2006 IEEE

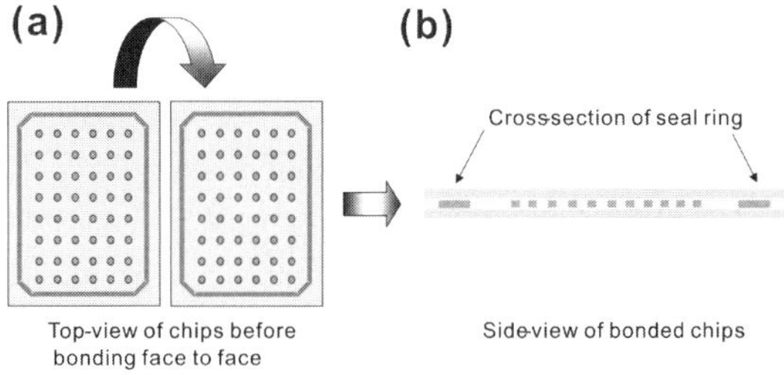

Fig. 6.6 Chip seal design concept. Reprinted with permission from [10] © 2006 IMIC

3. *Seal design:* As shown in Fig. 6.6, a seal design, which has extra Cu bond area around electrical interconnects, chip edge and wafer edge, can prevent corrosion and provide extra mechanical support [10].

6.4.2 Copper Pad Fabrication

A well-fabricated copper pad can guarantee full contact between two Cu surfaces during the bonding process without interference from contamination, native oxide, or other materials. Important factors include copper pad fabrication process and the surface treatment prior to bonding:

1. *Copper fabrication:* In order to make sure the two damascene Cu bond pads can fully contact during bonding, it is necessary to recess the surrounding material, usually oxide. It has been shown that a combination oxide recessing process, using a SiO_2 chemical–mechanical planarization (CMP) process and dilute hydrofluoric acid (HF) wet etching, gives the best bond quality [10].
2. *Surface cleanliness:* The surface cleanness is directly related to the difficulty of the bonding process if there are limitations on other parameters. To remove native oxides and other contaminants, an HCl clean followed by DI wafer rinse and dry has been widely used prior to bonding process [5].

6.4.3 Bond Parameters

Bond parameters include bond temperature, duration, pressure, chamber ambient condition, and anneal option. Some parameters are also limited by the device performance and cost considerations. Here are the guidelines for each parameter:

1. *Bond temperature*: It has been shown that higher bond temperature results in better bond quality [11]. However, in order to meet BEOL process requirements, the bond temperature is usually fixed at or below 400°C.
2. *Bond duration*:Although this effect is not as dominant as bond temperature, longer bond duration improves the bond quality [11]. However, for manufacturing cost consideration, shorter bonding process times are always preferred. In order to achieve a good bond quality, a minimum of at least 30 min is suggested at a bond temperature of 400°C [5].

3. *Bond pressure*: Sufficient bond pressure can hold two Cu surfaces close enough for bonding completion. The maximum bond pressures, 4000 N for 100-mm wafer in EVG AB1-PV bonder and 10 000 N for 200-mm wafer in EVG bonder, are both shown to provide good bond quality [5].
4. *Chamber ambient condition*: High-vacuum environment can effectively reduce the possibility of oxide formation and surface contamination, which will hinder high-quality bond, before and during the bonding process. The vacuum conditions, 10^{-3} torr for 100-mm wafer in EVG AB1-PV bonder and 10^{-4} torr for 200-mm wafer in EVG bonder, are both shown to provide good bonding quality [5, 9]. In addition, nitrogen or forming gas purges after loading wafers in bonder chamber are shown to be significant [5].
5. *Anneal:* Generally both microstructure and strength of the bonded layer show that further nitrogen anneal at the same bonding temperature improves the bond quality, especially for blanket copper films [5]. This criteria is not applicable for low-temperature bonding (below 300°C) since its bond strength is too low to withstand the thermal stress during anneal [12].
6. *Bond profile:* For patterned Cu pads on 200-mm wafers, the bond profile for the best bond quality, including strength, alignment, and electrical connectivity, is to apply a small bond pressure (~1000 N) prior to temperature ramping, followed by high bond pressure (~10 000 N). The temperature ramping rate should be kept low such as 6°C min^{-1} if possible [9].

6.5 Cu Bond Quality Characterization and Alignment Accuracy

The bond quality can be evaluated with image analysis, mechanical test, electrical connectivity, and thermal reliability. Although there are no concrete definitions toward Cu bond quality, researchcarried out at MIT and IBM have established several general guidelines. In addition to bond quality, the alignment accuracy is another requirement before the bonding technology can be used in real applications.

1. *Image analysis*: Electron microscopy, such as TEM and scanning electron microscopy (SEM), is a good method for examining the morphology of bonded layer. A TEM image of the bonded layer can clearly show the existence of an original interface. In addition, a bonded layer that has survived TEM sample preparation is a good indication of bond strength. Therefore, TEM observation results of the Cu-bonded layer can be categorized into three groups: (1) no interface but grain structure (identified as "Grain"), (2) interface structure (identified as "interface"), and (c) failure of TEM sample preparation (identified as "TEM failed") [11].
2. *Mechanical test*: Mechanical tests for Cu bond strength include quantitative and qualitative approaches, such as dicing test, tape test, pull test, and shear test. Dicing test is the most common approach by identifying a "failed piece" when the bonded piece separates due to the sawing stress. Figure 6.7 provides a map including the results from image analysis and dicing test under different bond temperatures and durations [12].

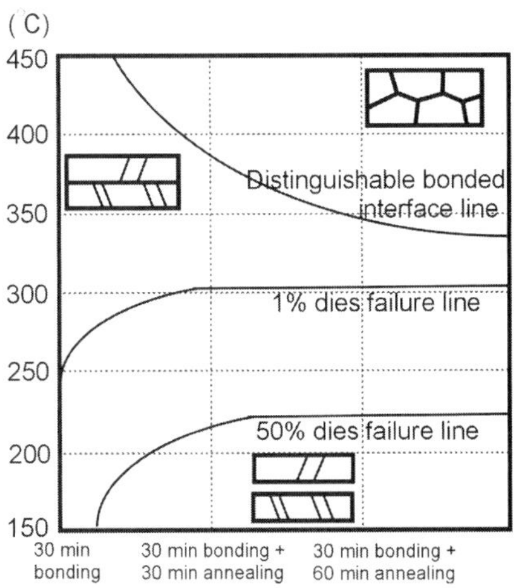

Fig. 6.7 Morphology and strength map for copper wafer bonding [12]. Reproduced by permission of ECS – The Electrochemical Society

Based on the results of pull and shear testing, the strength of Cu bond can be as high as 70 MPa and 80 Kgw, respectively. In addition, mechanical tests have shown that the post-bond thermal anneal improves bond strength only in high bond temperature samples [13].

3. *Electrical characterization:* Generally, the specific contact resistance of bonded Cu interconnects with different sizes is approximately 10^{-8} ohm-cm^2 [9, 14], which is consistent with measurements taken from similar structures build by depositing and patterning metal lines [15, 16]. In addition, chain measurements have shown excellent performance and stability of bonded interface across wafers after multiple stress cycles [9], as shown in Fig. 6.8.
4. *Thermal reliability:* Thermal reliability testing is an important method of evaluation for subsequent process steps and life reliability of Cu-bonded structures. Cu-bonded structures across wafers successfully passed thermal reliability tests which subjected the samples to 1000+ deep thermal cycles between –55°C and +125°C [9].

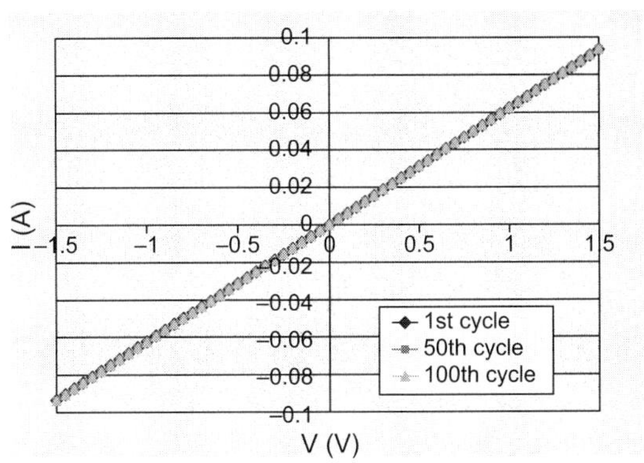

Fig. 6.8 Stable *I–V* characteristics of Cu-bonded chain after multiple current stressing. From [9] © 2006 IEEE

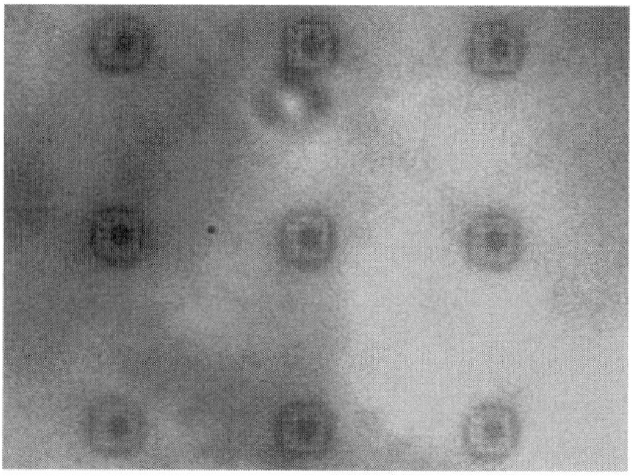

Fig. 6.9 Infrared image showing typical misalignment results after bonding [9] © 2006 IEEE

5. *Alignment accuracy:* Alignment accuracy of bonded structures can be inspected by a microscope directly if one wafer/chip is transparent. If both sides are not transparent, the accuracy can be inspected by cross-sectional microscopy analysis, removal of one substrate, or by infrared (IR) inspection if both sides of substrates are polished. An IR image of bonded Cu circular-to-pad structure is shown in Fig. 6.9. The misalignment after bonding can be kept within the 0.5-μm range using the best bonding conditions [9].

6.6 Reliable Cu Bonding and Multilayer Stacking

Based on previous Cu-bonding development, one can achieve a good Cu bond structure which passes the quality characterization requirements with high accuracy alignment. Figure 6.10 shows a high-quality Cu-pad-to-Cu-pad bonding result [9].

Multiple silicon layer stacking based on Cu bonding, grind back, and etch back was applied to demonstrate a strong four-layer-stack structure, as shown in Fig. 6.11 [17]. All bonded Cu layers in this structure became homogeneous layers and did not show original bonding interfaces, as shown in Fig. 6.12 [17]. This means that the bonded structure is strong enough to withstand the applied stresses during grind

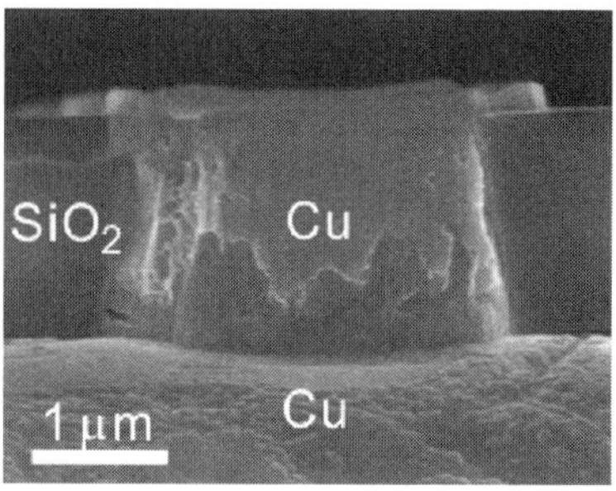

Fig. 6.10 Scanning electron microscopy image of a 4-μm Cu interconnect bonded to Cu pad showing a high-quality Cu-to-Cu bonding interface [9] © 2006 IEEE

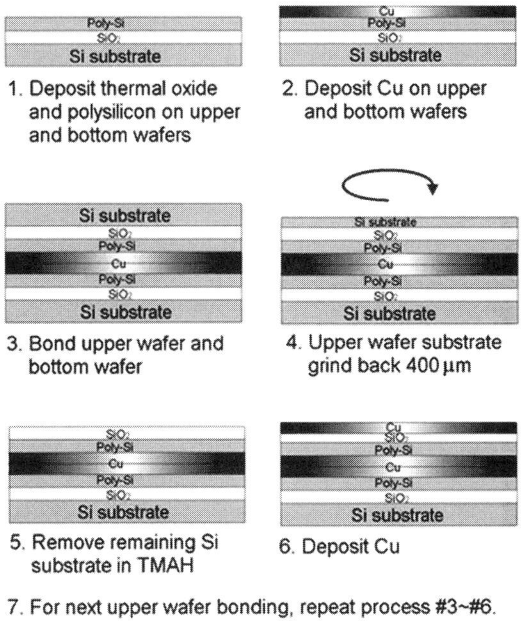

Fig. 6.11 Fabrication flow for the multiple silicon layer stacking based on Cu bonding. Reprinted with permission from [17] © 2005, American Institute of Physics

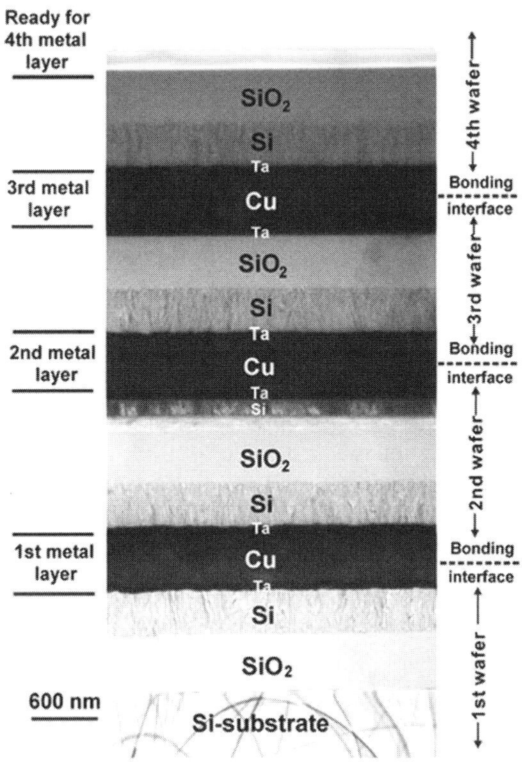

Fig. 6.12 Multiple silicon layer stacking based on Cu bonding, grind back, and etch back. Reprinted with permission from [17] © 2005, American Institute of Physics

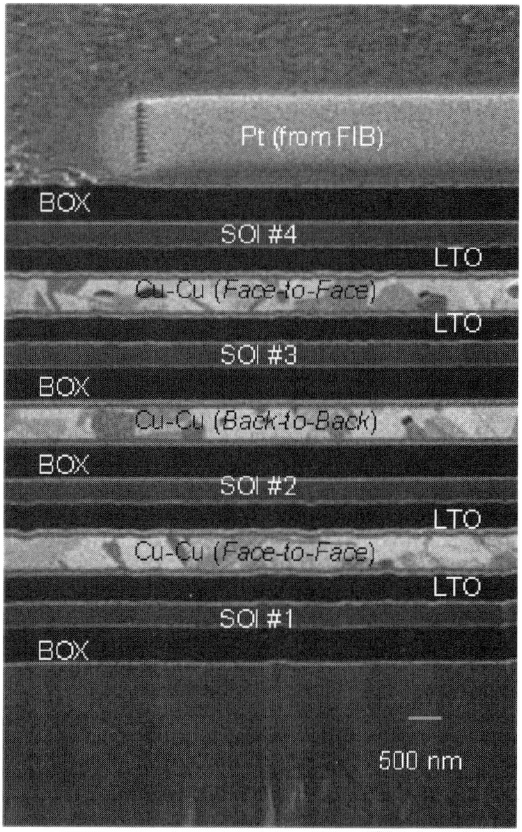

Fig. 6.13 Focus ion beam (FIB) image clearly shows a silicon four-layer stack achieved by stacking two silicon bilayer stacks. This paves a promising path to multilayer and multifunctionality silicon stack [18]. Reproduced by permission of ECS – The Electrochemical Society

back and etch back. The successful development of multiple silicon layers stacking based on Cu bonding means 3D integration using Cu bonding is feasible [17].

The four-layer stack in Fig. 6.12 is a sequential process. It begins with a face-to-face bonding orientation and subsequent face-to-back bonding orientation. It is also possible to achieve a four-layer stack in a multiplicative approach. One can begin by bonding two wafers face-to-face and perform wafer thinning to achieve a bilayer stack. By bonding two bilayer stacks one can achieve a four-layer stack. Subsequent multiplication is also possible as needed. A process flow of this method is described in [18]. Figure 6.13 shows a four-layer stack achieved using this flow.

6.7 Applications of Cu Wafer Bonding

Copper thermocompression bonding can be applied to fabricate 3D ICs and a number of proposed process flows can be found in [19]–[21]. IBM and other companies have announced to use the technology taking the advantages of reduced wire length in 3D ICs for both processor-on-processor and memory-on-processor chip designs [22]. These applications can be achieved based on Cu bonding with through-Si via (TSV) technology. Morrow et al. [23] at Intel reported the first demonstration of integrating 300-mm wafer stacking via Cu bonding with strained-Si/low-K 65-nm complementary metal-oxide semiconductor (CMOS) technology. Sets of active devices such as 65-nm metal-oxide semiconductor field-effect transistors (MOSFETs) and 4-MB static random access memory (SRAMs) were bonded face-to-face using copper pads. Patti et al. [24] reported Cu-bonding technology developed at Tezzaron. A number of 3D devices have been demonstrated using this stacking process such as stand-alone memory, CMOS sensor, 3D FPGA, mixed-signal ASIC, and processor/memory stack.

6.8 Summary

In this chapter, a method of wafer bonding using metallic copper is described. Bonding is achieved using thermocompression of wafers coated with Cu layer. Both blanket and patterned wafers are studied. Fundamental aspects of bonding mechanism as well as process development of Cu wafer bonding are examined. Multilayer stacking of silicon layers using Cu as the bonding medium is also presented. Applications of Cu wafer bonding for 3D ICs are described briefly in the final section.

Acknowledgments The authors wish to thank the contributions and support for the work presented in this chapter from MIT, IBM, RPI, and EV Group. This project was partially funded by DARPA under contract number N66001-04-C-8032.

References

1. Kim TH et al (2003) Room temperature Cu–Cu direct bonding using surface activated bonding method. J Vacuum Sci Technol A21(2):449–453
2. Shigetou A et al (2003) Room temperature bonding of ultra-fine pitch and low-profiled Cu electrodes for bump-less interconnect. IEEE, Electronic Components and Technology Conf. (ECTC), 53rd, Proceedings., New Orleans, Louisiana, USA, May 27–30, 848–852
3. Suga T, Takahashi Y, Takagi H et al (1993) Ceram Trans 33:323
4. Chen KN et al (2006), Bonding parameters of blanket copper wafer bonding. J Electron Mater 35(2):230–234
5. Chen KN et al (2001) Microstructure examination of copper wafer bonding. J Electron Mater 30:331–335
6. Chen KN et al (2002) Microstructure evolution and abnormal grain growth during copper wafer bonding. Appl Phys Lett 81(20):3774–3776
7. Thompson CV, Carel R (1996) J Mech Phys Solids 44:657
8. Sanchez JE, Artz E (1992) Scripta Metall Mater 27:285
9. Chen KN et al (2006) Structure design and process control for Cu bonded interconnects in 3D integrated circuits, 2006 International Electron Devices Meeting (IEDM), pp 367–370
10. Chen KN et al (2006) Improved manufacturability of Cu bond pads and implementation of seal design in 3D integrated circuits and packages, 23rd International VLSI Multilevel Interconnection (VMIC) Conference, Fremont CA, Sep pp 25–28
11. Chen KN et al (2003) Temperature and duration effect on microstructure evolution during copper wafer bonding. J Electron Mater 32(12):1371–1374

12. Chen KN et al (2004) Morphology and bond strength of copper wafer bonding. Electrochem Solid St Lett 7(1):G14–G16
13. Chen KN et al (2006) Investigations of strength of copper bonded wafers with several quantitative and qualitative tests. J Electron Mater 35(5):1082–1086
14. Chen KN et al (2004) Contact resistance measurement of bonded copper interconnects for three-dimensional integration technology. IEEE Electron Devices Lett 25(1):10–12
15. Schwartz GC (1998) Handbook of semiconductor interconnection technology. Marcel Dekker, New York, p 187
16. Rabaey J (1996) Digital integrated circuits. Prentice Hall, Englewood Cliffs, NJ, p 465
17. Chen KN et al (2005) Processing development and bonding quality investigations of silicon layer stack using copper wafer bonding. Appl Phys Lett 87(3):031909-1–031909-3
18. Tan CS et al (2005) Multi-layer silicon layer stacking based on copper wafer bonding. Electrochem Solid St Lett 8(6):G147–G149
19. Reif R et al (2002) Fabrication technologies for three-dimensional integrated circuits. In: Proceedings of the International Symposium on Quality Electronic Design
20. Reif R et al (2002) 3-D interconnects using Cu wafer bonding: technology and applications. In: Advanced Metallization Conference
21. Reif R et al (2004) Technology and applications of three-dimensional integration. In: 206th Electrochemical Society Fall Meeting
22. Morrow PR et al (2006) Design and fabrication of 3D microprocessors. In: MRS Fall Meeting, Boston
23. Morrow PR et al (2006) Three-dimensional wafer stacking via Cu–Cu bonding integrated with 65 nm strained-Si/Low-k CMOS technology. Electron Device Lett 2(5):335–337
24. Patti R (2006) Three-dimensional integrated circuits and the future of system-on-chip designs. Proc IEEE 94(6):1214–1224

Chapter 7
Cu/Sn Solid–Liquid Interdiffusion Bonding

A. Munding, H. Hübner, A. Kaiser, S. Penka, P. Benkart, and E. Kohn

The concept of solid–liquid interdiffusion (SLID) is known as a metallurgical joining technique since the 1960s [7] and has gained significance in semiconductor bonding during the last decades. It has raised particular interest in three-dimensional (3D)-integration technology because of its attractive properties, mainly its iterative applicability in chip stacking, which it owes to the irreversible nature of diffusion. The concept is also known as isothermal solidification and transient liquid phase (TLP) bonding [30], and its beneficial properties are particularly prominent in the Cu–Sn system.

Figure 7.1 visualizes the future shrinking of solder ball dimensions. Balls with a diameter of 0.5 mm and a pitch of 1 mm are currently in use for ball grid arrays (BGAs) for the connection of the package to the board. Micro bumps, which are used inside the packages, have diameters below 100 μm. Irrespective of the solder volume, the thickness of the intermetallic compound (IMC) layers stays constant. Thus, with shrinking ball diameters, the brittle IMCs will become more and more prominent and will dominate the thermodynamical behavior of the contacts. Below 10 μm the plastic stress response will completely vanish. This fact constitutes a motivation for the investigation of micro contacts consisting purely of IMCs.

This chapter first describes the physical and metallurgical principles of forming bonds based on isothermal solidification (cf. Section 7.1), and introduces parameters that define the bonding process window. In Section 7.2 a production-ready chip-stacking technology with results on reliability is presented. Section 7.3 lists present-day uses of SLID in selected exemplified 3D-integration concepts. Section 7.4 contains a brief comparison of the assembly alternatives, namely wafer-to-wafer (W2W) versus chip-to-wafer (C2W), and the concluding Section 7.5 closes the chapter.

A. Munding
Institute of Electron Devices and Circuits, University of Ulm, Ulm, Germany. Now with Liebherr-Elektronik GmbH, Lindau, Germany
e-mail: andreas.munding@liebherr.com

C.S. Tan et al. (eds.), *Wafer-Level 3D ICs Process Technology*,
DOI: 10.1007/978-0-387-76534-1_7, © Springer Science+Business Media, LLC 2008

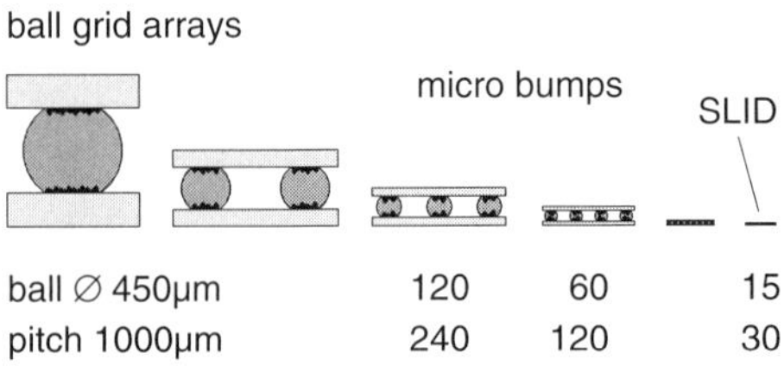

Fig. 7.1 As solder volume shrinks in future bump technologies, intermetallic compounds are still present to the same extent, starting to dominate the joint characteristics. solid–liquid interdiffusion contacts are completely made up by intermetallics

7.1 The Principle of Solid–Liquid Interdiffusion

Soft soldering is a technique widely used in the electronics industry in order to reliably join two conductors. The mechanical properties of the contacts are given by the solder material, the characteristics of which are a low melting point and plastic deformation when subjected to mechanical stress. A second class of soldering – hard soldering or brazing – is characterized by high-melting solder alloys and a hard, elastic response to mechanical stress [27, 15]. Brazing usually is not used for electronic applications.

Solid–liquid interdiffusion is something in-between, since the process by definition is soft soldering in terms of the low melting temperature of the filler alloy, but the result is a high-melting and nonductile joint, which has the characteristics of a hard soldered connection. This extreme difference between the low processing temperature and the final high melting temperature is one of the most interesting benefits of SLID. It allows for sequential processing at the same process temperature, a property which makes it well suited for multilevel chip stacking.

There are two prerequisites for the SLID-process:

1. At least two metals are involved, which are able to form intermetallic phases – a low-melting metal (solder) and a high-melting metal.
2. The solder thickness must be in the range of a few microns in order to solidify within reasonable processing time. This indicates that SLID preferentially is suited for planar technologies and small geometries.

There are many metal systems, showing an isothermal solidification behavior, with different characteristics concerning the complexity of the phase equilibrium, the reaction rates, or the processing temperatures. The reflow temperature is chosen at least 20 C above the melting temperature of the solder. A higher temperature favors wetting, while a lower value is crucial with respect to the homogeneity of the processing oven. Table 7.1 shows a selection of suiting low-melting metals and alloys together with their melting points.

The high-melting material is usually one of the noble metals like gold, silver, or copper. In combination with Sn, the reaction rates are decreasing from Au to Ag to Cu. In principle, other metals can also be used for SLID; however, the chemical

Table 7.1 Examples of Pb-free solders and solder alloys

Low-melting solders and alloys	Composition	Melting temperature (°C)
Sn	Pure	232
In	Pure	156
Ga	Pure	27
Sn-3.5Ag	Eutectic	221
In-48Sn	Eutectic	120
In-22Bi	Eutectic	73
BiIn	IMC	110

reactions between the two metals may be severely hindered by the existence of a stable metal oxide at the interface. Nickel (Ni) often is used as a barrier material to Sn because of its rather slow reaction kinetics with Sn.

Few low-melting metals have technical relevance, but the choice is broadened by alloys and low-melting intermetallic phases thereof. The special SLID systems containing Hg as low-melting component are called amalgams, and have broadly been used in dentistry. They are not used in the electronic industry. Likewise ruled out are alloys containing Pb.

In the following, the basics of SLID will be demonstrated based on the metal systems Cu/Sn, which is one of the best-investigated and well-established metal systems.

7.1.1 Liquefaction and Liquid Phase Properties

For an implementation of a SLID bonding process for chip stacking it is necessary to understand and gain sufficient control over two regimes. The first one will be treated in the course of this section, and the latter one in the succeeding section.

1. The liquefaction of solid solder material, its wetting, and its behavior in the liquid state, and
2. the phase growth kinetics and metallurgy at the Cu/Sn interface.

7.1.1.1 Characteristics of Soldering with Thin Solder Layers

When the two surfaces to be joined are brought into contact, the uniformity of the solder flow is critical, which is illustrated in Fig. 7.2 where the footprint of a faulty solder joint is shown. The thickness of the solder initially was 3 μm, and the joint area of the chips was 1 cm^2. Both chips had been pressed together by a force of 30 N.

Severe lateral solder flow leads to solder squeeze-out at the lower edge and to solder depletion in the opposite region. The critical moment is when the solder

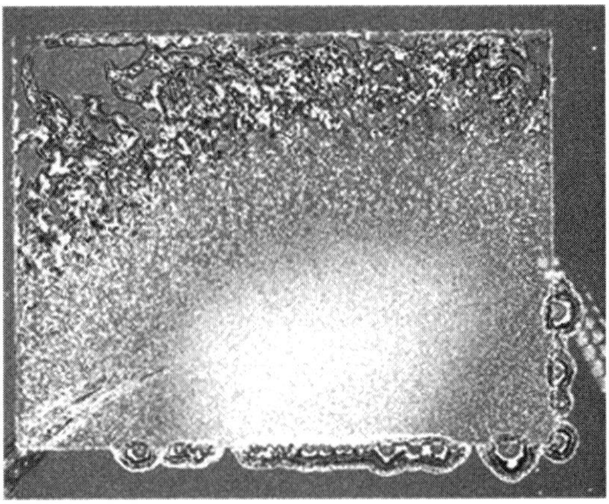

Fig. 7.2 Footprint of a faulty solder joint of two silicon chips, bonded using a 3 μm thick layer of solder at 30 N of down force on an area of 1 cm^2

liquefies, because this does not happen instantly; liquefication starts at a few points, from where the fronts spread radially at approximately 1 cm s^{-1}. Depending on the distribution of the liquid areas, the force may not act precisely in the center *S* of the area of the liquid (cf. Fig. 7.3a). The resulting torque causes a lateral flow of the solder. The system collapses, resulting in a wedge-shaped solder joint. Moreover, the strong flow disturbs the phase growth and a cross-linking of the crystallites cannot be established. The time constant associated with these processes is in the range of only 0.1 s.

A practical alternative is sketched in Fig. 7.3b. Patterning of the area into several pads impedes the lateral flow and each of the pads can act as a local spring, giving feedback to the system (cf. next section). A stable chip configuration is reached, when the sum of the restoring forces F_i reaches equilibrium with the external force *F* in point A, which need not be the center of the area *S*. Of course, the stable position in general will not be perfectly parallel but slightly tilted.

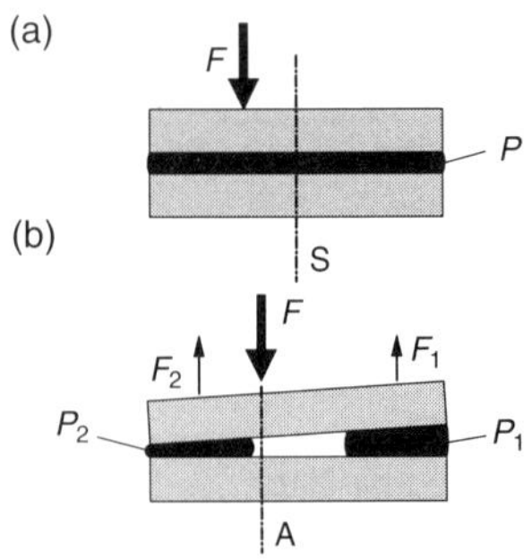

Fig. 7.3 Forces acting on a chip during the liquefaction of a thin solder layer with (**a**) unpatterned metallization and (**b**) patterned metallization

7.1.1.2 Hydrostatics of the Solder Pad

As has been shown above, the short moment, during which the solder is in the liquid state and the phases grow rapidly, has the largest impact on the quality of the solder joint. Thus, it seems reasonable to take a closer look on the hydrostatic behavior of a liquid metal film.

In order to simplify the situation, we will assume two circular pads with a cylindrically shaped volume of liquid solder in between (Fig. 7.4). The second pad on top of the solder defines a boundary condition, which is different from conventional solder reflow, where the free surface of liquid solder is allowed to contract into a solder ball. Here, the cylinder-shaped solder is brought into contact with the opposing pad prior to melting.

The hydrostatic pressure p inside of the liquid is given by [1].

$$p = \gamma\left(\frac{1}{R_1} + \frac{1}{R_2}\right), \tag{7.1}$$

where γ is the surface tension, and R_1 and R_2 are the two radii of curvature in any two planes, which are perpendicular to each other. For reason of convenience, R_1 will be chosen as the constant radius of the pad, given by the pattern. R_2 is the variable curvature of the free surface, seen in the vertical cut. With no external force, the hydrostatic pressure balances the surface tension of the liquid, which acts on the circumference of the pad. The weight of the chip and solder can be neglected. If an additional orthogonal force F is acting in the center of the pad, the following equation holds [19]:

$$F = pA + F_{\text{circ}} = \pi R_1^2\gamma\left(\frac{1}{R_1} + \frac{1}{R_2}\right) - 2\pi R_1\gamma\cos(\beta), \tag{7.2}$$

where F_{circ} is the circumferential force resulting from the surface tension, and β is the surface angle at the interface between solder and pad. When the gap becomes smaller, the solder is pressed outwards, R_2 is positive, becomes smaller, and the hydrostatic pressure increases. The resulting external force F is diminished by the value of F_{circ}, which always acts downwards and decreases as the distance from the initial position increases. When the pad is pulled apart, the surface profile bends inwards, resulting in a negative, that is, R_2 and a negative tensile pressure p, which then acts in the same direction as F_{circ}.

The maximum external force F is given by the condition $R_2 = d/2$, where d denotes the variable solder-filled gap distance between the pads. Further

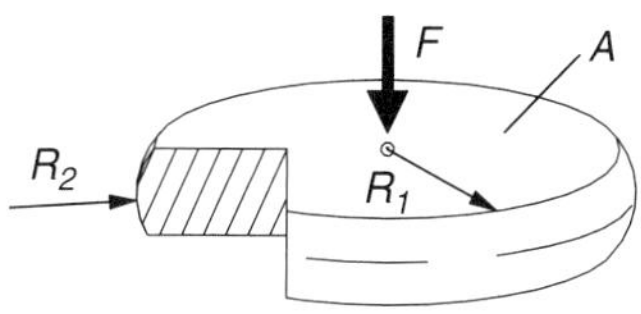

Fig. 7.4 Simplified model for evaluating the hydrostatic pressure in a single joint with liquid metal

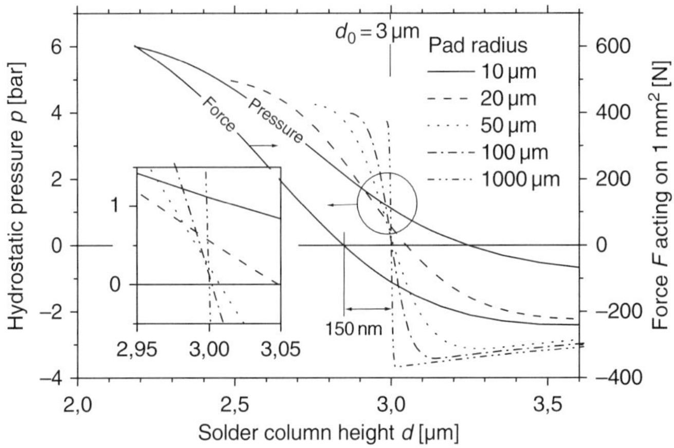

Fig. 7.5 Response of a single liquid metal joint to displacement for a solder thickness $d_0 = 3\ \mu m$

compression would increase R_2 further and decrease the internal pressure of the liquid, which no longer could balance the acting force F. At this moment the contact collapses by squeezing out solder.

With $\gamma_{Sn} = 0.56\,N\,m^{-1}$, the surface tension of tin [4], the hydrostatic pressure p inside a solder joint of bond gap d can be calculated. Figure 7.5 shows $p(d)$-curves for joints with a pad diameter from 10 μm to 1000 μm and an initial solder thickness $d_0 = 3$ μm. At high pressure, the curves end when the joints collapse due to solder squeeze-out. Large pads react very stiff on displacement Δd, that is, with a large Δp, whereas smaller pads display a soft characteristic. Especially for small contact pads, the force-free equilibrium ($p = 0$) is obtained at a smaller gap distance d than defined by the original solder thickness d_0. For example, given a contact with a radius of 10 μm, the surface tension has the tendency to compress the gap by 150 nm, thus raising the hydrostatic pressure to approximately 2 bar.

In contrast, a 100 μm large pad develops the same pressure at a displacement of only a few 10 nm. Thus, an optimized number of large dummy pads surrounding the small electrical contact pads is able to stabilize the joints at a given gap distance. The high pressure inside the pads favors wetting and eliminates any need to apply external pressure or bonding force. Such optimized design also enables free-floating chips to self-align within 1 μm (cf. Section 7.2.3). At the same time, the large dummy pads reinforce the contact planes and shield the small electrical contacts from shear stress.

7.1.2 Metallurgy and Intermetallic Growth

As mentioned earlier, besides the static behavior of a liquid solder metal it is essential to understand the dynamic phase transition reactions of the metal system in the relevant temperature and pressure ranges.

The IMC does not show the compliant plastic deformation behavior known from solder balls, but nevertheless the thin and planar interconnect layer shows outstanding thermomechanical properties and is more suitable for C2C rather than for chip-to-board connections.

7.1.2.1 Phase Front Propagation in Cu/Sn Diffusion Couples

A thin film of low-melting solder Sn is placed in between two sheets of high-melting metal (Cu) – for example, by covering one of the Cu sheets with Sn.

With reference to the binary phase diagram in Fig. 7.6, the situation can be described as follows: At first, there are two phases, Cu on the left-hand side and Sn on the right-hand side of the diagram. After heating, the molten Sn dissolves Cu, until it is saturated. Now, the high-melting intermetallic phases start to precipitate and accumulate at a high rate on the Cu surfaces. In a first step, all Sn is transformed into the metastable IMC Cu_6Sn_5 (η-phase), whose calotte-shaped crystallites expand into the Sn melt and finally close the gap (cf. Fig. 7.7a).

The η-phase in turn reacts with Cu to the stable and planar IMC Cu_3Sn (ε-phase). Figure 7.7b shows a cross-section of such a sample in an early stage of this process (the Cu layers are nearly consumed already). All four phases involved are still present: Cu, Cu_6Sn_5, Cu_3Sn, and Sn.

This process terminates when all Sn is reacted to Cu_3Sn. Now the system consists again of only two phases, the Cu on the left of the phase diagram, which is in contact with the ε-phase in the middle of the diagram. Due to Gibbs' phase rule, the system is now thermodynamically stable and no nonreacted Sn remains. Therefore, there will be no further phase growth even at high temperatures.

The reactions are diffusion controlled [3]. Below the melting temperature, the phase growth is slow; but when the tin melts the reaction immediately is accelerated by two orders of magnitude. The first step is a fast, convection-assisted diffusion

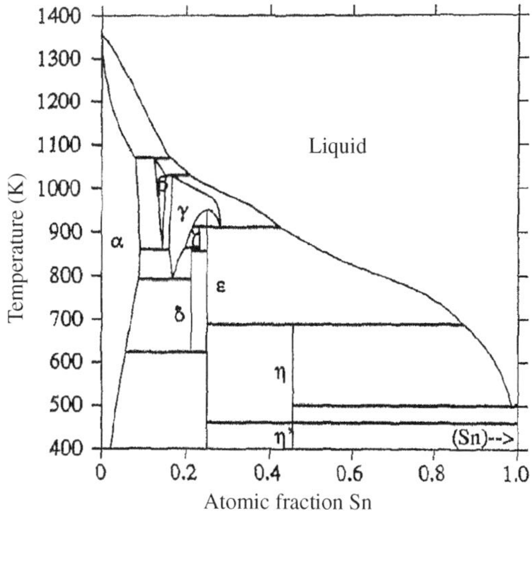

Fig. 7.6 Equilibrium phase diagram of the Cu–Sn system [32], © 1996 Carl Hanser Verlag

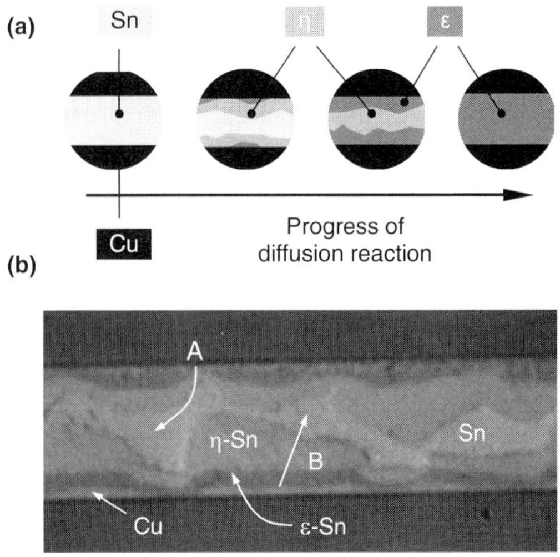

Fig. 7.7 (**a**) Progress of IMC growth, and (**b**) Microscope picture of a prepared cross-section of a Cu/Sn couple after reflow at 250°C for < 30 s

of the copper via the grain boundaries into the liquid Sn (path A in Fig. 7.7), while the second step, which requires the solid state diffusion through the Cu_3Sn layer (path B), will progressively be slowed down by the growing thickness of the ε-phase. Figure 7.8 shows this time dependence for both IMCs with data from [2] and [3]. Whereas the solidification of a 3 μm thick Sn layer will be terminated within approximately 10 s, the complete transformation to the stable Cu_3Sn needs more than 10 min. Closing the gap by the calottes of the η-phase is the mechanism,

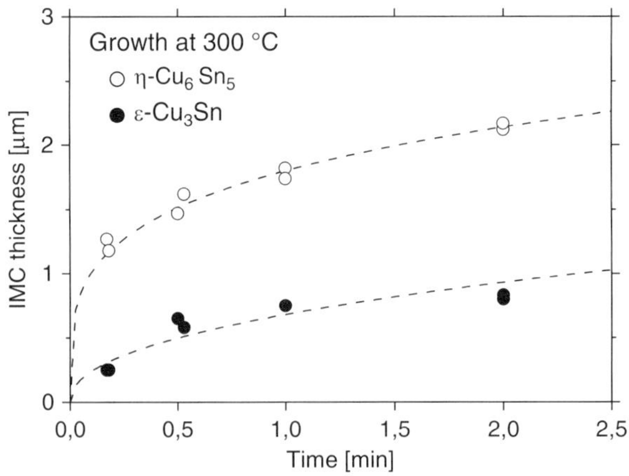

Fig. 7.8 Growth of the intermetallic phases Cu_6Sn_5 and Cu_3Sn at 300°C, as evaluated by Bader et al. [3]. Data used with permission, © 1995 Elsevier Ltd

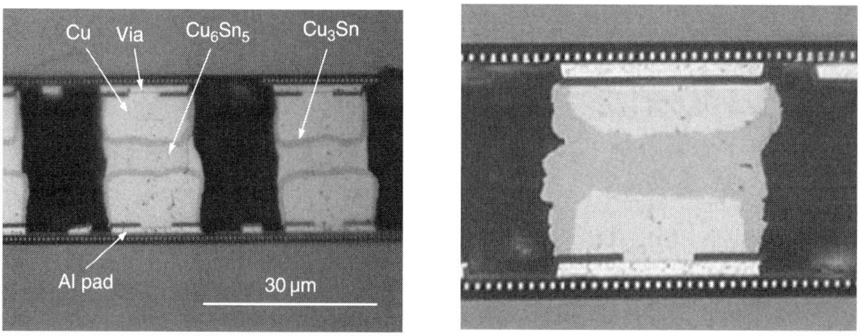

(a) (b)

Fig. 7.9 Cross-section of an electrical contact (**a**) after reflow, and (**b**) close-up of the completely alloyed joint

which already provides the joint with mechanical strength, whereas the final step can be done economically in a subsequent batch process.

Figure 7.9a depicts a cross-section of electrical contacts after melting and wetting. Figure 7.9b shows one of the contacts after complete alloying in a batch oven. The close-up demonstrates an ideal metallurgy and the simplicity of the metallurgical system.

7.1.2.2 Calculation of the Metal Pad Thickness

For a binary intermetallic phase A_aB_b, the relation of the film thicknesses d_x is given by

$$\frac{d_\mathrm{A}}{d_\mathrm{B}} = \frac{a}{b}\frac{M_\mathrm{A}}{\rho_\mathrm{A}}\frac{\rho_\mathrm{B}}{M_\mathrm{B}}, \tag{7.3}$$

where M_x is the molecular weight and ρ_x the respective density. As a rule of thumb, for the Cu/Sn system

$$d_\mathrm{Cu} : d_{\mathrm{Cu_6Sn_5}} : d_{\mathrm{Cu_3Sn}} = 1 : \frac{3}{2} : \frac{7}{3} \tag{7.4}$$

and the initially present copper and tin thicknesses should be related as

$$d_\mathrm{Cu} \geq 1.5 \cdot d_\mathrm{Sn}, \tag{7.5}$$

distributed to each side of the contact. In an asymmetric initial configuration with only one pad covered by Sn, more Cu needs to be provided in the Sn-capped pad in order to account for the earlier onset of intermetallic growth at that interface (cf. Section 7.1.3 for details).

7.1.3 Process Conditions and Characteristics

All occurring phase transitions (solid–liquid, liquid–solid, solid–solid) introduce time constants, which may depend on local conditions but need to be maintained on a global scale across the entire surface. Therefore, a few implications exist, which constrain both the metal system and the usable equipment, thus narrowing the parametric window of SLID-based bonding processes.

7.1.3.1 Surface Metal Oxides

Typically, the metal pads to be bonded are exposed to ambient atmosphere at room temperature, and natural oxide layers of a few Å in thickness form on their surfaces. Reported experimental results [20] show that these oxides present a barrier to Cu/Sn interdiffusion and inhibit intermetallic growth at the respective interface, whereas most Sn is already consumed by its underlayer. After Cu oxide removal, a successful bond was achieved. In addition, a dramatic increase in shear strength was observed after pretreatment in a 5% HCl solution [37].

However, most important is the prevention of the formation of new oxides during the bonding process at the temperatures involved. This is achieved either by bonding in vacuum, in an inert gas ambient, or using underfill material, which can act as sealing encapsulant.

Bonding in a reactive gas environment offers the advantage of both removal and prevention of new oxide growth in one step, using a simple set-up. When heated above the activation temperature of the reactive gas, the presence of oxygen can be tolerated to a certain extent, and the chamber used for bonding can be a half-open recipient, which keeps the partial pressure of oxygen low and the reactive gas concentration near the sample high enough. Such a fluxless soldering technique has been successfully demonstrated with formic acid vapor (HCOOH in N_2) [23].

7.1.3.2 Parasitic Consumption of Reactive Solder

The metal system for SLID needs to be designed to provide sufficient unreacted Sn at all surface locations, so as to avoid terminating the bonding reaction prematurely. Such abortion will likely result in freezing of the chip in a tilted position before hydrostatic forces allow leveling (cf. Section 7.1.1) or, worse, result in unbonded areas.

The consumption of reactive Sn occurs during various processing steps, and needs to be accounted for in the design by means of allowances:

1. *Storage at room temperature*: The activation energy for the growth of η-Cu_6Sn_5 is considerably low. At room temperature this IMC can grow at a rate of 6 nm per day, thus consuming 200 nm of reactive Sn in approximately 35 days [36].
2. *Preheating during fluxless reactive gas activation*: Simulation of different scenarios showed little intermetallic growth at preheating temperatures below 150°C

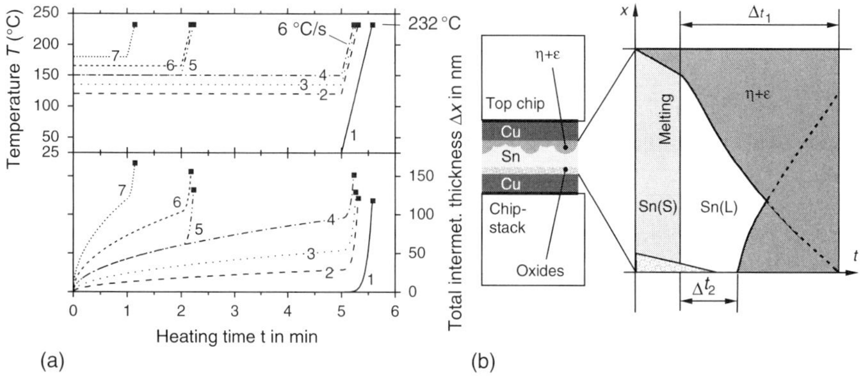

Fig. 7.10 Growth of overall IMC in various stages of the bonding process: (**a**) solid state heating scenarios [6], © 2007 Elsevier Ltd, and (**b**) asymmetric conditions in the soldering gap [25]

and a preheating time below 5 min [6, 25], as can be verified by Fig. 7.10a. Still, the model results in up to 100 nm of overall intermetallic growth.

3. *Heating ramp to melting point of Sn*: During this time, another 50 nm–100 nm of intermetallic growth and respective Sn consumption are estimated (cf. Fig. 7.10a). Fast heating at a high rate above 6°C s^{-1} is beneficial to preserve most of reactive Sn.
4. *Delay until the molten Sn has wetted the Cu pad surface*: The least control by means of process parameters is given over the dynamics of wetting and the time of onset of intermetallic growth at the joining interface. Consequently, this stage governs the design of the metal system and determines the amount of Sn necessary in the process.

The scenarios depicted in Fig. 7.10 are modeled on the base of the equation

$$x = \left(\int_0^t k_{\mathrm{sqr}}^2 (T(\tau)) \mathrm{d}\tau \right)^{\frac{1}{2}} \tag{7.6}$$

for nonisothermal diffusional growth [18], where the temperature- and time-dependent growth constant is given by

$$k_{\mathrm{sqr}}(T,t) = k_{\mathrm{sqr},0} \cdot \exp\left(\frac{-Q_0}{k_{\mathrm{B}} T(t)} \right), \tag{7.7}$$

where k_{B} is the Boltzmann constant, $k_{\mathrm{sqr},0} = 35 \cdot 10^6$ nm·s$^{-0.5}$, and the activation energy $Q_0 = 0.57$ eV, both obtained from diffusion experiments [25].

A minimum growth of intermetallics in the solid state can be assumed when heating directly from room temperature, as shown in scenario 1 in Fig. 7.10. However, for high-precision alignment accuracy and reactive gas activation, preheating of the part with the Cu surface is recommended. Growth in scenarios 2 and 3 at temperatures below ~150°C occurs at a low rate, and results in an overall intermetallic thickness similar to scenario 1. Scenarios 4 and 5 reveal that a shortening of the preheating time is significantly beneficial to avoid excessive parasitic reactive solder consumption, whereas a preheating temperature above 150°C, as in scenarios 6 and 7, accelerates growth to an extent that cannot be compensated by a shortened preheating time.

7.1.3.3 Wetting versus Intermetallic Growth

While in a face-to-face (F2F) chip-stacking process (cf. Section 7.2) the metallization can be designed symmetrical with both Cu and Sn layers on either side, in a multichip 3D-integration scheme, the reactive solder can only be supplied by the previously unheated new chip, which will be placed on top of the already soldered lower chip. The resulting asymmetric metallization introduces a time delay between the onset of intermetallic growth on the upper and lower interface.

Figure 7.10b shows the solder-filled gap between the pads in a cross-section and the solidification progress on an uncalibrated timescale. Intermetallic compound growth starts readily from interface A, where Cu and Sn have been deposited sequentially and thus are in direct contact. If this interface remains the sole consumer, all molten Sn will be solidified after Δt_1. When the adjoining Cu surface B can be wetted after $\Delta t_2 < \Delta t_1$, the phase fronts from both interfaces will meet in the middle. This is the basic requirement for stable bonds, which has to be fulfilled on a global scale at any point of the bonding area.

In order to fulfill this criterion, the time constant Δt_1 can be extended by increasing the solder layer thickness. Thereby, the total bonding time needed to transform all η into ε-IMC will also increase, and the risk to produce shorts by solder squeeze-out will be higher. Since more Cu needs to be provided (cf. Eq. 7.5), the resulting joint will also be larger, and the achievable interconnect density will be lower.

It is therefore desirable to shorten the time-to-wetting Δt_2. However, the wetting time depends on many conditions: the wave-like propagation of liquefaction (causing long travel times over large pad areas), the bonding temperature (which is more reactive at higher temperatures), the solder composition (i. e., alloys with Ag for lower surface tension), the solder and underlayer morphology, and the type of fluxless approach, to name but a few. Because of this high number of factors the time-to-wetting Δt_2 is difficult to control.

7.1.3.4 Electromigration

Future projections not only predict a higher input/output (I/O) connection density per chip, but also an increase of current density from presently typical 0.2 A up to 0.4 A per I/O in the near future [12]. This has direct consequences for the lifetime of miniaturized solder joints in terms of heat dissipation, thermal stress fatigue, and electromigration (EM), which in solder connections is already relevant at current densities two orders of magnitude lower than those associated with today's on-chip copper interconnects [35]. As a measure for the stability of a specific material against EM the critical product $j\Delta x$ can be used [35]:

Table 7.2 Summary of electrical mechanical and thermal properties of intermetallics in the (Cu/Ni)–Sn system [10, 11, 32]: η-Cu_6Sn_5, ε-Cu_3Sn, and Ni_3Sn_4

Property	Units	Cu	Sn	Ni	Cu_6Sn_5	Cu_3Sn	Ni_3Sn_4
Density	g cm^{-3}	8.9	7.3	8.9	8.28	8.9	8.65
Young's modulus	GPa	117	41	213	85.56	108.3	133.3
Shear modulus	GPa				50.21	42.41	45.0
Vicker's hardness	kg mm^{-2}	30	10	15	378	343	365
Electrical resistivity	μΩ·cm	1.7	11.5	7.8	17.5	8.93	28.5
Thermal conductivity	W/(m·K)	3.98	0.67	0.905	34.1	10.4	19.6
Thermal expansion	10^{-6}/K	17.1	23	12.9	16.3	19.0	13.7
(Re-)Melting point	°C	1083	232		415	676	

$$j\Delta x = \frac{Y\Delta\varepsilon\Omega}{Z^{*}e\rho}, \tag{7.8}$$

where j is the current density, Δx the conductor length, Y the Young's modulus, Z^* the effective charge number, Ω the atomic volume, ρ the resistivity, e the electron charge, and ε the elastic limit of the solder material. As can be seen in Table 7.2, the IMCs formed here display a higher Young's modulus and a lower electrical resistivity than soft Sn-based solders, and therefore will result in a higher critical product and a higher stability against EM. In addition, the high homologous temperature (difference of operating temperature and melting point) of IMCs also contributes to the joint stability.

7.2 Chip-Stacking: The SOLID-F2F Process

From 2002 till 2005 Infineon Technologies AG developed a chip-stacking process, called SOLID-F2F, to join two chips in F2F orientation, based on SLID soldering [13, 14]. This development is based on a feasibility study[1], carried out in cooperation with Fraunhofer IZM (Institute for Reliability and Microintegration, Munich, cf. Section 7.3.2).

The typical elements of the interface of this process are shown in Fig. 7.11. The top chip A and the bottom chip B are joined together in F2F orientation. The inter-chip communication is established with contacts, which are placed within the area of the top chip. The chip stack will be connected to the package by wirebonds or solder balls, using the pads C. The image gives a close-up view of the interface, with the top chip sketched open for a better view.

Both top and bottom chips are coated with a mirrored pattern of metal layers, each 5 μm thick. The small contacts D are for C2C interconnection, surrounded by insulation trenches. They can be stand-alone or clustered and their placement as well as their dimension and shape can be designed individually. The vias to the last metal layer are even smaller, similar in diameter to vias within this last metal layer. Thus,

[1]BMBF Grant Number 01 M 2999A

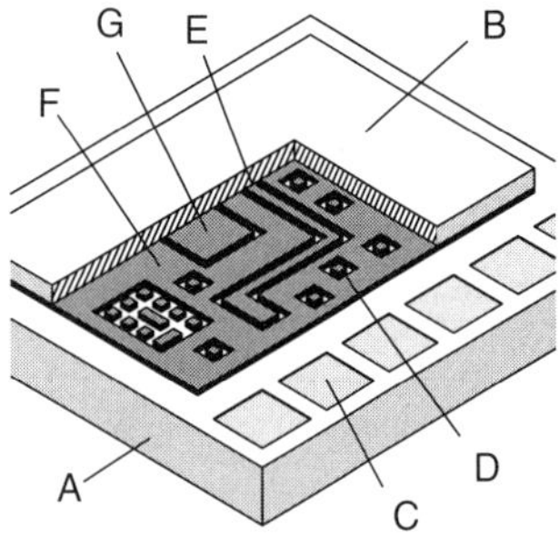

Fig. 7.11 The chip configuration for the SOLID-F2F technology

only a minimum area is consumed by the landing pads. The metallization can also be used for rewiring E. The large area F, which is not used for interconnects, can optionally be metallized, too. This metal layer is located on top of the passivation and could be used for mechanical reinforcement of the small electrical point contacts (cf. Section 7.1.1) and to favor heat dissipation. Optionally, it can electrically be connected as an electrical shield to reduce cross-talk between the two circuits. In this plane, large test pads G can be designed, which also connect to the last metal layer, and allow for testing prior to assembly. Thus, large test pads will no longer consume a major amount of space. In order to vent the insulation gaps and to control the soldering process, the plane F is subdivided into small squares by fine trenches (not shown in the drawing, but cf. Fig. 7.14). Underfill is not needed for mechanical stabilization, unless the metal coverage of the interface is small, say below 10% of the chip area.

7.2.1 Process Flow

The starting point is a wafer, completely processed in the frontend in any technology and wafer size, with an opened passivation (Fig. 7.12, a) and tested. The last metal layer B is either an Al or a Cu metallization.

Both chips of the future twin stack will be preprocessed in parallel with the same process flow on wafer scale: the wafer will be sputter-coated with 50 nm of TiW (10/90 by weight) as barrier and adhesion layer C, followed in situ by 100 nm of Cu

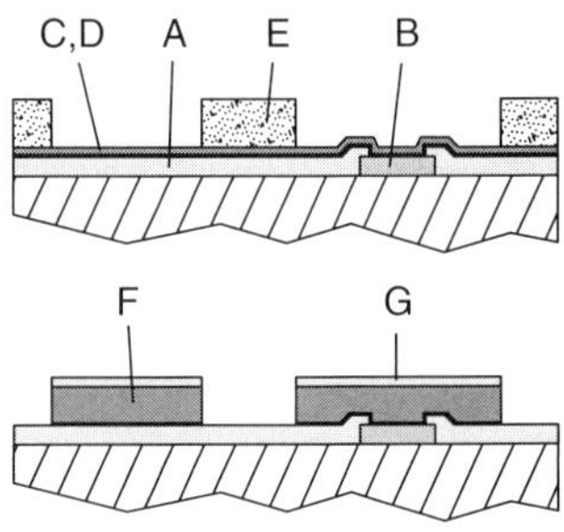

Fig. 7.12 Process flow for SOLID-F2F. The parts of the stack are prepared in parallel on wafer scale, then assembled on a C2W scale

as seed layer D. Then, a wafer-scale proximity lithography step with a 10 μm thick resist E is applied in order to define the future insulation trenches. Inside the open areas of the resist pattern a 5 μm-thick copper layer F is plated. Additionally, 3 μm of Sn or SnAg solder alloy G is plated in situ. Finally, the resist is stripped and the seed layer and barrier layer in the openings are wet-chemically etched in order to remove the shorts.

After metallization, the top wafer is thinned and diced. Then, the bottom wafer is populated with top chips using a flip-chip bonder [22]. An advantage of this C2W processing scheme is the possibility to test the chips prior to their attachment and to pick and place only known good dies (KGDs).

The populated wafer is transferred into a vacuum oven and soldered in inert gas or vacuum. Vacuum conditions are not essential, but for optimum results it is crucial to have an oxygen-free ambient, as discussed above. In order to prevent contamination or corrosion by residues, in this soldering process no solid or liquid fluxes are used to fix the chips onto the wafer. Instead, the chips are held in place by a low-melting (55°C) organic (bibenzyl), which vaporizes completely during the heating period of the soldering process. This agent is inert with respect to the metal surfaces and does not impair the soldering. During pick and place it is dispensed onto the surface of the top chip by a jet dispenser. Nozzle and chip are heated to moderate 70°C. When the chips are placed onto the cold wafer, the agent freezes immediately and fixes the chip. In contrast to adhesives, there is no time delay to wait for chemical cross-linking or evaporation of solvents, and there is no volume shrinking, which possibly could displace the chips.

After soldering, the wafer carrying the stacks is diced. For final packaging, the chip stack is treated like a monolithic single die: it is attached to a substrate and wirebonded. Also, a flip-chip package, using solder bumps, would be possible.

7.2.2 Behavior of Underfill in Small Gaps

A typical method for underfilling is to dispense the viscous filler material alongside the chip edges and let capillary forces draw the material into the interchip air gap. Since SLID is a planar technology, the gap between two soldered chips is much smaller as in the case of solder bumps. The distance between the pads may therefore also be less. The question to be answered is "Can a capillary with a cross-section of $10 \times 10\ \mu m^2$ and a length of several millimeters be filled by organic underfill". In the following, this filling capability is estimated in terms of viscosity and geometrical dimensions.

For reason of simplicity, let us assume a cylinder-shaped capillary of diameter D, which on the left-hand side is in contact with a reservoir of underfill (Fig. 7.13). The weight of the underfill material shall be neglected. Then, the capillary force generates a negative pressure of

$$\Delta p = 2\frac{\gamma}{R}, \tag{7.9}$$

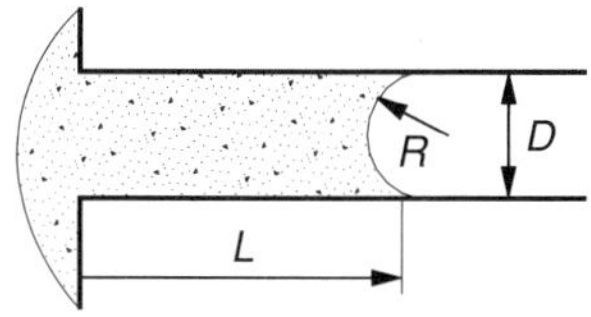

Fig. 7.13 Capillary model for estimating the underfilling conditions in a SOLID-F2F interchip gap

with γ the surface tension of the underfill, and R its radius of curvature (cf. Eq. 7.1). This low pressure draws the liquid into the capillary. The volume flow $\dot{V}$ of a liquid column of length L and with viscosity η in the capillary is given by (Hagen–Poiseuille)

$$\dot{V} = \frac{\pi \Delta p D^4 t}{128 \eta L}. \tag{7.10}$$

With $R = \frac{D}{2}$ and $\mathrm{d}V = \pi t \left(\frac{D}{2}\right)^2 \mathrm{d}x$, the velocity v of the liquid in the capillary is

$$v = \frac{\gamma D}{8 \eta L}, \tag{7.11}$$

and the time t for filling of the capillary to the length L is given by

$$t = 8 \int_0^L \frac{\eta x}{\gamma D} \mathrm{d}x = 4 \frac{\eta L^2}{\gamma D}. \tag{7.12}$$

Thus, large diameters shorten the time for underfilling, and long traces impede it with a square dependence. For example, glycerine with η = 1.5 Pa·s and γ = 0.06 N m^{-1} will fill a 1 mm long and 10 μm wide capillary in 9 s. Also, a conventional nonfilled underfill and even-filled underfill is able to perform the same process; however, the filler material is prone to agglutination.

Figure 7.14 is an infrared image of a soldered interface with a 15 μm high underfilled gap. The filler has reliably filled the 10 μm wide trenches to a length of 500 μm, but when they widen to a cavity, the decreasing capillary forces are no

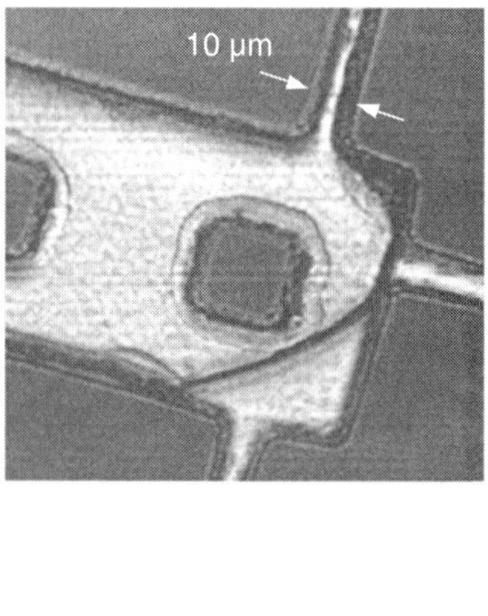

Fig. 7.14 Infrared image of underfilled SOLID-F2F interface

longer able to drive the material fast enough. Thus, the curing begins before the fill of the cavity is completed. Small remaining voids are not critical to reliability, but too large voids should be avoided by means of design rules. The close-up shows a detail of a partly filled cavity.

7.2.3 Self-Alignment

The patterning of the interface must guarantee electrical contact and a secure insulation distance to the adjacent metal areas even at maximum misalignment. But electrical contact alone is not sufficient with respect to reliability, as can be seen in Fig. 7.15. If the contact pads are misaligned severely, the copper will be eroded asymmetrically and the IMC can even extend to the via. Moreover, the current density in the solder layer rises and the electric breakdown strength drops due to the reduced distance to the adjacent metal structures. Thus, perfect alignment is necessary, especially for small contact diameters and high contact densities, despite the fact, that accurate bonding tools are expensive or slow.

Again, the surface tension, which tries to minimize the free surface area of the melt, can here be used as the driving force for self-alignment. But another fact has to be considered: the wetting of the copper surface by the molten tin takes a time interval similar to that needed for the solidification of the gap (cf. Section 7.1.2). In case of the F2F bonding process, it has shown to be beneficial to cover both sides of the pattern with tin in order to shorten the wetting time; otherwise, the tin melt would freeze before the chip could be properly aligned.

Figure 7.16 shows a chain of accurately aligned contacts. Starting with an initial displacement of 10 μm, all chips of the wafer have been self-aligned to within 1 μm. At the same time, plating both sides with Sn guarantees a reliable protection of the copper by a continuous coverage with the inert ε-phase (cf. next section).

7.2.4 Reliability Results

Based on the SOLID-F2F process, a set of reliability tests have been performed on dedicated test structures. Here, the investigated reliability risks were corrosion, cracking, stress migration, and EM. In order to avoid additional phase

Fig. 7.15 View of a soldered SOLID-F2F joint with a displacement introduced during alignment

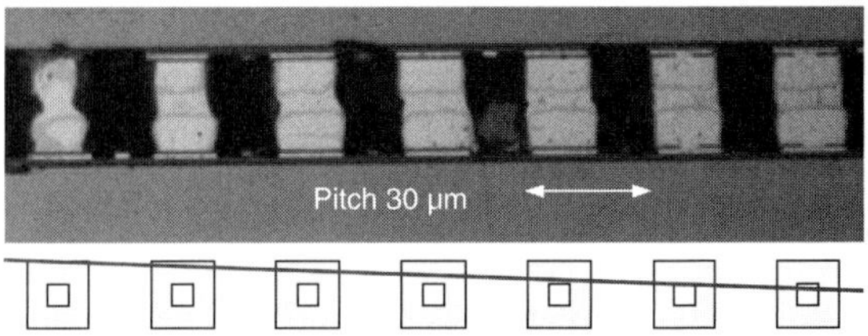

Fig. 7.16 Cross-section of self-aligned joint structures. The initial accuracy was within 10 μm; after surface energy minimization, accuracy was better than 1 μm

transformation during testing the thermal budget of the process was chosen such that phase transformation was completed before testing.

7.2.4.1 Thermal Cycling

During thermal cycling (TC) tests cracking can occur due to thermomechanical mismatch of the different materials used for the SOLID-F2F process. The cracks could either lead to openings in a contact chain or to an increase of leakage currents, if copper diffuses into cracks formed within the isolation material. This is detected by measuring the isolation current between comb meanders within the SLID metallization, as well as between meander structures in other metallization levels.

Thermal cycling tests between –65°C and 150°C were performed on contact chains with 500 contacts with a footprint of of $15 \times 15\ \mu m^2$ at a rate of 1 cycle per hour (cf. Fig. 7.17a). Readouts after 100, 300, 500, and 1000 cycles were carried out and no indication of crack formation was seen. The morphology of the contacts did not change during TC (cf. Fig. 7.18) and the electrical resistance of the chains was very stable (Fig. 7.17b). All measured leakage currents did not increase over cycle time.

The only risk of failure was detected for contacts with very large misalignment. The contact area was too small to withstand the thermomechanical stress (see Fig. 7.19) and cracking within the brittle Cu_3Sn phase occurred. With the self-aligning process described earlier in Section 7.2.3, the contact area was large enough to overcome the reliability risk of cracking.

7.2.4.2 High-Temperature Storage

High-temperature storage (HTS) was performed on fully alloyed contacts at 175°C and at 200°C for a maximum period of 1000 h in nitrogen atmosphere. There was no underfill material used for the high-temperature storage tests. The underfill material is based on an organic matrix and starts to degrade above 175°C. All phase transformations of the solder contacts were completed before starting the tests. The expected failure mechanism is therefore not resistance increase due to further alloying of the solder contacts but stress migration-induced voiding. The contacts proved to be stable, and no resistance increase on any test structure was detected.

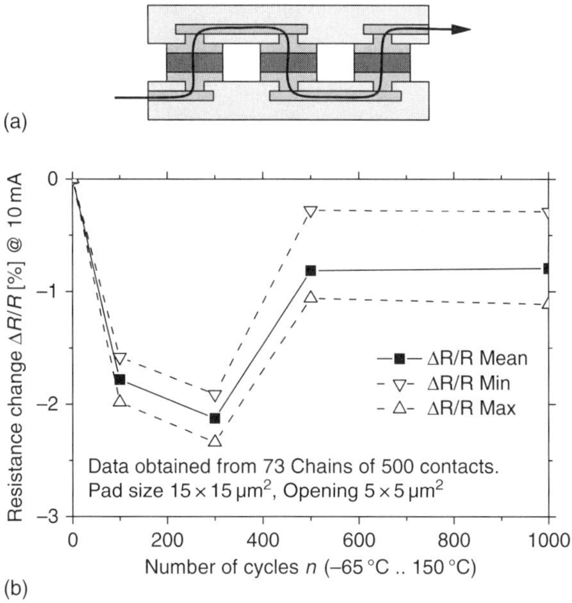

Fig. 7.17 (**a**) Chain of SOLID-F2F joints, and (**b**) change of resistance during TC

7.2.4.3 Corrosion

Generally, corrosion is a serious problem for unprotected solder contacts. Without any underfill or a metal ring at the edge of the chip, oxygen and moisture can attack the copper and the alloy surface of the contacts. Therefore, an autoclave test (AC)

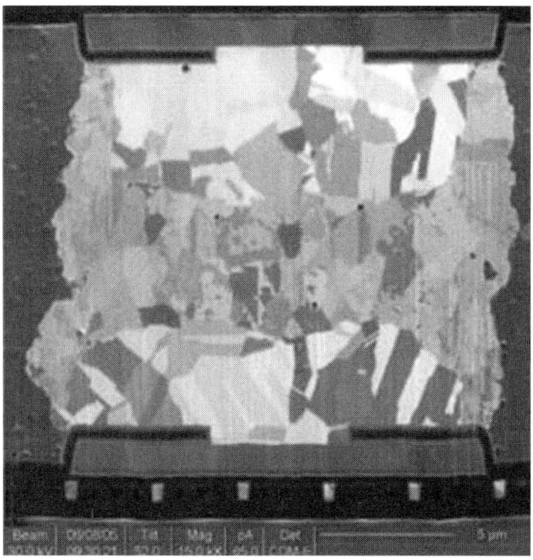

Fig. 7.18 Joint microstructure after 500 cycles from –65°C to –150°C

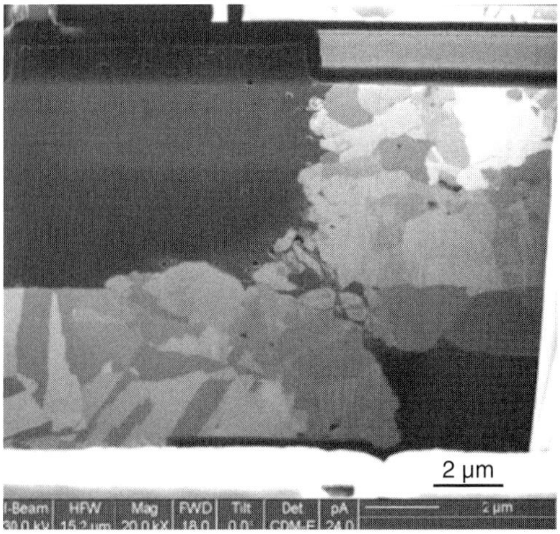

Fig. 7.19 Failure in a strongly misaligned joint after TC

was performed on partly alloyed contacts at 121°C and 100% relative humidity (RH) for 96 h. Attack of the metastable Cu_6Sn_5 phase was detected by means of a focused ion beam (FIB)-milled sample cross-section in both cases, but the Cu_3Sn phase was stable and did not show any corrosion. Since the plated Cu is completely covered with alloy, only little Cu oxidation is detected (cf. Fig. 7.20). When operated at moderate temperature, underfill is sufficient to reliably inhibit all corrosion.

The AC test and an additional humidity test, unbiased highly accelerated stress test (UHAST: 130°C, 85% RH, 96 h), were performed on underfilled metal comb structures (15 μm spacing, 2.5 mm total length, Fig. 7.21a) and chains of $15 \times 15\ \mu m^2$ contacts at a pitch of 30 μm. The resistance of the contact chains remained unchanged. Physical failure analysis did not reveal any attack of moisture on the contact material. However, an increase in leakage current was detected. The reason was humidity absorption of the underfill material, which reduced the electrical isolation. This is, however, not a problem assigned only to SOLID-F2F,

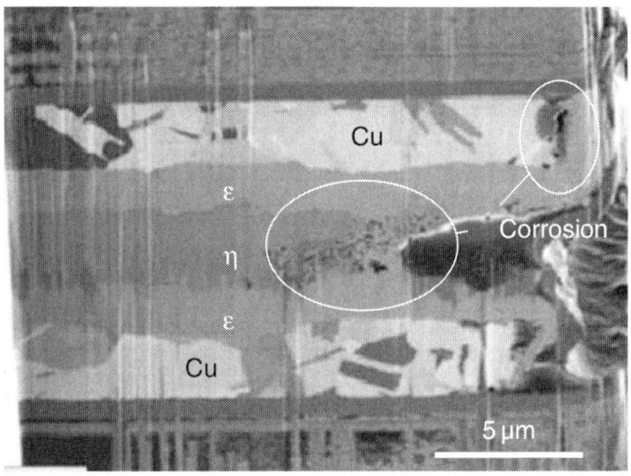

Fig. 7.20 FIB cross section of a sample stored at 121°C at 100% RH for 96 h

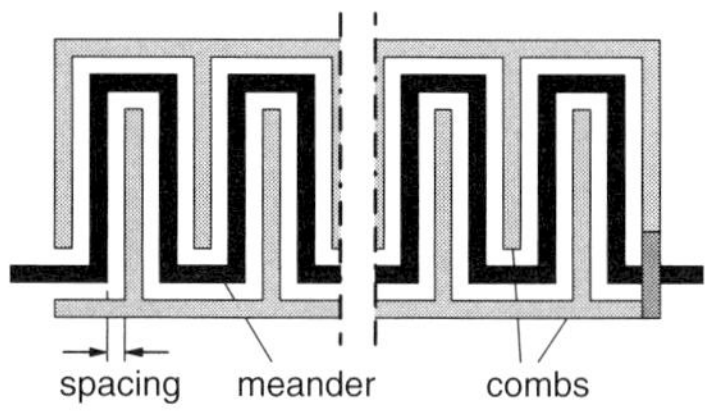

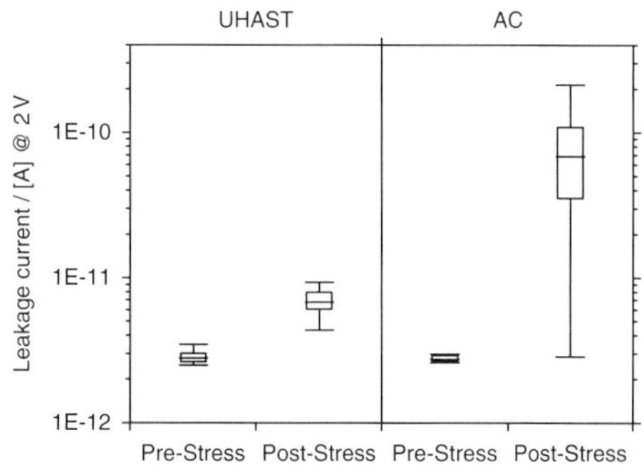

Fig. 7.21 (**a**) Comb structure and (**b**) measured leakage current in humid conditions

but to all 3D-integration approaches, which use high-density contact arrays, as well as for micro bumps. The increase in leakage current can be seen in Fig. 7.21b.

7.2.4.4 Electromigration

For this experiment, a single contact structure was designed with uncritical current feeding lines. The current was routed through the single contact, connecting the top and bottom chips. Since no under fill material was applied due to a metallic seal ring around the tested contact, oven temperatures up to 250°C could be applied. The increase of electrical resistance was measured directly at the contact by means of four-point-probing, with a resolution of the test equipment in the milliohm range.

Different stress current levels were chosen (cf. Table 7.3) and the increase of resistance within 1500 h of stress time was measured. The morphology of the contact remained stable during this EM test. In comparison to other bump technologies no further phase change was detected, indicating that the intermetallic phases remained stable (cf. Fig. 7.22). However, a significant resistance increase of up to 20% was measured within 1000 h of stress time, applying 100 mA at 200°C.

Root cause of this resistance increase is not void formation in the contact or the feeding lines. Due to a nonfunctional barrier system (TiW) between Cu and Al, interdiffusion was detected with element analysis. This interdiffusion is driven by the applied current and not seen in any high-temperature storage experiments. It leads to a prominent formation of Cu–Al intermetallic phases in the lower Al level. In addition, a significant material build-up is found (Fig. 7.22). In general, Al–Cu compounds have a higher resistivity when compared to pure Cu or Al lines.

Table 7.3. Stress conditions

Temperature	200°C – 250°C
Current	50 mA – 100 mA
Via current density	2 mA μm^{-2} –4 mA μm^{-2}

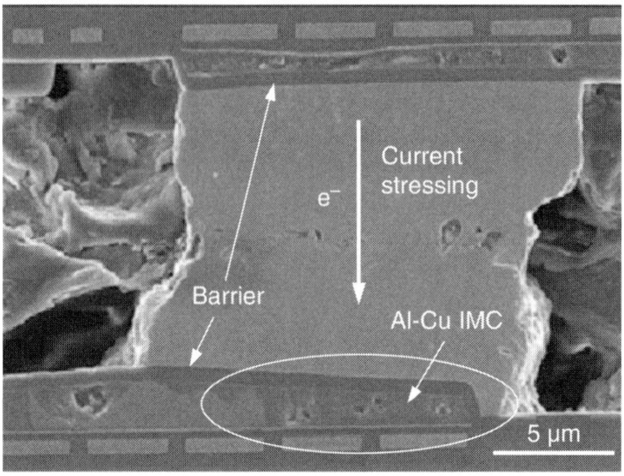

Fig. 7.22 Electromigration-induced voiding (upper interface) and compound formation (lower interface) at the Cu/Al interface. The barrier appears unchanged

It is assumed that a stable barrier will lead to superior EM results. Kinetic studies to determine activation energy and current density exponent should be performed with a functional barrier system between Cu and Al.

7.3 Three-Dimensional Integration: SLID for Multichip Stacking

In the past years, SLID has been investigated in a variety of different 3D-integration approaches, a selection of which is represented in Fig. 7.23. In Germany, joint efforts have been made in the frame of governmentally co-sponsored projects involving Infineon, Fraunhofer IZM, and the University of Ulm, and in Japan by ASET, a consortium of 18 companies with researchers from Toshiba, NEC, Oki, Elpida, etc.

Most of the approaches (Fraunhofer, ASET) typically follow a "vias-first" processing scheme, where the vias are etched and filled from the front side of a wafer through the intermetal dielectric (IMD) of the on-chip interconnect system. This has the advantage that after thinning, the vias are open at the backside and ready to be connected. However, the routing of on-chip interconnections is impaired more,

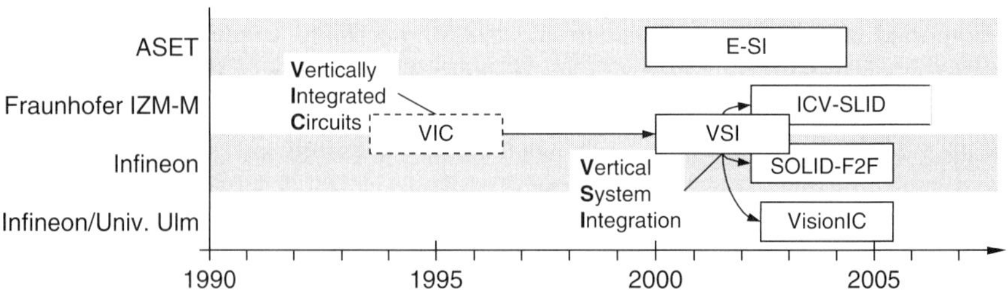

Fig. 7.23 Cu/Sn SLID bonding in 3D integration. A brief historical overview of selected research activities

when the density of the through-chip vias increases. Moreover, the vias are typically longer than the shortest path through the chip, because the via length is increased by the thickness of the IMD layers, which are typically as thick as 10 μm. When using thinned chips with a remaining substrate thickness of 10 μm, the via length is doubled and performance is reduced as compared to the shortest through-chip path.

Other approaches exist (University of Ulm/Infineon), where the via length is kept short by backside etching. Then, only the first metal layer needs to provide the landing pads, the other metal layers are not affected and can be designed freely. At similar via diameters, the benefit will be a lower resistivity and relaxed filling conditions, because the influence of the geometry decreases with the aspect ratio of the structure. At similar aspect ratios, the backside-etched through-chip vias occupy less area. The cost for these benefits are additional process steps, which ensure proper back-to-front alignment of the vias, and a good electrical contact to the first metal layer.

7.3.1 Scheme with SLID and Backside Vias

This section introduces a processing scheme, which combines a "thinning-first" process module and a through-silicon via process module with SLID stack building. It has been developed at the University of Ulm from 2002 to 2005 during a multipartner project (BMBF Grant Number 01 M 3127B) and is based on the SOLID-F2F process of Infineon, described earlier in Section 7.2. Firstly, the full concept is described, with its potential to integrate complementary metal-oxide semiconductor (CMOS)-based as well as other technologies on a variety of substrates in a C2W assembly process, where full wafer processing can be used just until before stacking the diced chips onto the base wafer.

Then the feasibility of this concept is demonstrated on the base of C2C-bonded experimental samples incorporating bare Si substrates without active CMOS circuitry. Finally, the 3D-integration of a diamond DUV sensor is presented as an example for the heterogeneous integration of different materials.

7.3.1.1 Basic Considerations

When starting to think about a 3D-integration technology, one has to balance the advantages and disadvantages of frontend or backend approaches. Generally a pure backend approach seems to provide benefits, as a change of the standard processing cycles is not necessary. Nevertheless, sometimes, significant drawbacks are encountered due to constraints in circuit design, when using a sole backend technology. The success of a 3D-integration technology depends on how much of the benefits of both, frontend and backend approaches can be realized at a reasonable cost and, at the same time, provide a high interconnect density, short interconnect length, and reduced parasitics. In order to achieve this, the via footprint needs to be in the range of a few microns (micro vias).

In the proposed processing scheme only substrate-related changes prior to the CMOS circuit fabrication process are introduced. These are in detail the etching of markers into the silicon substrate by a dry process, and standard silicon epitaxy in order to produce a highly boron-doped buried layer, which can be used as an etch stop during substrate thinning. Thus, the device fabrication process is not significantly affected.

The proposed 3D-integration scheme can be divided into three stages (cf. Fig. 7.24), namely the preprocessing (which also includes the active device fabrication), the wafer processes (thinning and vias), and the stack-building.

7.3.1.2 Preprocessing

Stage 1 starts with a bare silicon wafer with no particular prerequisites, except for (100) orientation. Because the vias are etched after the active device fabrication and after thinning (vias-last approach), the issue of back-to-front alignment needs to be addressed. The alignment concept described here is based on markers etched into the front-side surface of the bare wafer by means of reactive-ion etching (RIE). These markers are used as a reference to align the circuit on the front side. After thinning, the markers again become visible at the backside and can be used to align the vias. The markers can be placed in the periphery of the active chip area, that is, within the dicing grooves where they do not disturb device fabrication.

After the patterning of the markers a highly boron-doped silicon epitaxial layer (p^{++}-silicon) is grown, which is used as an etch stop layer during chemical thinning [31]. To obtain an acceptable etch stop, a boron doping level of approximately 8×10^{19} cm^{-3} is sufficient, which is low enough to avoid dislocations and lattice mismatch in the silicon substrate. The thickness of this etch stop layer can be customized according to the precision and homogeneity of available thinning tools. Here, a thickness range between 0.5 μm and 1.0 μm was found suitable.

On top of the etch stop layer, the active layer of the device is grown, whereby the doping concentration may be adjusted according to the specification of the circuit fabrication process. The thickness of this functional epitaxial layer defines the final thickness after the wafer-thinning step. With this thinning process, a high precision in reaching the target thickness of approximately 10 μm is achieved.

Any wafer, which has been prepared in such a way, can now undergo a specific circuit fabrication process, based on the technology of choice. The process flow is visualized in Fig. 7.24b (Stage 1).

7.3.1.3 Processing

Stage 2 of this 3D-integration scheme starts with the deposition of the front-side part of the SLID metal system. This is performed by using conventional lithography and electroplating processes.

Before thinning, the wafers are attached to a carrier wafer for mechanical support. Since attachment involves a temperature-curing step, both wafers should be thermally matched (similar coefficients of thermal expansion). Since carrier release

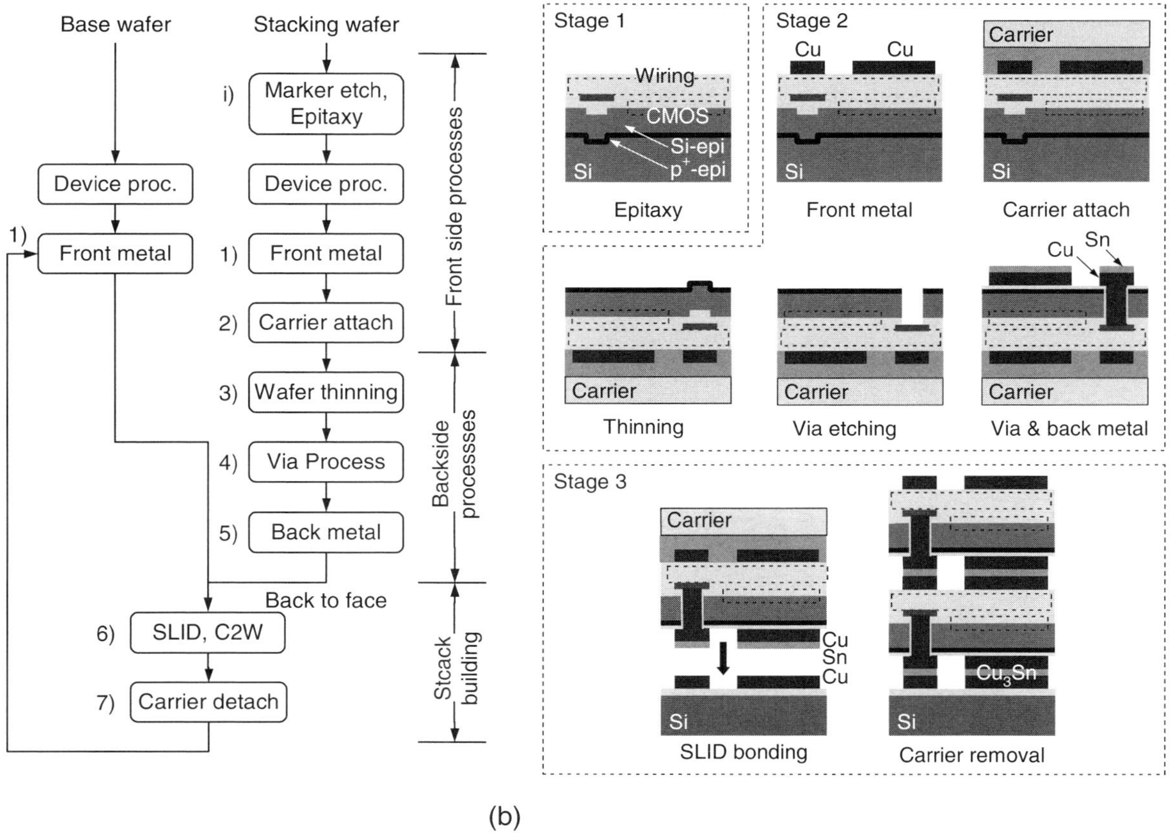

Fig. 7.24 (**a**) 3D-integration process flow. (**b**) The concept includes controlled backside thinning, backside micro vias, and SLID bonding technology [6], © 2007 Elsevier

is needed after stacking, selection of a carrier wafer must consider release processing requirements (see Fig. 7.24). In the attempt discussed here, a gluing process in combination with a release layer technique was used (cf. Fig. 7.25). The adhesive layer itself withstands the entire processing sequence easily, but cannot be dissolved in the final step. Thus, a specific release layer was introduced, allowing easy removal of the carrier (supporting wafer) at the end of the processing flow [16]. The thinning itself is a two-step sequence. First, the wafers are mechanically ground down to a wafer thickness of approximately 50 μm. In a second alkaline spin-etch step, crystal defects are removed and the previously mentioned buried alignment marker topography is exposed at a typical residual wafer thickness of 10 μm [5]. The chemical thinning results in a mirror-like surface, and thus no further polishing is necessary.

As etching mask for the via fabrication, and as electrical insulator on the backside of the wafer, silicon dioxide is deposited by a dedicated plasma-enhanced chemical vapor deposition (PECVD) process, which had been optimized for a low deposition temperature of 200°C or below [16]. Thus, the overall thermal budget and thermal stresses in the thinned circuitry are kept at a minimum. The oxide mask itself is patterned by photolithography and dry etching in fluorine plasma.

For the etching of the micro vias an anisotropic silicon etching (ASE) process (also known as "Bosch process" [24]) was used, which results in vertical side walls of the etched geometry [8, 40]. Thus, small vias with $5 \times 5\ \mu m^2$ cross-section and a pitch of 15 μm were realized, leading to an average interconnect density of about $4400\ mm^{-2}$. However, this density can only be achieved locally, because in a practical application there will be a tradeoff between perforated and active chip area. The ASE process stops when reaching the dielectric underneath the landing pad in the metal 1 layer (lowest metallization level). This insulation layer has to be removed in a subsequent anisotropic dry etching step, also referred to as contact opening (cf. Fig. 7.24b).

The via process continues with the deposition of the sidewall insulation by PECVD silicon dioxide, followed by sputter-deposition of TiW as adhesion and diffusion barrier, and Cu as seed layer for the following electroplating step. A composition of 10 wt.% Ti in TiW was chosen to obtain good adhesion properties, as

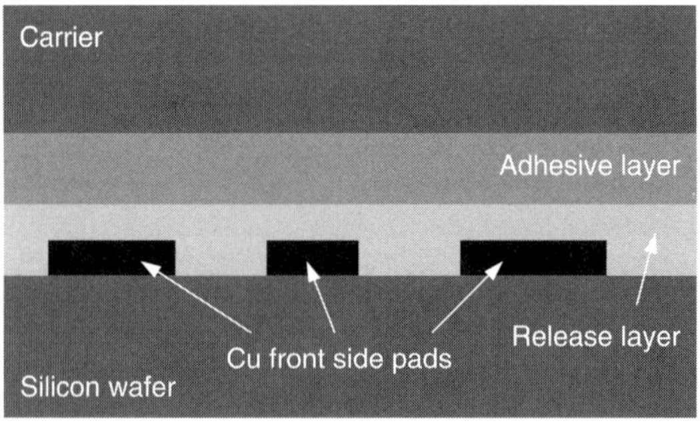

Fig. 7.25 Handling concept for thinned wafers. The permanent adhesive layer bonds the carrier to the release layer on the front side of the device wafer

well as sufficient barrier properties up to nearly 800°C [38]. Electroplating was not only used for via filling, but also for the deposition of the SLID metal stack in a combined through-mask plating step of Cu, followed by a top layer of Sn. Fig. 7.24b, Stage 2 shows the schematic cross-section of a wafer with filled vias, which is now ready for the chip-stacking process. The wafer can now be diced and tests can be performed to in order to preselect functional chips (KGDs).

7.3.1.4 Stack Building

In order to stack the chip layers with the SLID process, two different geometries of metal pads are used. These are electrical joints with an area of $9 \times 9\ \mu m^2$ and much larger mechanical joints with an area of about $80 \times 80\ \mu m^2$, to improve the stability of the stack. Thus, proper alignment of the different layers to each other has to be assured. This can be done by a flip-chip bonder or die-attach system with sufficient accuracy.

During alignment of the levels to be stacked, a formic acid purging (HCOOH in N_2) of both the base wafer and the die is performed. Subsequently, soldering for 10 min at 300°C is performed in formic acid atmosphere to prevent thermal-induced oxidation of the metal surfaces. A down force of 10 N on a chip size of $5 \times 5\ mm^2$ in this case yielded satisfying soldering results. After completed soldering, cool-down time seems not to be critical; therefore, compressed air cooling can be used.

To allow further chip soldering on top of the already soldered stack, carrier wafer removal is done by immersion of the soldered stack in suitable solvents. Remaining residues of the glue can easily be removed by an oxygen plasma treatment. Figure 7.26 shows a chip stack after such a treatment.

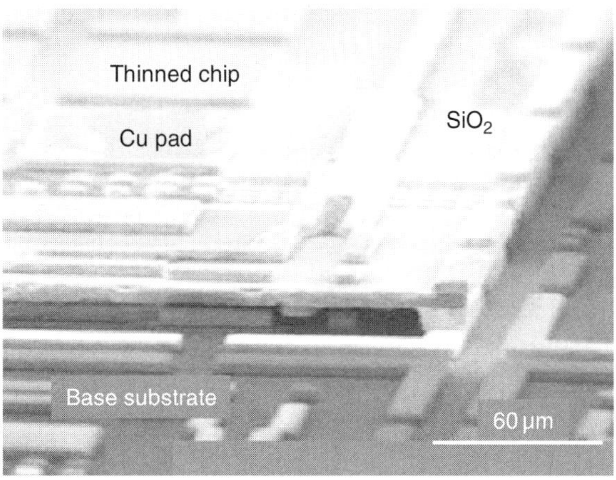

Fig. 7.26 Thinned chip, back-to-face SLID bonded onto a base substrate [5], © 2005 IEEE

7.3.1.5 Results

After the removal of the carrier from the soldered chip, optical inspection and electrical measurements are possible. Figure 7.27 and Fig. 7.28 show chip stacks of different heights, which were realized by processing of bare silicon samples with sizes of $5 \times 5\,cm^2$ and an initial thickness of 675 μm. Mechanical stacks with a height of up to seven levels were demonstrated. No mechanical degradation was observed after six solder sequences on the base substrate as well as on the thinned chips.

Electrical characterization of the vias and solder joints revealed a joint resistivity of less than 0.5 Ω per interconnect [5]. The overall resistance of the interconnects

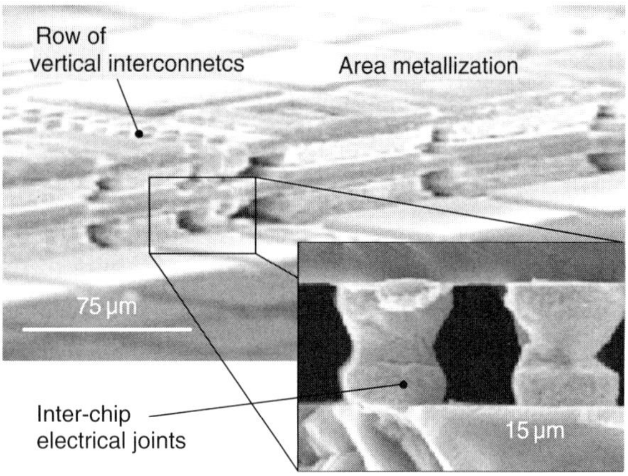

Fig. 7.27 Chip-stack with two thinned chips on top of a base chip. The inset shows an example of C2C electrical connections [26], © 2006 IEEE

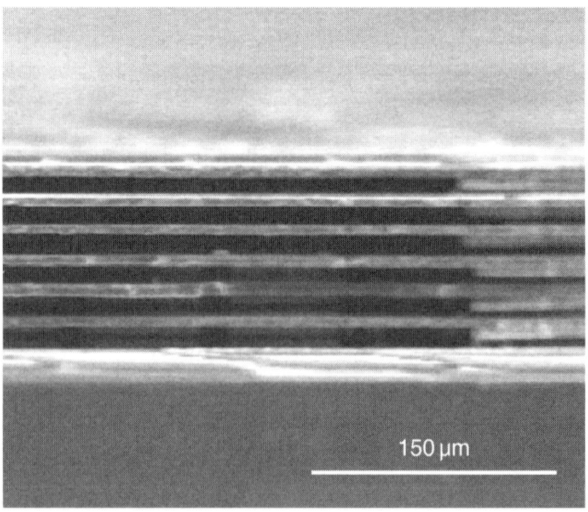

Fig. 7.28 Chip stack with seven levels of passive metallization [5], © 2005 IEEE

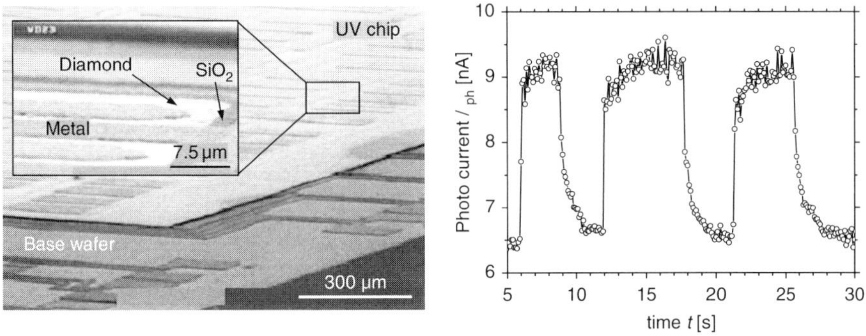

Fig. 7.29 Three-dimensional integrated diamond-based UV sensor array. (**a**) SEM picture [16], and (**b**) response to pulsed exposure at a wavelength of 193 nm [17], © 2006 Elsevier

seems to be dominated by the interface resistances between the metal on the front side of the wafer and by the via filling. Thus, great care must be taken to optimize this part of the process.

Figure 7.29 shows a first attempt of heterogeneous 3D integration incorporating a diamond-on-silicon substrate as top sensor level. The same processes are used as for the other silicon chips, except for the specific fabrication steps of the diamond UV sensor arrays. This example should be seen as proof-of-concept experiment for the integration of a different materials configuration with, for example, CMOS circuit technology. Another example may be the integration of a GaAs or GaN submillimeter wave antenna system.

Figure 7.29a depicts this configuration, with a base chip and a SLID-soldered thinned diamond-on-silicon UV sensor layer on top. The inlay shows the planar MSM detector geometry of the sensors. The DUV response of a through-connected sensor element to a chopped input signal is shown in Fig. 7.29b. An electrical field of 1 kV cm^{-1} was applied between the sensor fingers. Although the photocurrent of such an element was only in the nanoampere range, it could be clearly detected against the leakage current of the micro vias [17].

7.3.2 *Scheme with ICV-SLID Technology*

The development of 3D-integration technology at the Fraunhofer Institute for Reliability and Microintegration (IZM) in Munich started in the 1990s, and a variety of process modules with interchip vias (ICVs) have been introduced, spanning from an adhesive-based wafer-stacking scheme [28] to a SLID-based C2W scheme [29].

7.3.2.1 Processes

First presented was a W2W stacking technology, where the wafers are bonded onto each other using polyimide as adhesive. In this technology, the ICVs are etched from

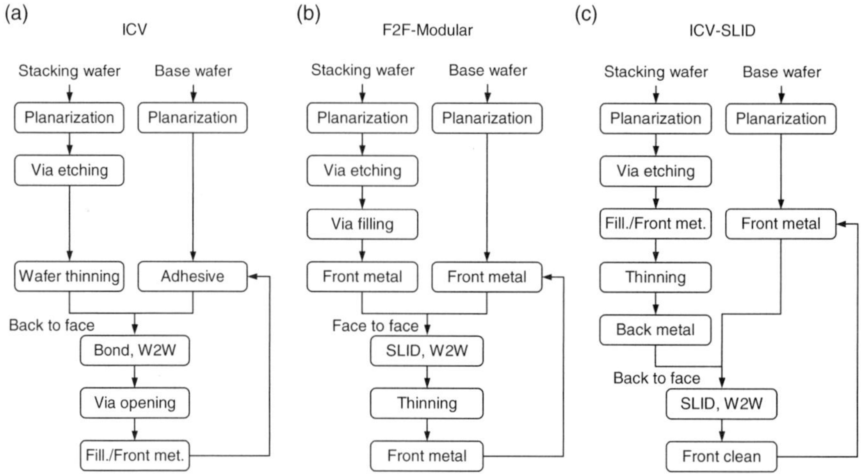

Fig. 7.30 Fraunhofer IZM-M process flow for 3D integration with ICV-SLID technology (ICVs, solid–liquid interdiffusion) [21]

the front side through the dielectric and the Si of a CMOS wafer (cf. Fig. 7.30a). The wafer is then thinned in a three-step grinding, wet-etching, and polishing sequence. After bonding onto the stack with polyimide at a final curing temperature of 400°C the vias are isolated with conformal CVD of O_3/TEOS at approximately 400°C, then filled with CVD-tungsten. Before filling, an anisotropic dielectric etching process was used to open the contact through the polyimide layer, the passivation and the IMD of the lower wafer [28]. Finally, standard TiN/AlSiCu metallization was used to connect the W-plugs with the top-level metallization of the upper wafer.

During subsequent research (cf. Fig. 7.23) this processing scheme was extended by modules, which also include SLID of Cu and Sn, namely the F2F modular process and the ICV SLID processes depicted in Fig. 7.30b and 7.30c.

With ICV-SLID, a concept was developed which requires as little on-the-stack processing as possible by combining ICV and F2F-modular. In the resulting stacking sequence, via opening and filling, as well as the critical on-the-stack thinning step are not necessary, since they are carried out as batch processes before stacking (cf. Fig. 7.30c).

In ICV-SLID, micro vias are etched from the front side into fully processed CMOS wafers. Via isolation, via filling, and solder metallization, as well as via connection to the top metal layer are followed by wafer thinning, down to a thickness of 10 μm–50 μm. The through-chip vias are opened during the thinning process. They are then contacted by the backside metallization step, which also forms the solder metallization. The following stacking sequence uses SLID to solder the thinned chips onto the wafer by means of SLID. After carrier removal and front-side cleaning, the stack is ready for placing further chips on top.

Other concepts under development are mixed C2W/W2W handling schemes, which allow population of handling wafers with KGDs from various donor wafers, which are then thinned and stacked on a wafer level [21].

7.3.2.2 Results

Figure 7.31 shows the cross-sectional structures resulting from the ICV-SLID concept described above. As can be seen in the FIB picture, the joint interface solely consists of the Cu_3Sn IMC [29]. An average resistance of 1.0 Ω in a $2.0 \times 2.0\ \mu m^2$ tungsten-filled via of 10 μm depth, and 0.43 Ω per SLID contact (including diffusion barriers) has been measured. Chains with 240 vias have been realized with this technology [21], and the influence of the ICV-SLID process steps on the functionality of CMOS transistors in a thinned layer has been studied [39].

7.3.3 Scheme with Copper Bump Bonding

From 1999 to 2004 research within the Association of Super-Advanced Electronic Technologies (ASET) in Japan was directed toward the development of ultrahigh-density electronic system integration (E-SI, cf. Fig. 7.23). The project work covered three areas, namely optoelectronic packaging, circuit design and probing, and high-density 3D-LSI chip-integration technology, the latter of which will be discussed here. Stack building in this technology was performed by copper bump bonding (CBB), a technique which involves SLID of Cu with Sn-covered Cu bumps.

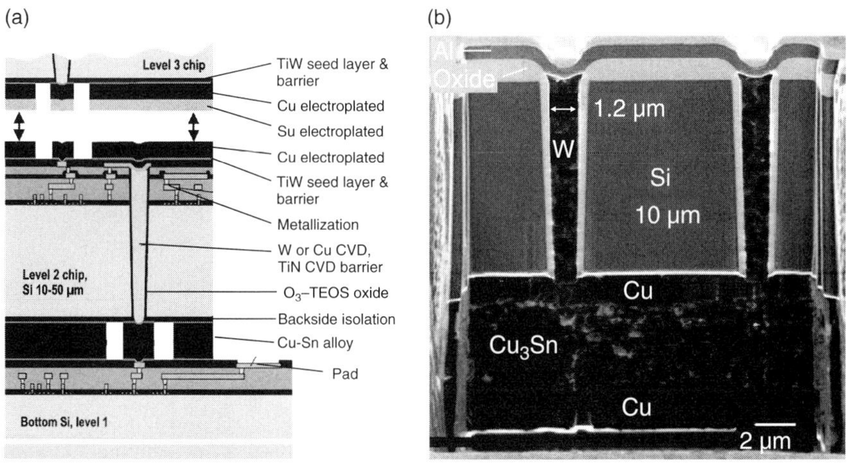

Fig. 7.31 ICVs and SLID metal system developed at Fraunhofer IZM. (**a**) Schematic representation of a stacked circuit [39], © 2005 Elsevier, and (**b**) FIB cross-section of the bonded interface [29], © 2003 MRS

7.3.3.1 Process Scheme

The process scheme developed by ASET is outlined in Fig. 7.32 and follows a "vias-first" concept. The vias are filled by means of Cu electroplating, and special focus was put on the development of a fast and economic plating process with a complex chemistry of leveling, accelerating, and inhibiting additives [33]. As a result, 70 μm deep vias of 10×10 μm^2 were filled within 1 h by chemically controlled preferential deposition at the bottom of confined high-aspect-ratio geometries. Excess Cu on the surface was removed by chemical–mechanical polishing (CMP).

After bonding onto a carrier, thinning, and thereby exposing the Cu at the backside, the Si substrate is selectively etched back, so as to form protruding Cu posts, which are then covered by a 1.5 μm thin layer of electroless plated Sn. This is done in preparation of the following CBB process, which incorporates SLID and is used for stacking of the devices (cf. Fig. 7.33).

7.3.3.2 CBB Process

In the CBB process mechanical stability is provided by underfill resins. For highest reliability these should be thermally matched with the substrate material, which usually is achieved by filler particles in the resin. Studies with filled and unfilled resins using various bonding techniques revealed highest reliability for CBB joints consisting of IMCs, even when the thermal mismatch of the underfill was highest. This is attributed to the low equivalent plastic strain in joints that are solely made up of IMCs [33]. Figure 7.34 shows the plastic strain as a function of the underfill coefficient of thermal expansion (CTE) for Au-bumped, IMC, and SnAg (2.5 wt.% Ag) joints. Joints with high plastic deformation in unfilled resin failed TC from –40°C to 125°C, and for the SLID-based CBB joints a lifetime of at least 1000 cycles was verified [33].

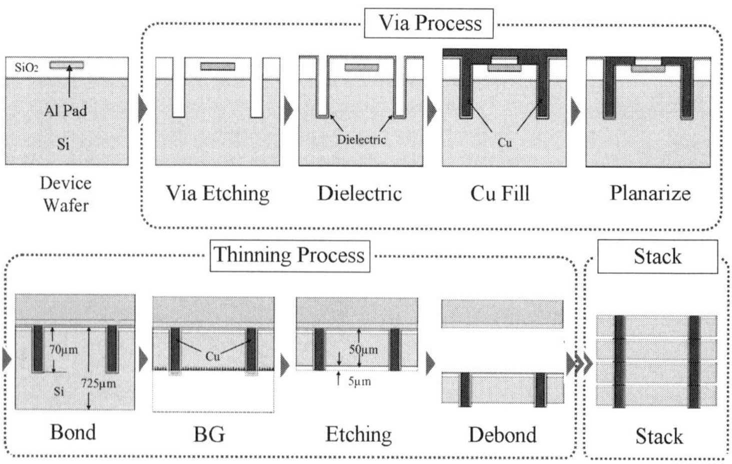

Fig. 7.32 Process flow for 3D integration with CBB technology [9], © 2005 Wiley Periodicals Inc

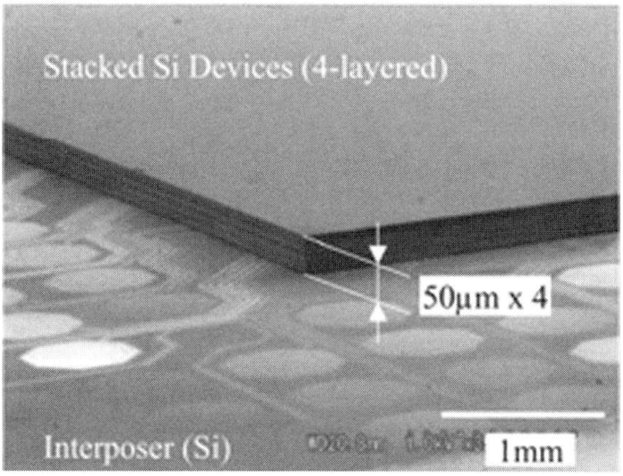

Fig. 7.33 CBB bonded stack with four thinned chips on top of an interposer [33], © 2005 Wiley Periodicals Inc

Bond tests at a constant bonding time of 60 s and varying temperatures ranging from 240°C to 450°C were performed, thus obtaining joints at various stages of the SLID reaction. It was found that the bond strength was highest at a temperature of 350°C [34]. Under these conditions only ε-Cu_3Sn is present in the joint, and the superior strength is attributed to the ε-phase, whereas the presence of the η-phase resulted in inferior bond strength. Figure 7.35 shows such a joint after bonding at 350°C for 60 s.

With this technology, $10 \times 10\ \mu m^2$ vias with a length of 50 μm and a resistance of 0.03 Ω per via were achieved at a pitch of 20 μm. The electrical path through a CBB joint, including Cu, UBM (under bump metallurgy, barrier and adhesion layers), and Al pad wiring was 0.45 Ω [34].

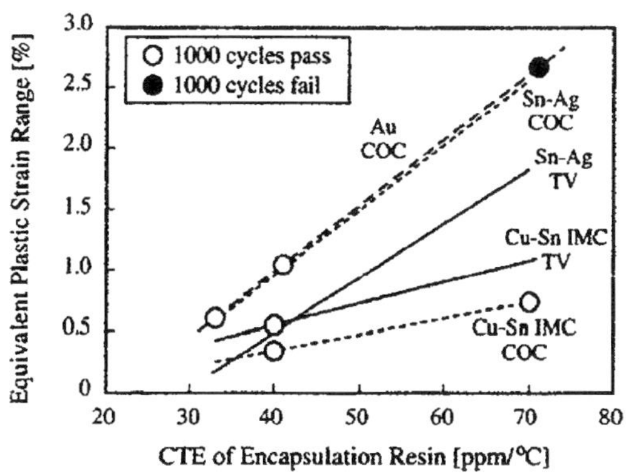

Fig. 7.34 Plastic strain in microjoints under various conditions. Lowest deformation is observed in joints consisting of pure IMC [33], © 2005 Wiley Periodicals Inc

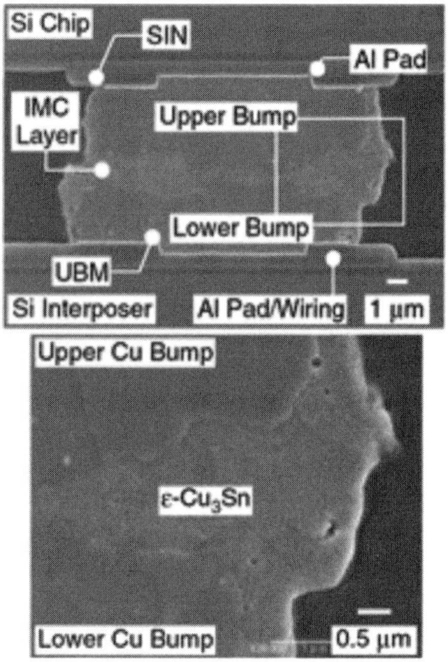

Fig. 7.35 Copper bump bonding joint bonded with initially 1.5 μm thick electroless Sn-2.5Ag solder [34], © 2004 J Jpn Appl Phys

7.4 Brief Discussion on W2W Versus C2W Schemes

The chip-stacking process can be laid out either as W2W or as C2W process (a third variant, C2C can be considered a subform of the C2W procedure). Both alternatives have their specific advantages and disadvantages, which will be shortly contrasted in the following sections.

Obviously, the W2W process is a highly parallel process. All chips on the wafers are aligned at the same time. Thus the effort spent for the alignment procedure can be much higher than in the serial chip-stacking approach of C2W. Although recent flip-chip bonding equipment can place chips with an alignment accuracy of 10 μm (3σ) at a speed of 5000 UPH (UPH, units per hour, e.g. 8800 FC Quantum, Datacon Technology GmbH), the population of a wafer typically takes some 10 min. A wafer aligner is able to align two wafers with an accuracy of 1 μm in a fraction of that time.

High-precision bonding technologies, as, for example, thermocompression Cu–Cu bonding, Au–Au bonding, diffusion bonding, and adhesive bonding, which need long temperature cycles, preferentially are often used in W2W approaches for reasons of cost-effectiveness.

However, one of the most striking arguments for chip stacking is the possibility to combine chips of different technologies in one package. This system-in-package (SiP) concept was the driver for the early development of wire-bonded multichip packages. However, technology mixing on the wafer-level necessitates not only a very precisely matched chip layout, but also the complete wafer map has to be compatible. Since the lithography tools of different fabrication sites usually are not calibrated to each other, the reticle dimension depends on the stepper tool, and moreover the geometry of the kerf depends on the technology. Thus, in practice, this approach can generate a large overhead of complications.

In the meantime, also the memory community realizes the demand for chip stacking in order to provide for packages with densities far beyond those of planar single dies. Even though stacking of a multitude of similar memory chips seems to be the ideal application for a W2W procedure, here the yield loss is in general fatal. Assuming a wafer yield of 90%, an eight-fold stack would end up with a yield loss of more than 50%. Moreover, DRAM memory chips after test only tolerate a very limited temperature budget.

To overcome the yield loss, only pretested chips, KGDs), should be stacked. Unfortunately, not all methods of chip testing can be performed on a wafer level. But this is a concern for the C2W procedure too, when the chips are to be picked from the diced wafer.

Last, but not least, wafer stress is a challenge for W2W processes, most notably after wafer thinning. The silicon substrate with the multitude of inter metal oxides and metallization layers in interaction with the high-temperature processes during manufacturing is a complex thermomechanical system, which balances at a certain stress equilibrium. Now, when the silicon carrier is thinned substantially, a new stress equilibrium establishes itself, leading to a bending of the wafer. Stress release even can result in a change of the wafer diameter.

In summary, chip manufacturers presently favor C2W processing, especially when flip-chip bonders are already available on the fabrication sites, which then also can be used for C2W attaching. The soldering process is not a cost driver, with the completely populated wafer soldered once.

7.5 Conclusion

Over the past 10 years various key players of 3D integration in Europe and Asia have chosen to evaluate a solder-based bonding method for stacking of multiple chip layers. These approaches are based on different concepts, but – as a common feature – mainly employ the principle of solidification by interdiffusion of Cu with thin layers of Sn-based solders. The resulting joints are rigid, and are therefore suitable for bonding substrates with similar coefficients of thermal expansion, as in chip stacking and 3D integration.

The associated physical and chemical reactions comprise two interacting regimes, which determine the processing window: firstly, the liquefaction and the hydrostatic behavior of liquid solder, and secondly, the diffusion-driven intermetallic phase growth in both liquid and solid states. Therefore, the following aspects of the bonding process need to be considered:

- The Sn-based solder layers need to be in the range of a few microns.
- Rapid heating preserves more Sn for wetting and allows thinner solder layers.
- The phase growth starts already at room temperature and terminates when all Sn is converted to Cu_3Sn.
- Chips of typical size require a patterning of the metal layers.
- The pattern geometry influences the balance of forces during bonding and enhances the underfill capabilities and the self-alignment properties.

Four selected applications of SLID as part of a complete integration scheme for semiconductor devices and technology have been introduced: one in the field of two-layer chip stacking without through-silicon micro vias, and three applications for SLID with micro vias and thinned chips.

For the application of SLID in a two-layer F2F chip-stacking scheme results of reliability tests have been presented. The following tests have been performed:

- Thermal cycling (TC): up to 1000 cycles from –65°C to 150°C,
- High-temperature storage (HTS) up to 200°C and 1000 h
- *Humidity tests*: Autoclave at 121°C, 96 h, 100% (AC) and unbiased highly accelerated stress test at 175°C, 96 h, 85% (UHAST)
- *Electromigration*: Current stressing at up to 250°C during 1500 h

The SLID joints exhibited a high degree of stability. The metallurgy has proven to be very stable against the tested stress conditions. However, current-driven diffusion of Cu through the TiW barrier layer has been detected. Moreover, leakage currents at low pad distances due to moisture uptake of the underfill material have been identified as a general concern in micro contacts. The SLID bonding process can be scaled to large-volume production by employing an intermediate evaporable glue during pick and place. Bonding itself is then carried out in a batch oven, where the chips can align themselves if allowed to move freely.

The presented integration schemes with micro vias can be divided into front-side approaches and backside approaches, which refers to the faces of a processed circuit wafer, from which the micro vias are formed. Both types bear advantages and disadvantages, which are summarized here:

- *Front-side vias*: The micro vias reach through the dielectric of the metallization layers and impose an obstacle to metal line routing. The vias need to be longer at the same substrate thickness and are therefore a challenge to metallization.
- *Backside vias*: The micro vias do not affect the metallization layers and provide the shortest through-chip path. But back-to-front alignment and via conductance versus via isolation are challenging issues that need to be solved then.

One example of a backside approach and two examples of the front-side approach are presented in detail, where a further distinction can be made with regards to the level of processing. In all schemes, most processing can be carried out on a wafer level; thus, it is the assembly or stacking processes that determine this classification. While W2W stacking provides a high throughput, C2W stacking provides a high yield. However, methods exist, which can enhance the throughput of a C2W approach, that is, by splitting the SLID bonding process into a quick pick-and-place step and a longer batch-curing step.

An overview of work on SLID in 3D integration has been presented in this chapter, which shows the potential and maturity of this bonding technique.

Acknowledgments This work consists of contributions of several authors, which are listed here according to their affiliations:

- Institute of Electron Devices and Circuits, University of Ulm, Germany
- Andreas Munding, now with Liebherr-Elektronik GmbH, Lindau, Germany
- Alexander Kaiser, now with Reinhardt Microtech GmbH, Ulm, Germany
- Peter Benkart, now with MicroGaN GmbH, Ulm, Germany
- Erhard Kohn
- Qimonda AG, Munich, Germany
- Holger Hübner
- Infineon Technologies AG, Munich, Germany
- Sabine Penka

References

1. Adamson AW (1990) Physical chemistry of surfaces. ISBN 0-471-61019-4, John Wiley & Sons Inc, New York
2. Bader S (1990) Erzeugung thermisch stabiler Mikroverbindungen durch isotherme Erstarrung (in German). Dissertation, Fakultät für Chemie, Universität Stuttgart
3. Bader S, Gust W, Hieber H (1995) Rapid formation of intermetallic compounds interdiffusion in the Cu–Sn and Ni–Sn systems. Acta Met et Mat 43:329–337
4. Beer SZ (1972) Liquid metals, chemistry and physics. ISBN 0-8247-1032-0, Marcel Dekker Inc, New York
5. Benkart P, Kaiser A, Munding A, Bschorr M, Pfleiderer H, Kohn E, Heittmann A, Hübner H, Ramacher U (2005) 3D chip stack technology using through-chip interconnects. IEEE Des Test Comput 22:512–518
6. Benkart P, Munding A, Kaiser A, Kohn E, Heittmann A, Hübner H, Ramacher U (2007) 3-Dimensional integration scheme with a thermal budget below 300°C. Sens Act A 139: 350–355
7. Bernstein L, Bartholomew H (1966) Applications of solid–liquid interdiffusion (SLID) bonding in integrated-circuit fabrication. Trans Met Soc AIME 236:405–412
8. Bhardwaj J, Ashraf H, McQuarrie A (1997) Dry silicon etching for MEMS. In: Proc Symp on Microstructures and Microfabricated Systems. ECS Annual Meeting, Montreal, Quebec, Canada, May 4–9
9. Bonkohara M, Takahashi K, Ishino M (2005) Trends and opportunities of system-in-a-package and three-dimensional integration. Electron Comm Jpn Pt 2 88:37–49
10. Fields RJ, Low SR (2002) Physical and mechanical properties of intermetallic compounds commonly found in solder joints. Research Report, NIST, Metallurgy Division
11. Frear DR, Burchett SN, Morgan HS, Lau JH (1994) The mechanics of solder alloy interconnects. Van Nostrand Reinhold, New York, p 60 ff

12. Frear D (2007) Issues related to the implementation of Pb-free electronic solders in consumer electronics. J Mat Sci: Mat Elec 18:319–330
13. Hübner H, Ehrmann O, Eigner M, Gruber W, Klumpp A, Merkel R, Ramm P, Roth M, Weber J, Wieland R (2002) Face-to-face chip integration with full metal interface. In: Melnick BM, Cale TS, Zaima S, Ohta T (eds) Advanced Metallization Conference, San Diego, V18:53–58
14. Hübner H, Penka S, Eigner M, Gruber W, Nobis M, Kristen G, Schneegans M, Barchmann B, Janka S (2006) Micro contacts with sub-30 μm pitch for 3D chip-on-chip integration. MAM, Genoble
15. Humpston G, Jacobson DM (1993) Principles of soldering and brazing. ISBN 0-87170-462-5, ASM International
16. Kaiser A (2007) Dreidimensionale Systemintegration: Technologische Entwicklung und Anwendung (in German). Dissertation, Fakultät für Ingenieurwissenschaften und Informatik, Universität Ulm
17. Kaiser A, Kück D, Benkart P, Munding A, Prinz GM, Heittmann A, Hübner H, Sauer R, Kohn E (2006) Concept for diamond 3-D integrated UV sensor. Diam Relat Mater 15: 1967–1971
18. Kempen ATW (2001) Solid State Phase Transformation Kinetics. Dissertation, Faculty of Chemistry, University of Stuttgart
19. Kim J-M, Fujimoto K (2003) Parametric design for resin self-alignment capability. IEICE Trans Electron E86(10):2129–2136
20. Klumpp A, Merkel R, Wieland R, Ramm P (2003) Chip-to-wafer stacking technology for 3D system integration. In: Proc 53rd ECTC, pp 1080–1083
21. Klumpp A, Ramm P, Wieland R, Merkel R (2003) Integration Technologies for 3D Systems. Int Workshop 3D System Integration, Munich
22. Kostner H, Hübner H (2003) New Flip-Chip on Chip Process Supercedes Embedded Technologies. EMPC 2003, Friedrichshafen
23. Lin W, Lee YC (1999) Study of fluxless soldering using formic acid vapor. IEEE Trans Adv Pack 22:592–601
24. Lärmer F, Schilp A (1994) Method of Anisotropically Etching Silicon. German Patent DE 4241045.
25. Munding A (2007) Interconnect technology for three-dimensional chip integration. Dissertation, faculty of engineering and computer science, University of Ulm. ISBN 978-3-86727-406-7, Cuvillier-Verlag, Göttingen, Germany
26. Munding A, Kaiser A, Benkart P, Kohn E, Heittmann A, Hübner H, Ramacher U (2006) Scaling aspects of microjoints for 3D chip interconnects. In: Proc 36th ESSDERC, pp 262–265
27. Rahn A (1993) The basics of soldering. ISBN 0-471-58471-1, John Wiley & Sons Inc, New York
28. Ramm P, Bollmann D, Braun R, Buchner R, Cao-Minh U, Engelhardt M, Ehrmann G, Grassl T, Hieber K, Hübner H (1997) Three dimensional metallization for vertically integrated circuits. Microelectron Eng 37/38:39–47
29. Ramm P, Klumpp A, Merkel R, Weber J, Wieland R, Ostmann A, Wolf J (2003) 3D System Integration Technologies. In: Proc MRS Symp 766, pp 3–14
30. Schmid-Fetzer R (1995) Fundamentals of bonding by isothermal solidification for high temperature semiconductor applications. In: Design Fundamentals of High Temperature Composites, Intermetallics, and Metal-Ceramics Systems, ISBN 0-87339-310-4, The Minerals, Metals and Materials Society, 75–98
31. Seidel H, Csepregi L, Heuberger A, Baumgärtel H (1990) Anisotropic etching of crystalline silicon in alkaline solutions. J Electrochem Soc 137:3626–3632
32. Shim J, Oh C, Lee B, Lee DN (1996) Thermodynamic assessment of the Cu–Sn system. Z Metallk (Int J Mat Res) 87:205–212

33. Takahashi K, Taguchi Y, Hoshino M, Tanida K, Umemoto M, Yonezawa T, Kondo K (2005) Development of less expensive process technologies for three-dimensional chip stacking with through-vias. Electron Comm Jpn Pt 2, 88:50–60
34. Tanida K, Umemoto M, Tanaka N, Tomita Y, Takahashi K (2004) Micro Cu bump interconnection on 3D chip stacking technology. Jpn J App Phys 43:2264–2270
35. Tu KN (2003) Recent advances on electromigration in very-large-scale-integration of interconnects. J App Phys 94:5451–5473
36. Tu KN, Thompson RD (1982) Kinetics of interfacial reaction in bimetallic Cu–Sn thin films. Acta Met 30:947–952
37. Umemoto M, Tanida K, Nemoto Y, Hoshino M, Kojima K, Shirai Y, Takahashi K (2004) High-performance vertical interconnection for high-density 3D chip stacking package. In: Proc 54th ECTC, pp 616–623
38. Wang SQ, Suthar S, Hoeflich C, Burrow, BJ (1993) Diffusion barrier properties of TiW between Si and Cu. J App Phys 73:2301–2320
39. Wieland R, Bonfert D, Klumpp A, Merkel R, Nebrich L, Weber J, Ramm P (2005) 3D Integration of CMOS transistors with ICV-SLID technology. Microelectron Eng 82:529–533
40. Xie H, Erdmann L, Zhu X, Gabriel KJ, Fedder GK (2002) Post-CMOS processing for high-aspect-ratio integrated silicon microstructures. IEEE J MEMS 11:93–101

Chapter 8
An SOI-Based 3D Circuit Integration Technology

James Burns, Brian Aull, Robert Berger, Nisha Checka, Chang-Lee Chen, Chenson Chen, Pascale Gouker, Craig Keast, Jeffrey Knecht, Antonio Soares, Vyshnavi Suntharalingam, Brian Tyrrell, Keith Warner, Bruce Wheeler, Peter Wyatt, and Donna Yost

8.1 Introduction

Lincoln Laboratory has developed a wafer-scale three-dimensional (3D) integrated circuit (IC) technology whereby 3D chips are constructed by transferring, bonding together, and electrically connecting the active sections of ICs that were fabricated on silicon-on-insulator (SOI) substrates [1]. Layer transfer techniques were originally developed at Lincoln Laboratory to transfer thin GaAs strips fabricated on a GaAs template for the fabrication of solar cell devices [2]. This concept was later applied to the fabrication of displays by Kopin Corp. [3], who transferred ICs fabricated on SOI substrates; the transferred layer consisted of the buried oxide (BOX), the thin SOI film, and the multilevel interconnect. Further work at Kopin, Northeastern University, and Lincoln Laboratory led to an enhanced wafer-scale 3D IC technology. The building blocks of that 3D technology were SOI circuit fabrication, low-temperature wafer–wafer adhesive bonding, transfer of SOI analog and digital circuits to photodiode imaging circuits, and electrical connection of the circuit structures with through-silicon vias (TSVs) [4]. This technology led to the first demonstration of a "true" 3D circuit with the successful operation of a 3D 64×64 visible imager [5]. More recently, numerous institutions have also used a variety of layer transfer techniques to develop 3D integration technologies [6–11].

The technology used to construct the 3D imager had serious limitations. The adhesive bond limited subsequent processing temperatures to less than 200°C. This prevented using tungsten as an electrical connection and a post-processing anneal at 400°C to repair oxide damage from plasma processes. The adhesive bond also limited a 3D IC to two functional layers since an additional bonding and transfer process would disturb the initial adhesive bond. The TSV design required deep silicon etch and oxide refill processes, and outgassing from the adhesive

J. Burns
Lincoln Laboratory, Massachusetts Institute of Technology, Lexington, MA, USA
e-mail: jab@LL.MIT.EDU

C.S. Tan et al. (eds.), *Wafer-Level 3D ICs Process Technology*,
DOI: 10.1007/978-0-387-76534-1_8, © Springer Science+Business Media, LLC 2008

bond prevented scaling the TSV size to less than 6 μm. Finally, the wafer–wafer alignment equipment had an overlay error of 2 μm, which set the pitch of the TSVs to 10-μm minimum. As a result, a new 3D IC technology was developed at Lincoln Laboratory to overcome these limitations.

8.2 Lincoln Laboratory's Wafer-Scale 3D Circuit Integration Technology

8.2.1 Three-Dimensional Fabrication Process

Three-dimensional circuits are fabricated on 150-mm SOI substrates using a 180-nm fully depleted SOI process that includes mesa isolation of transistors and three levels of metal interconnect. A new term, "tier", was adopted to distinguish among design layers, physical layers, and transferred layers of a 3D IC and is the functional section of a wafer that consists of the active silicon, the interconnect, and for an SOI wafer, the BOX. A tier is approximately 10-μm thick. The 3D assembly process and a 3D chip consisting of three tiers are illustrated in Fig. 8.1. The process begins by transferring tier 2 to the base tier, tier 1, after face-to-face infrared alignment, oxide–oxide bonding at 275°C, and a wet etch of the handle silicon to expose the BOX of tier 2. The BOX is used as an etch stop for the silicon etch to produce a uniformly thin active layer and is an essential step in the 3D assembly technology. For this reason all circuits to be transferred must be fabricated with SOI substrates. The handle silicon of a transferred tier is removed by grinding the silicon to a thickness of about 70 μm followed by a silicon etch in a 10% tetramethylammonium hydroxide (TMAH) solution at 90°C. Since the ratio of silicon-to-BOX etch rates in TMAH is 1000:1, the handle silicon is removed without attacking the BOX and without introducing a thickness variation in the transferred tier, a factor that is essential when forming the vertical connections between tiers, the 3D vias. In both etches the edge is protected to ensure that the wafer can be handled by cassette-to-cassette equipment and that the silicon removal process does not attack the oxide–oxide bond. Three-dimensional vias are designed to lie in the mesa-isolated regions of the tiers so that lining the vias with a deposited dielectric is not required to achieve insulation between the vertical connections.

The 3D vias are patterned and etched through the BOX and deposited oxides to expose metal contacts in both tiers. The 3D vias are then filled with tungsten that is planarized by chemical–mechanical polishing (CMP) to electrically connect the two tiers. The metal contact in the upper tier is an annulus with a 1.5-μm opening that also functions as a self-aligned hard mask during the plasma etch of the oxide beneath it to reach the metal land in the lower tier. In order to fully land the 3D via, the size of the metal pad, and thus the pitch of the vertical interconnect, are proportional to twice the wafer–wafer misalignment.

A third tier, tier 3, can then be added to the tier 1–2 assembly using the same processes, except that the front side of tier 3 is bonded to the BOX of tier 2, and 3D vias connect the top-level metal of tier 3 to metal pads on the BOX of tier 2.

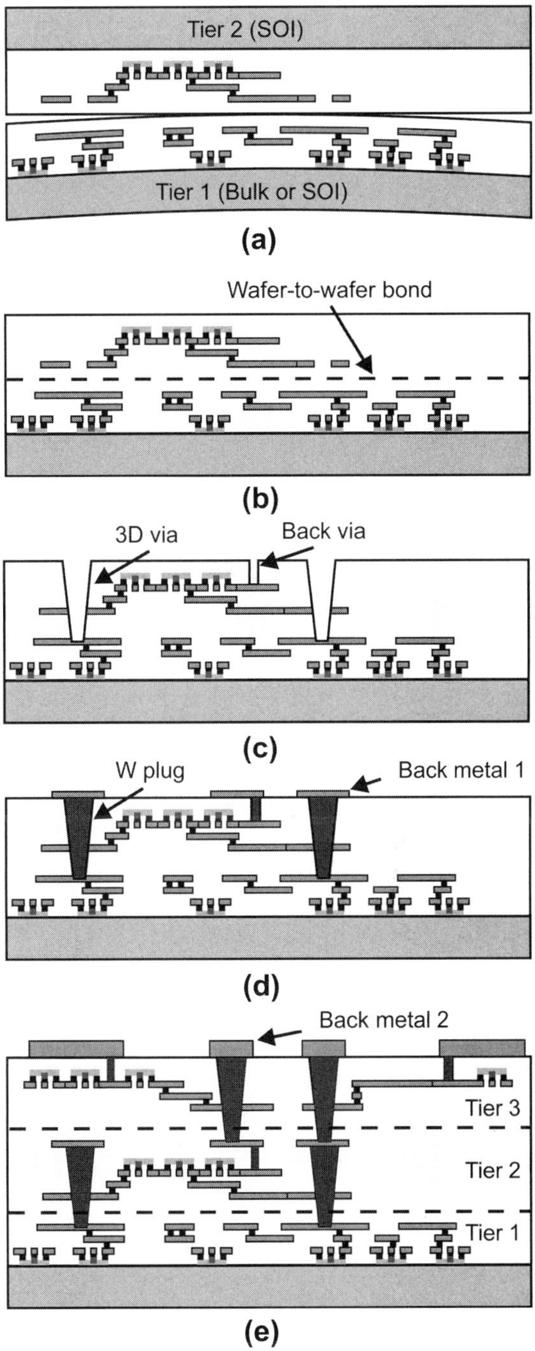

Fig. 8.1 Assembly process for a 3D chip: (**a**) Two completed circuit wafers are planarized, aligned, and bonded face to face; (**b**) the handle silicon is removed; (**c**) 3D vias are etched through the deposited BOX and the field oxides; (**d**) tungsten plugs are formed to connect circuits in both tiers; and (**e**) after tier 3 is transferred, bond pads are etched through the BOX for testing and packaging. From [1] © 2006 IEEE

The 3D chip is shown after bond pads are etched to expose the back of the first-level metal for probing and wire bonding. If the 3D chip is a digital circuit, the bond pads are etched through the BOX and deposited oxides of tier 3. If it is a backside-illuminated imager, tier 1 is a bulk silicon detector wafer in which photodiodes were fabricated. An additional transfer to a carrier wafer is then required, and bond pads are etched after thinning the backside of the detector wafer to tune the silicon to the required optical absorption. The cross-sectional scanning electron micrograph (SEM) of a three-tier ring oscillator shown in Fig. 8.2 illustrates interconnections between tiers and the compactness possible with the 3D technology.

8.2.2 Three-Dimensional Enabling Technologies

The building blocks of Lincoln Laboratory's 3D circuit integration technology are SOI circuit fabrication, precision wafer–wafer alignment, low-temperature wafer–wafer oxide bonding, and electrical connection of the circuit structures with dense vertical interconnections.

An SOI technology is required for all tiers that are to be transferred, but a FDSOI technology [12] is preferred since FDSOI is a low-power technology and is particularly advantageous for 3D circuits to minimize heat generation within the transferred tiers. The principal technology features are a 400-nm BOX thickness, 40-nm SOI channel thickness, mesa isolation, elimination of sidewall leakage with sidewall implants, 4-nm gate oxide, dual-doped polysilicon gates, oxide–nitride spacers, cobalt salicide, three levels of aluminum-based interconnect with

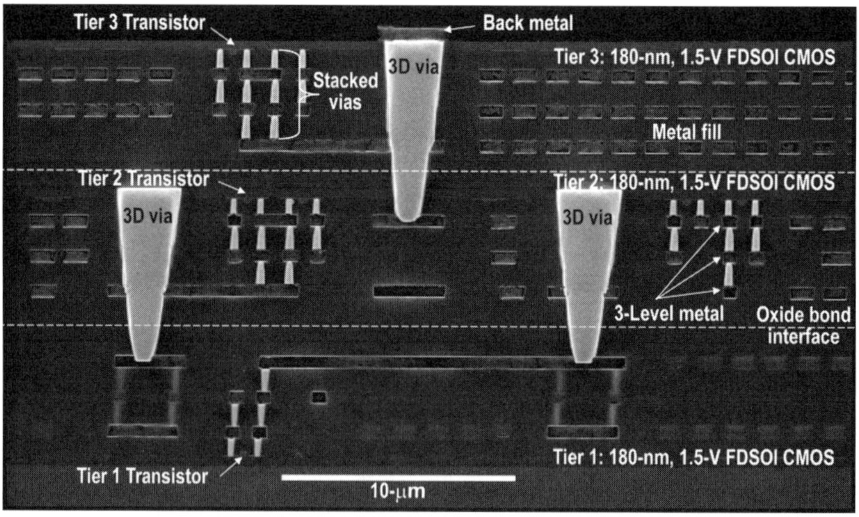

Fig. 8.2 Cross-sectional SEM of a three-tier ring oscillator in which inverters in each fully depleted SOI (FDSOI) tier are electrically connected with 3D vias. Note that the 3D vias are located in the isolation (field) region between transistors

planarized dielectrics, and tungsten plug contacts and via connections. All features are defined using a 248-nm Canon stepper with a 22 × 22-mm field of view. Mesa isolation simplifies the placement and fabrication of vertical connections between tiers since the connections are formed in the field-isolation regions. This 180-nm technology has an unloaded inverter delay of 30 ps per stage at 1.5 V; f_T for the n- and p-type field-effect transistors (FETs) are 65 GHz and 40 GHz, respectively.

The 3D via pitch is a critical factor in the viability of the 3D technology since for optimal circuit density the minimum pitch of the 3D vias should be comparable to that of the 2D vias that connect multilevel metal layers. The 3D via pitch is determined by the 3D via size, which is dependent upon the achievable oxide-etch aspect ratio, by the design rule that the 3D vias be fully landed, and by the wafer–wafer alignment tolerance, which is primarily constrained by the overlay capability of the wafer–wafer alignment system. The pitch, P, can be written as

$$P = 2 \times \mathrm{WA} + 3\mathrm{DV} + \mathrm{MS}, \tag{8.1}$$

where WA is the wafer–wafer alignment capability, 3DV is the diameter of the 3D via at the landing pad, and MS is the minimum metal–metal spacing. Initially, a modified mask aligner with an overlay tolerance of ± 2 μm was used, which, combined with a 1.5-μm opening in the annulus, led to a 5.5-μm square landing pad and a 6-μm pitch. To scale the inter–tier connections, modern wafer stepper technologies were used to design and fabricate a precision wafer–wafer alignment system whose overlay goal was ± 0.25 μm [13]. The basic components of the precision alignment system are two InGaAs cameras with subpixel edge detection for infrared alignment; a six-axis piezoelectric stage with nanometer-scale resolution to provide X, Y, Z, Θ, tilt, and tip motions while tier 2 is aligned to a fixed tier 1; and a precision XY air-bearing stage controlled by an environmentally compensated laser interferometer to map wafers before alignment. The latter is intended to measure the grid distortion of each wafer and select the best match between wafers and, eventually, to compensate for distortion between wafers by substrate heating. The system is shown in Fig. 8.3a, and the alignment repeatability data of Fig. 8.3b show that an overlay of ± 0.4 μm has been achieved. Alignment is measured using box-in-box alignment marks fabricated in the same metal layers that contain the 2D alignment marks, and the features are measured after substrate removal using a commercial overlay tool [14]. The overlay goal of ± 0.25 μm will require wafer mapping before 3D integration in order to bond wafers with equivalent grid distortion.

The wafer–wafer bonding process in 3D integration has three requirements. The first is that the room-temperature bond be sufficiently strong to prevent wafer slippage between the wafer alignment and wafer bonding processes, since the alignment and a 275°C heat treatment take place in two separate instruments. Second, the bond temperatures must not exceed 500°C, the upper limit of the aluminum-based interconnect. Finally, the bond must be sufficiently strong to withstand the 3D fabrication process. Complementary metal-oxide semiconductor (CMOS) wafers to be bonded are coated with 1500 nm of a low-temperature oxide (LTO) deposited by low-pressure chemical vapor deposition (LPCVD) at a temperature of 430°C.

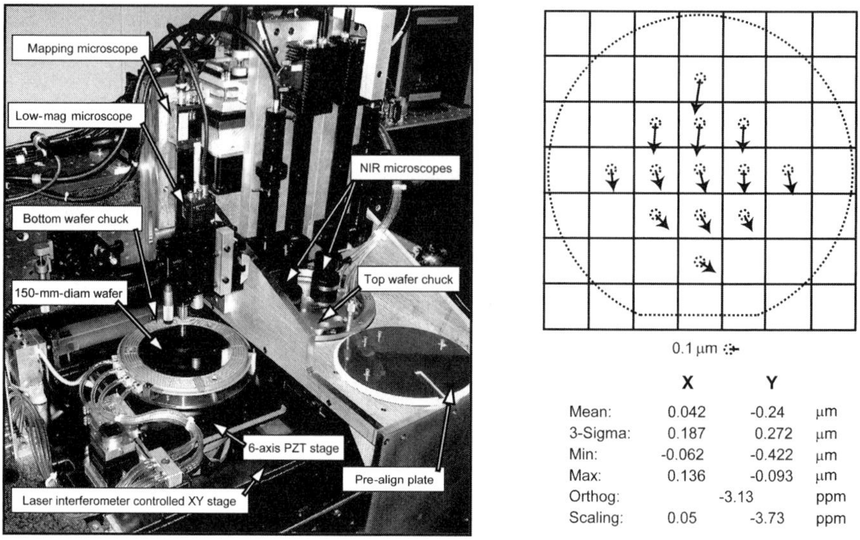

(a) (b)

Fig. 8.3 (**a**) Precision wafer–wafer alignment system and (**b**) a vector plot of wafer–wafer alignment using the precision aligner for two 150-mm-diameter bonded wafers with a 22-mm die size. The alignment error was modeled in terms of displacement, orthogonality, and scaling. Successive alignment measurements show a 3σ repeatability of <0.4 μm. From [1, 15] © 2006 IEEE

A thickness of 1000 nm of the oxide is removed by CMP to planarize and smooth the surfaces to a roughness of <0.4 nm RMS (root mean square)(as measured by atomic force microscopy using a 10-μm square scan). The wafers are immersed in H_2O_2 at 80°C for 10 min to remove any organic contaminants and to activate the surfaces with a high density of hydroxyl groups [16], after which the wafers are rinsed and spun dry in nitrogen in a standard rinse/dryer. The wafers are aligned and bonded by initiating contact at the center of the top wafer. When the surfaces are brought into contact, weak (~0.45 eV) hydrogen bonds are created at the interface (Si–OH:HO–Si), the bond interface propagates radially within 2–5 s to the edge of the wafer pair, and after 30 s the wafer pair can be removed from the aligner without disturbing the bond and wafer alignment. The bond strength is increased by a thermal cycle that creates covalent bonds at the interface from the reaction Si-OH:HO–Si → Si–O–Si + H_2O, with the Si–O bond having a bond energy of 4.5 eV.

Optimal thermal cycle parameters for this particular bonding technique were determined by measuring bond strengths in the temperature range 150–500°C. Fig. 8.4a is an Arrhenius plot of the measured surface energy for 1- and 10-h cycles. Surface energies were measured using the blade insertion technique [17], and crack lengths were measured using IR inspection at four points along the edge of the wafer pair. A 1-mm measurement error was assumed and the maximum and minimum bond energies were determined after subtracting the 1-mm error from the shortest

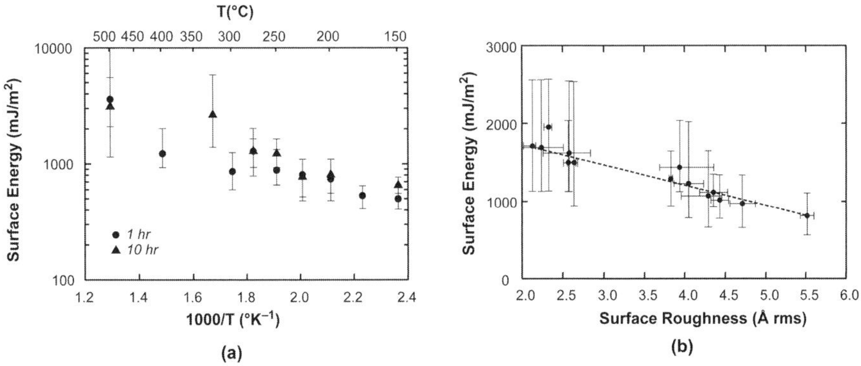

Fig. 8.4 A thermal cycle (**a**) is required to increase the wafer–wafer bond strength of the low-temperature bond process to at least 1000 mJ m^{-2}; heat treatments greater than 300°C may cause bubbles in the bond interface. The measured roughness in (**a**) was 0.4–0.5 nm but improved CMP processes have led to a reduction in surface roughness (**b**) and increased bond strengths. From [15] © 2007 IEEE

and adding 1 mm to the longest cracks, respectively. A 275°C, 10-h thermal cycle was chosen as optimum since a surface energy of about 1000 mJ m^{-2} is required to allow removal of the handle substrate and subsequent 3D via processing without disturbing the aligned pair; infrared inspection determined that voids were formed for bonding temperatures above 300°C as a consequence of water being formed and trapped within the bonding regions. Removal of the handle silicon of a transferred tier removes a barrier to water diffusion, and as a result voids are not formed during back-end-of-the-line (BEOL) processes that reach 475°C. The effect of surface roughness on bond strength in Fig. 8.4b indicates that low-temperature bonding requires careful control of the CMP polish process [15]. An additional factor to be controlled in the bonding process is surface planarity, which restricts surface steps to less than 50 nm.

The unrestricted placement of 3D vias is an integral part of the 3D technology. A 3D via, shown in Fig. 8.5, consists of a metal annulus in the upper tier, a metal land in the lower tier, and a tungsten plug that electrically connects the two features. The plug is formed in an oxide hole that is plasma etched with a resist mask that is aligned to the metal annulus with a mean and 3σ overlay error less than 100 nm and 300 nm, respectively. The annulus is a unique feature of the design since it forms the top electrical contact and masks the oxide etch to the metal land. After resist removal, tungsten is deposited and planarized by CMP leaving approximately 8 μm of tungsten in the hole. The present 3D via design consists of a 3-μm square landing pad, a 1.5-μm square annulus opening, and a 1.75-μm square resist window called 3D cut. The design of these intertier connections significantly reduced the pitch of 3D vias to 6 μm from 26 μm used in the 3D imager first reported [5].

The oxide via is etched in a Trikon Technologies low-pressure, high-density, helicon-based cluster tool in order to achieve a high-aspect-ratio etch [18]. The etch conditions are a balance between polymer deposition used to achieve a vertical

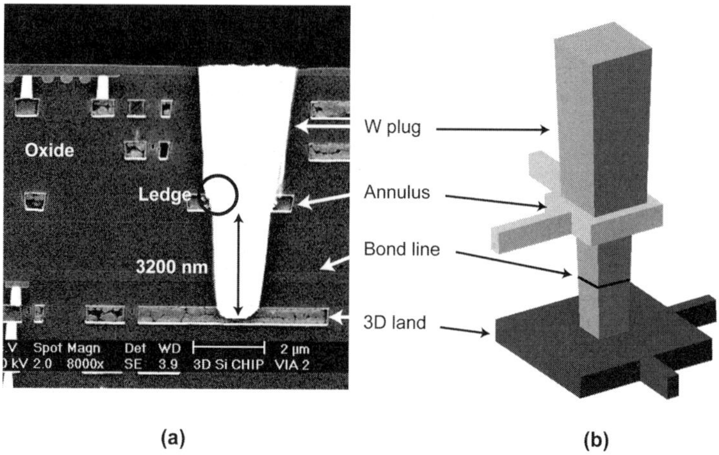

Fig. 8.5 (**a**) Cross-sectional SEM and (**b**) isometric drawing of a 3D via. The essential features shown are the metal annulus in metal 3 of the upper tier, the 3D land in metal 1 of the lower tier, and the tungsten plug that electrically connects the two tiers. Not shown is the 2.5-µm resist pattern that masks the oxide etch and determines the size of the via to the plane of the annulus. The ledge, the overlap of the plug and the metal annulus, is circled. From [1] © 2006 IEEE

sidewall and wafer bias to sputter the deposited polymer from the base of the via during etching. A section of the metal annulus, called a ledge, is exposed during the oxide etch, and the interaction of the plasma with the ledge affects the etch profile below the plane of the annulus. The metal interconnect is a deposited stack of titanium/aluminum–silicon/titanium/titanium nitride, and after tier transfer the metal stack is inverted and titanium is exposed to the plasma. Three-dimensional via chains were fabricated to optimize the process. Three-dimensional via resistance and yield were analyzed as a function of etch chemistry, system pressure, RF (radio frequency)bias, and 3D via design features. Initial 3D via chain tests had high yield and low resistance to the annulus but low yield to the landing pad on the lower tier. The cross-sectional SEM photos of 3D vias showed that the width of the metal ledge had been substantially reduced. The vias were pinched off below the annulus metal, but structures etched without the annulus displayed no via pinch-off. The metal ledge was eroded as a result of chemical and physical etching since titanium and titanium nitride are volatile in fluorine-based chemistries, but aluminum fluoride, a plasma etch reactant, is not and so a high wafer bias was necessary to remove it. In addition, the etch chemistry produces a polymer when the oxide etch reaches the metal annulus and interacts with titanium, leaving a metal organic deposit on the via sidewalls leading to via pinch-off. A yield model constructed from test devices with varying 3D cut and annulus sizes indicated the 3D via chain yield was proportional to the annulus opening and inversely proportional to the ledge width with the annulus opening being the major yield determinant factor. The yield of a 10 000 3D via chain increased from 35% to 100% by minimizing the exposed ledge, adding a Cl_2-based metal etch in-situ to etch the ledge, and adding a small amount of O_2 to

Table 8. 1 Three-dimensional via-etch chemistry

	3D Via etch	
	Oxide	Metal
Source power (W)	1750	2750
Bias power (W)	1300	250
Press (mTorr)	3.5	6
C_4F_8 (cc)	30	–
CH_2F_2 (cc)	40	–
CO(cc)	24	–
Ar (cc)	80	–
Cl_2 (cc)	–	75
BCl_3 (cc)	–	10
N_2 (cc)	–	15

From [1] © 2006 IEEE

the oxide etch to remove the etch polymer. The current process results in a 3D via with the median resistance of 0.75 Ω. The oxide-etch chemistry is summarized in Table 8.1.

8.2.3 Three-dimensional Technology Scaling

The overlay of features to be vertically connected between tiers places constraints beyond wafer–wafer alignment. In 3D technology, variations between wafers in *x*–*y* die placement and rotation add to the overlay error of the 3D via, and the error is doubled if one tier is flipped with respect to the other. The first constraint requires that the first lithographic level stepped on each wafer to be 3D integrated must place the center of the exposed die pattern coincident to the wafer's center to within 100 μm. This requirement ensures a uniform bond at the tiers' edge and aids wafer handling. The second constraint is to place the die parallel to the wafer flat to within 5 ppm. To assist the alignment process of the first level, features are included in the reticle to measure pattern placement and field rotation. The exposure stepper's alignment algorithm ensures that subsequent lithographic levels satisfy the die placement criteria. The third constraint is tier distortion that occurs during 2D and 3D fabrication. Distortion degrades the registration between two tiers and, if not controlled, can result in 3D vias that are not fully landed since the placement of the metal annulus in the upper tier cannot compensate for different grid distortions in the two tiers. Note that in 2D lithography grid distortion can be accommodated during level–level registration by the alignment algorithm. Registration between two tiers is measured with the overlay tool [14], and the total overlay error is deconvolved into translation, rotation, expansion, and orthogonality errors in the *x* and *y* axes. A viable 3D technology thus requires the control of tier distortion throughout 2D fabrication as well as during 3D fabrication. Wafer bow caused tier distortion and was minimized by depositing stress-compensation oxides on the backsides of wafers before wafer–wafer alignment. Tier distortion during wafer–wafer alignment also occurred because of wafer deformation by a vacuum hold-down chuck and was minimized by a redesign of the chuck. The 3D technology has been scaled as a result of the efforts to control overlay error. One example is the rule ST2 for spacing of 3D cuts in tier 2 in order to prevent shorting between tungsten plugs that connect separate 3D lands in tier 1:

$$\mathrm{ST2} = \mathrm{MS} + \mathrm{WA} + \mathrm{VS} - \mathrm{3Dcut}, \tag{8.2}$$

where MS is the minimum spacing between the 3D land metal features, 0.3 μm; WA is the wafer–wafer alignment error, 1 μm; VS is the required size of the tungsten plug in contact with the 3D land, 1 μm; and 3D cut is the resist window; 1.75 μm. The preceding work led to a reduction in spacing from 3.55 μm to 1.55 μm and a reduction in the size of the 3D land from 5.5 μm to 3 μm.

In 3D integration tier 3 is bonded to tier 2 face-to-back as seen in Fig. 8.1, and the 3D vias connecting those tiers penetrate the SOI and polysilicon planes. Exclusion zones that reduce the density of SOI active and polysilicon interconnect are required to prevent tungsten to SOI and polysilicon shorts. The spacing rule ST3 is dominated by the wafer–wafer alignment error:

$$\mathrm{ST3} = \mathrm{WA} + \mathrm{PS} + \mathrm{AE}. \tag{8.3}$$

Combining PS, the spacing between the tungsten plug and SOI, 0.175 μm, and AE, the accumulated alignment and feature size budget, 0.185 μm, with wafer–wafer alignment error leads to the spacing between 3D cut on tier 3 and SOI on tier 2 of 1.35 μm. The effect of the wafer–wafer alignment error on SOI and polysilicon

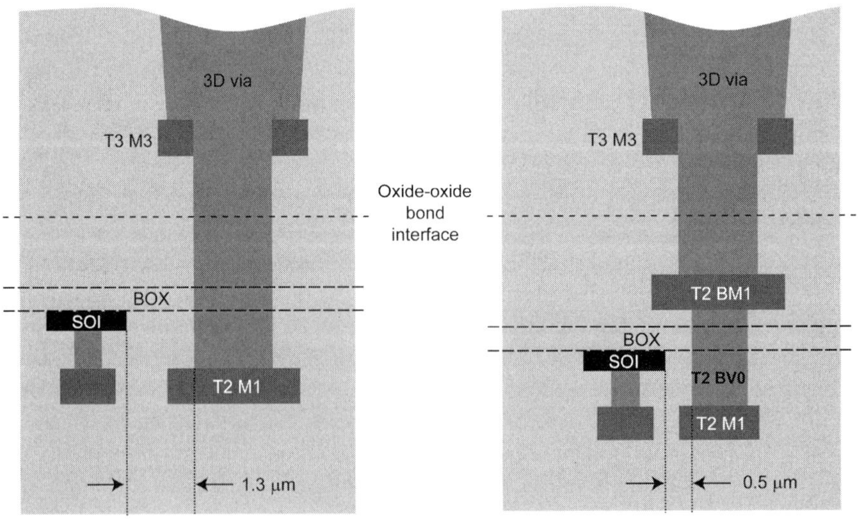

Fig. 8.6 The inclusion of back metal (BM1) and a back via (BV0) on tier 2 reduces the 3D via–SOI spacing from 1.3 μm to 0.5 μm

density on tier 2 can be removed by the addition of a metal land on the BOX of tier 2 that is connected to the metal 1 of tier 2 by the standard 300-nm via as shown in the drawing in Fig. 8.6. An additional stepper alignment is required but the spacing between the via, BV0, and SOI is reduced to 0.35 μm since the wafer–wafer alignment error has been removed from the overlay budget.

8.3 Transferred FDSOI Transistor and Device Properties

The 3D integration process creates FDSOI transistors in tiers 2 and 3 that differ from those in tier 1 as well as conventional SOI transistors since transistors in those tiers do not have a silicon substrate. As a result, the body potential of a transistor in a transferred tier is less well determined since the potential is a function of fringing electric fields from the source and drain, the field from the gate, the integrated charge in the body, the integrated electric field from oxide charge above the body, and the field from any electrodes above the transistor. Test transistors and circuits were designed for testing devices before and after 3D integration in order to determine the effect of 3D integration and the absence of a substrate electrode on their properties [19]. The relationship between the off-state current and drive current is shown in Fig. 8.7 after wafer fabrication and 3D integration for the same set of transistors in tier 3. The data indicate no change in transistor properties of the pFETs and a decrease of the nFET drive current of approximately 50 μA μm^{-1} for the equivalent off-state current. The latter is principally due to a 40-mV increase in nFET-threshold voltage. Ring oscillators and 8 × 8 multipliers were similarly characterized with the data indicating no change in delays per stage or operating frequency.

Of particular interest is the effect of radiation-induced charge trapped in isolation oxides, that is, field, trench, and BOXs, on circuits from which the substrate was removed as part of a 3D imager for space applications. In FDSOI transistors the main effect of total dose irradiation is a threshold shift due to charge trapping in the BOX because the front and back oxides are capacitively coupled. Removing the substrate after bonding a FDSOI tier to a handle wafer reduces the

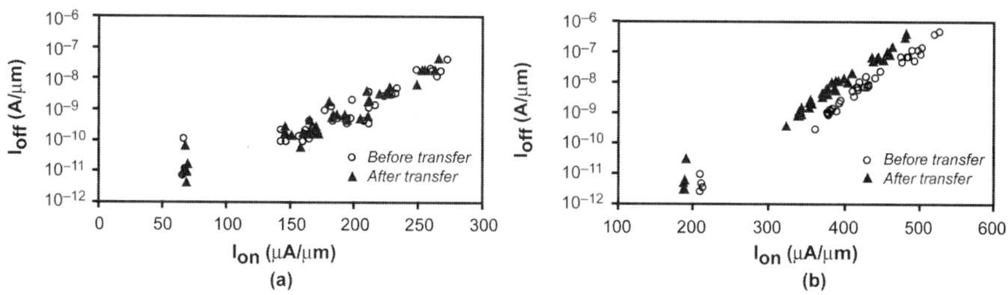

Fig. 8.7 Off-state vs. drive current for (**a**) pFETs and (**b**) nFETs in tier 3. The data show that 3D integration, including substrate removal, of transistors in tier 3 of a three-tier 3D wafer has an insignificant effect on transistor properties. The data were obtained from 40 transistors with 8-μm-wide transistors with gate lengths of 200, 250, 300, or 500 nm

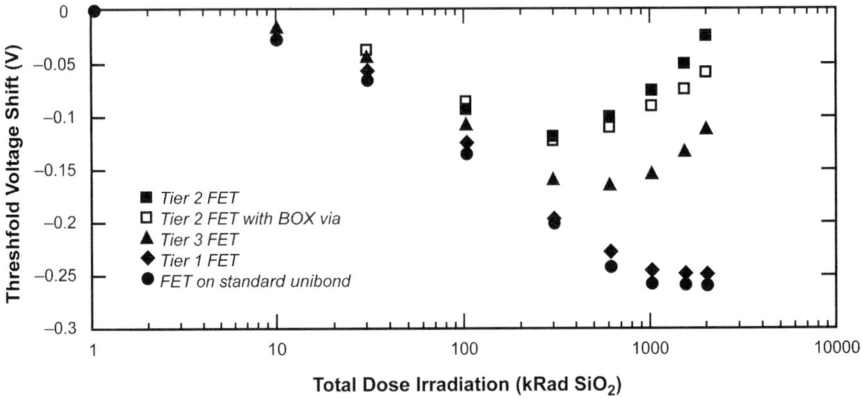

Fig. 8.8 Total ionizing dosage (TID)-induced threshold voltage shift for nFETs on tiers 1, 2, and 3 of a 3D chip with the transistors in the off-gate bias conditions during irradiation. Each transistor has 20 fingers with $L = 0.18\ \mu m$, $W = 10\ \mu m$. The *filled circles* are typical data for an nFET on a single standard SOI wafer. From [20] © 2003

radiation-induced threshold voltage shift and off-state leakage of FDSOI nFETs, and thinning the BOX further decreases the nFET sensitivity to irradiation [20]. The threshold shifts of transistors in the three tiers of a 3D chip were evaluated as a function of total integrated dose (TID) using 10-keV X-rays. As shown in Fig. 8.8, for nFETs in tier 1 the decrease in threshold voltage was similar in magnitude to that of nFETs in a conventional FDSOI wafer, while for nFETs in tiers 2 and 3 the shifts were less. Since both tiers 2 and 3 were flipped and bonded onto tier 1, these results are consistent with experiments of single-tier wafers bonded to handle wafers and the substrate removed; see Fig. 8.18 of [21]. These data indicate that the integration of multitier wafers reduces the radiation-induced threshold voltage shift for transistors in the upper two tiers and does not degrade the total dose tolerance of transistors in the first tier.

The effects of heat generation in tiers 2 and 3 of a 3D chip is an area of current research since devices in those tiers are embedded within dielectric layers and the heat generated is more difficult to remove than heat generated in tier 1. The effects of self-heating on transistors in a 3D circuit stack were compared by measuring the actual temperature of each tier as a function of power consumption using planar pn junction diodes as temperature sensors and SOI resistors as the heat source. The effectiveness of various heat-sinking techniques was also studied using vias that were etched through the BOX and filled with tungsten at the same time as the contacts were formed [21]. A diagram of the structure studied is shown in Fig. 8.9a along with heat transfer paths. Two types of heat-sinks were evaluated. One used metal pads at the top of the 3D stack to radiate the heat into air, and the other used tungsten-filled vias through the BOX to dissipate the heat into the substrate. Note that design rule constraints require the heat to be transferred through standard metal interconnects and vias, which have fairly high thermal resistance because of their

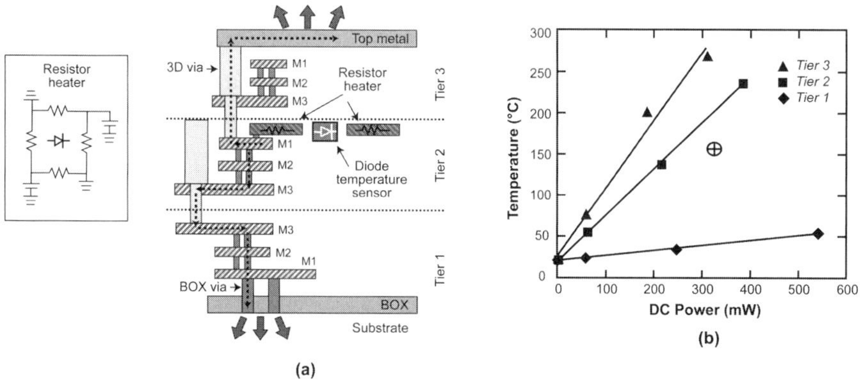

Fig. 8.9 (**a**) Heat flow was generated by resistors in each tier and (**b**) caused a temperature increase as measured by pn diodes in each tier. The symbol ⊕ shows temperature with metal pads on the 3D vias of tier 3. From [21] © 2007 IEEE

small size, before reaching the heat-sink. The measured temperatures of each tier are shown in Fig. 8.9b and indicate that the temperature of tier 3 is the highest and can exceed 250°C at 300 mW, equivalent to approximately 125 W mm^{-2} of surface area. Although both BOX vias and the top metal pads contributed to temperature lowering, the top metal is more effective in removing heat. Similar measurements with transistors showed that adding top metal or vias through the BOX can help dissipate the heat, but the improvement in drive current and transconductance is less than 10%. Fortunately, thermal impacts on FET performance are small and SOI 3D ICs are more forgiving with high-temperature operation than bulk silicon because of significantly reduced junction area. These results show that 3D ICs should be designed with circuits that dissipate the greatest power placed in tier 1 and the least power in tier 3.

8.4 Multiproject Circuit Design and Layout in Lincoln Laboratory's 3D Technology

The implementation of Lincoln Laboratory's 3D circuit design techniques followed closely the development of the 3D fabrication technology. The Defense Advanced Research Projects Agency (DARPA) funded work to make Lincoln's 3D circuit integration technology available to designers in universities and commercial institutions through a 3D Multiproject program. Three-dimensional circuit designs were to be submitted by those institutions and be integrated into a 3D Multiproject chip along with Lincoln Laboratory designs. It was expected that the experience gained by all the participants through circuit design and testing of their 3D chips would lead to a greater understanding of the advantages of 3D ICs. Initially 3D circuits and test devices were designed and laid out as if a 3D circuit was distinctly different from its 2D counter part. Experience gained in 3D circuit layout and 3D technology development made it clear that a more direct design approach was required to simplify the design and layout process, particularly for researchers who were not familiar with the details of the 3D technology. This required the development of 3D design rules, 3D design and layout enhancement to Mentor [22] and Cadence [23] computer-aided design (CAD) tools, and circuit design submission rules to ensure data compatibility when the designs were placed in the chip.

8.4.1 Three-Dimensional Design and Layout Practice

Three steps were adopted to aid the design and layout of 3D chips. The first was to view a 3D chip as a 2D chip with multiple SOI, polysilicon, and interconnect layers. The scanning electron micrograph (SEM) of the 3D chip of Fig. 8.2 was obtained from a 3D ring oscillator and can be viewed as a 2D chip with 3 SOI levels, 2 BOX layers, 3 polysilicon layers, and 10 metallization levels. The second step was to introduce a new term, tier. A tier is the section of a completed wafer that contains the active silicon, polysilicon, and interconnect, and if it is a SOI wafer, it also contains the BOX. As discussed in section 2, the 3D integration process bonds, transfers, and interconnects tiers to build a 3D chip. The term was chosen to avoid confusion with the various terms used to describe design and fabrication layers. The third step was to design and layout a 3D chip as it would be viewed in its completed state with the tiers numbered from 1 to *n* with tier 1 at the bottom and the bond pads located in the top tier, as shown in Fig. 8.1.

Three-dimensional design rules were developed and embedded into the existing design rules for the laboratory's FDSOI technology [12], which are the third revision and reflect technology and design advancements. The addition of the 3D rules increased the number of rules by 25%, a value that is consistent with the addition of the additional SOI, polysilicon, and interconnect layers. The major features controlled by the rules relate to the 3D via shown in Fig. 8.5. Those features are the size of the 3D cut which defines the resist window to etch oxide to the metal annulus, the metal annulus in the upper tier; the 3D land in the lower tiers which is contacted by the tungsten plug; the spacing between 3D vias; and spacing between 3D vias and other layers to avoid shorting of the tungsten plug to them. The present 3D via design consists of a 3-μm square landing pad, a 1.5-μm square annulus opening, and a 1.75-um 3D cut. The total etched oxide depth is 8 μm. The design of these inter–tier connections significantly reduced the pitch of 3D vias to 6 μm from 26 μm used in the 3D imager first reported [5]. In addition, metal fill densities were modified to ensure that the planarity achieved after CMP satisfies the requirements for low-temperature bonding.

8.4.2 Computer-Aided Design Tool Refinements

Experience from three previous 3D chip designs indicated that the CAD tools used for design and layout needed to be refined to reduce design errors and reticle compilation time. Some major problems in the past were misalignment of the tiers which led to nonfunctional 3D vias, 3D vias that lacked metal lands, and faulty designs due to misidentification of layers. To relieve some of these difficulties, Lincoln Laboratory created a Mentor Graphics-based 3D design kit and incorporated the 3D design rules in it. DARPA also funded North Carolina State University to develop a Cadence-based 3D design kit [24]. Modifications to the design kits included layer displays that were unique to each tier. For example, the polysilicon layers in each tier use the same color but different shading.

8.4.3 Three-Dimensional Design Optimization

A major challenge in 3D design is the placement of subcircuits such as an arithmetic logic unit (ALU) or a ring oscillator within a tier and its placement among the three tiers. A great deal of work has been expended to deal with this challenge [25]. An experimental architectural and topological design tool [26] was developed to optimize the 3D architectures as a function of the 3D technology which includes 3D via design rules, the number of tiers, and circuit area. A cost function was determined which could be performance, circuit area, or processing complexity, and a set of simulations was run to determine what process and design factors would minimize the function. The regularity of an imager design is a natural application for such a tool, and the tool was applied to the 3D laser radar (LADAR) imager discussed later, as shown in Fig. 8.10. For this analysis the cost function was the pixel size and the

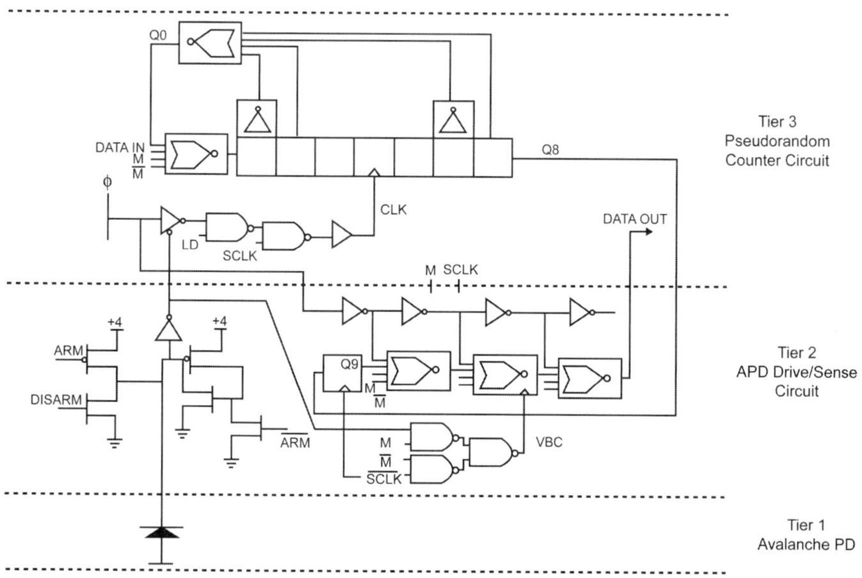

Fig. 8.10 The optimization analysis studied the pixel size of the 3D-LADAR chip as a function of portioning the subcircuits shown above among seven tiers. From [27] © 2005 IEEE

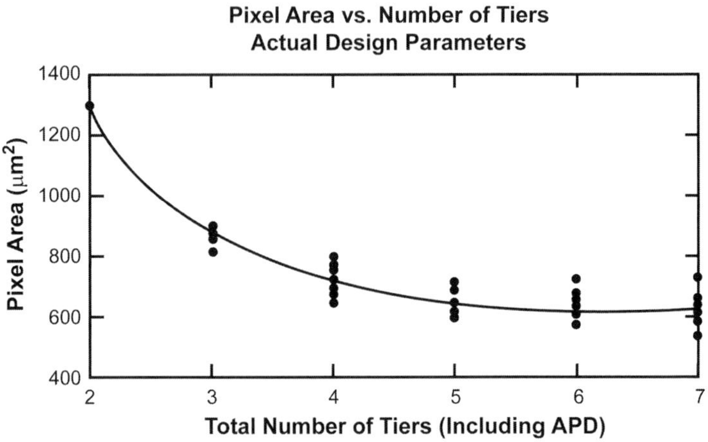

Fig. 8.11 The analysis indicates that partitioning the subcircuits among four tiers results in an optimum 3D architecture when pixel size is used as the criterion. From [26]

principal variable was the number of tiers for the design. The analysis considered placing the various subcircuits among seven tiers and the analysis was run. The results are shown in Fig. 8.11 and indicate the advantage of a three-tier design but that the optimum number of tiers for the 3D-LADAR application would be four. The actual 3D chip consisted of three tiers which was a compromise among processing complexity, fabrication throughput, and pixel size. However, this analysis demonstrated the utility of a 3D-specific architecture tool that would be used in the early stages of design to optimize the circuit architecture.

8.4.4 Wafer–Wafer Alignment Aids

The design of wafer–wafer alignment structures is a critical factor in achieving submicron wafer–wafer alignment. A set of alignment structures is placed in the center of each chip in every tier. The alignment structures achieve four functions. First, they are readily located features for visual inspection. Second, they can be used for manual wafer–wafer alignment and alignment measurements. Third, they are an essential part of automatic wafer–wafer alignment with the precision alignment system. Fourth, they are used for automatic wafer–wafer alignment measurements. The first wafer–wafer alignment consists of aligning features in metal 3 of tier 2 to features in metal 3 of tier 1, and the second wafer–wafer alignment consists of aligning features in metal 3 of tier 3 to features in BM1 of tier 2. In order to satisfy the alignment and measurement functions, metal from other layers, silicon, and polysilicon features are excluded from regions around the alignment structures in order that the structures on both tiers be clearly visible and exhibit high contrast in infrared reflection. The wafer–wafer alignment structure is shown in Figs. 8.12 and 8.13 as a composite image to illustrate the location of the tier 2–tier 1 and tier 3–tier

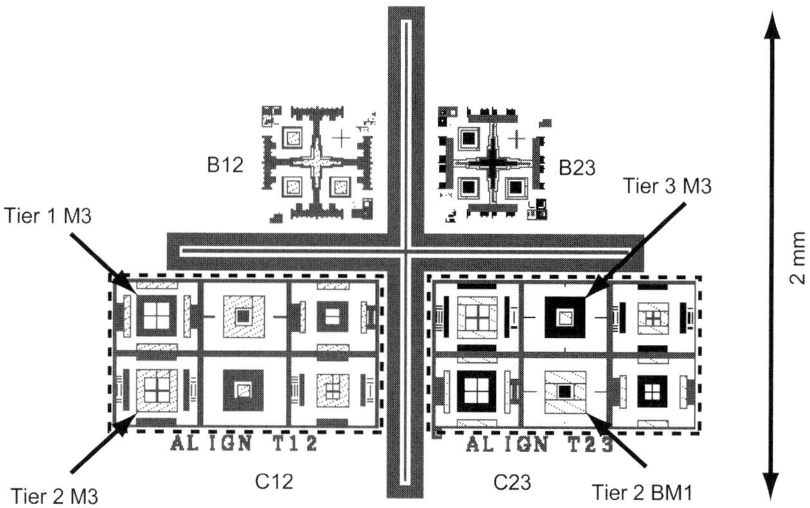

Fig. 8.12 A composite view of the two sets of alignment structures for tier 2–tier 1 and tier 3–tier 2 alignments and measurements. The features for alignment of tier 2–tier 1 and tier 3–tier 2 are on the left and right sides of the large cross, respectively. The features for the precision wafer–wafer alignment system are enclosed in the dashed boxes. The fill layers are suppressed for clarity

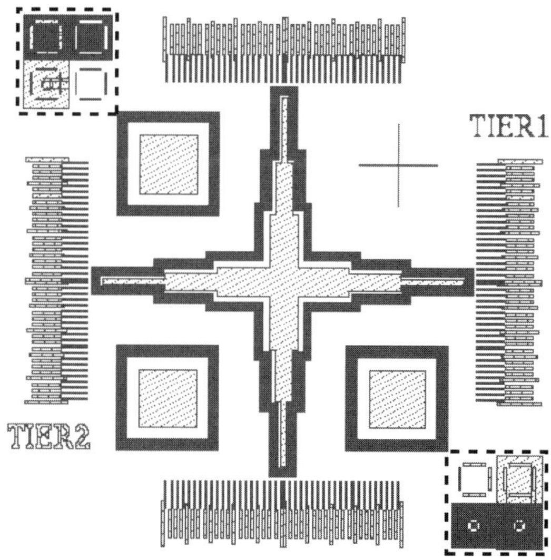

Fig. 8.13 The structures for automatic tier 2–tier 1 wafer–wafer alignment measurement are enclosed in the dashed boxes. Fill layers are suppressed for clarity

2 alignment features. The large cross is used as a target for wafer–wafer alignment and measurements. The two smaller crosses, B12 and B23, and their verniers are used for manual wafer–wafer alignment and manual alignment measurements. Until the development of the precision aligner, these were the features used throughout

the 3D program for alignment and measurements. The six rectangular features in C12 and C23 are used with the precision alignment system to achieve submicron overlay. In principle, only one of the six designs within C12 and C23 is required for alignment; however, the effectiveness of each design is being tested during 3D integration and eventually an optimum design will be selected and the remaining designs removed and the area made available for other 3D alignment investigations.

8.4.5 Three-Dimensional Design and Submission Procedures

The designer is requested to adhere to layout and submission requirements since the design will be integrated into a Multiproject 3D chip. A chip is designed as viewed in its completed state with the bond pads on the topmost tier using a layer-numbering procedure that appends the conventional layer numbers for 2D design to the tier number. For example, the numbers for polysilicon layers on tiers 1 and 3 are 209 and 409, respectively, while the metal 1 numbers for layers on tiers 2 and 3 are 327 and 427, respectively. These procedures ensure that the tiers have a common origin so that the rotation of tiers 2 and 3 about the *x*-axis can be performed automatically at Lincoln Laboratory during reticle compilation. Three-dimensional designs must be submitted as a gds file with the top cell of the design identified as 3DP_*org*_*Design*.gds, where 3DP is the Multiproject program name, *org* is the institution's name, and *Design* is the design name chosen by the designer. The bond pads must be placed in the top cell. The extent of the cell is defined as: X = ###, Y = ### in mm units and deliberate design rule violations, if permitted by Lincoln Laboratory, must be identified. All designs are analyzed and run through a post-processing procedure before the initiation of reticle compilation.

Upon receipt, the design is reviewed for obvious errors using a gds viewer. Then a batch process is used to add implant layers, bloat specific features, and add fill layers. Design rule checks (DRCs) are then run to verify that 3D-specific rules are satisfied. For example, a 3D land must exist for each 3D cut, the metal surround rules for a 3D cut in tier 2 must be satisfied in metal 3 of tier 1, and the metal surround rules for BM1 in tier 2 must be satisfied for a 3D cut in tier 3. A layout versus schematic (LVS) extraction is also run as part of the post-processing procedure. The DRC report, LVS netlist, and the post-processed files are then sent to the designer for review and correction if necessary. After a final review, all the 3D designs, test devices, alignment targets, and process control structures are integrated to create a master gds file that is sent to a mask vendor for reticle production.

8.4.6 Three-Dimensinal Circuit Design Examples

The design practices discussed in the previous section were used by 20 institutions to design 3D chips that were included in DARPA's first 3D Multiproject chip that is shown in Fig. 8.14. A variety of applications were investigated to study 3D design and performance issues. Three-dimensional-SRAM memories were designed by researchers at RPI [28]. The devices are the first three-tier memories fabricated and

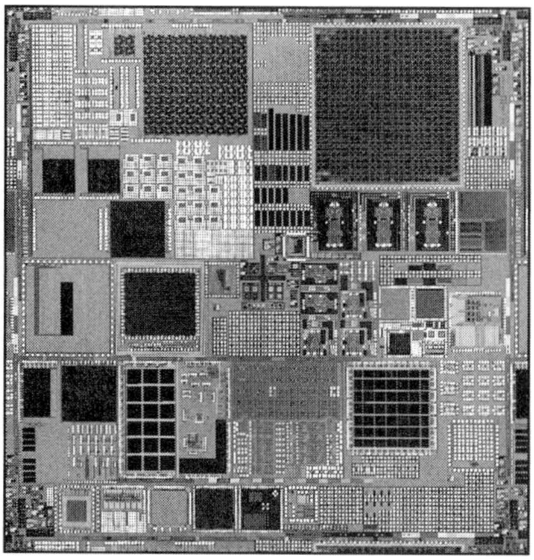

3D FPGAs, digital, and digital/mixed-signal/RF
ASICs exploiting parallelism of 3D interconnects
3D analog continuous-time processor
3D-integrated S-band digital beamformer
Stacked memory (SRAM, Flash, and CAM)
Self-powered CMOS logic (scavenging)
Integrated 3D nano-radio and RF tags
Intelligent 3D-interconnect evaluation circuits
DC and RF-coupled interconnect devices
Low-power multi-gigabit 3D data links
Noise coupling/cross-talk test structures and circuits
Thermal 3D test structures and circuits

(a) (b)

Fig. 8.14 (**a**) Photograph of the first 3D Multiproject chip, which contained integrated 3D chips designed by 20 institutions in addition to circuits designed internally and (**b**) a list of the 3D circuits and applications

lead to the promise that 3D will lead to greater design and programming flexibility. Three-dimensional designs from Cornell are shown in Fig. 8.15 and included functional field-programmable gate arrays (FPGAs) that promise higher bandwidth performance due to reduced interconnect delay. Also included in the 3D chip were

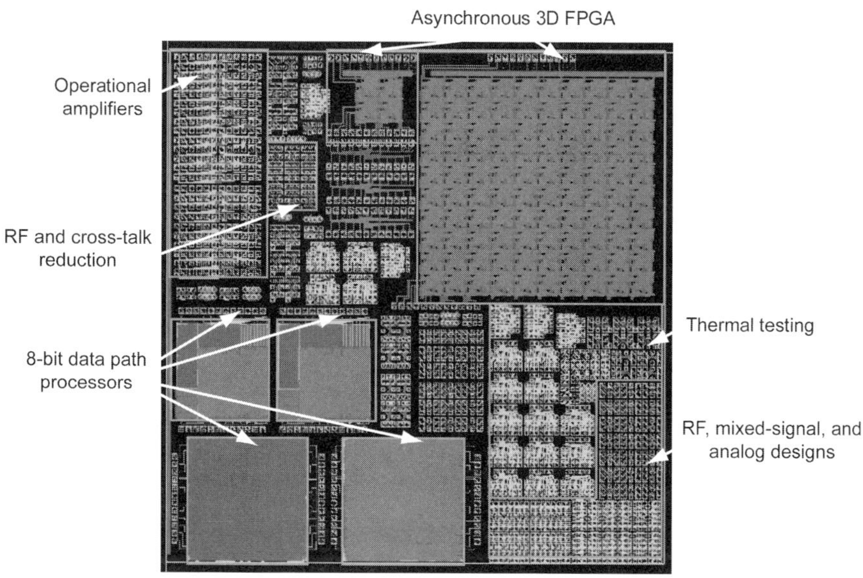

Fig. 8.15 Chip design incorporating a series of functional asynchronous 3D FPGAs and low-voltage adaptive analog circuits, with backgating and RF cross-talk reduction through 3D-integrated ground planes. (Courtesy of S. Tiwari of Cornell University)

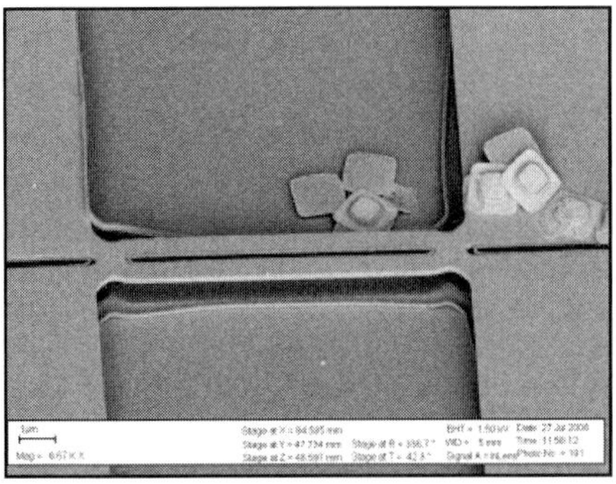

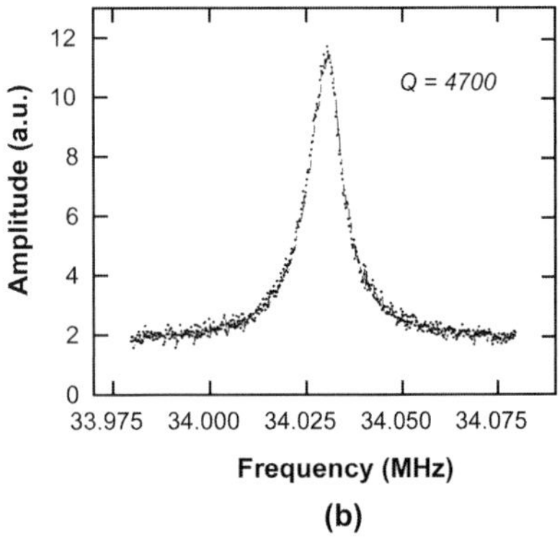

Fig. 8.16 (**a**) Scanning electron micrograph of a micromechanical resonator and (**b**) the frequency response. The resonator was 3D integrated with CMOS. Courtesy of M. Zalalutdinov of the Naval Research Laboratory

low-voltage adaptive analog circuitsachieved with backgated transistors and reduced RF cross-talk reduction through 3D-integrated ground planes. A 3D analog–digital converter for imaging applications was designed at Stanford [29] that simultaneously achieved a high frame rate and a high dynamic range with a 50-μm pitch.

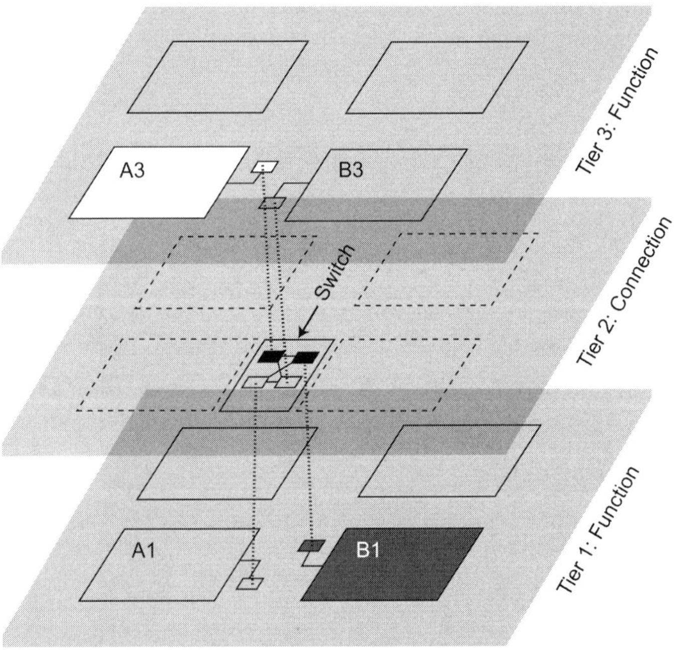

Fig. 8.17 A design concept being tested on the second 3D Multiproject chip is the use of tier 2 as a smart interconnection tier to achieve reduced signal delays and power supply droop. The tier can be part of a redundancy strategy

A 3D design by UCLA [30] investigated capacitively coupled interconnects and demonstrated a baseband impulse shaping interconnect and self-synchronized RF interconnect at >11 GHz with a bit error rate less than 1×10^{-14}. A 3D imager designed by Yale [31] demonstrated the feasibility of a three-color imager. A collaborative effort by the Naval Research Laboratory, BAE Systems, and Cornell University designed a 3D-integrated CMOS circuitry with micromechanical resonators that achieved a $Q = 4700$ at 34 MHz as shown in Fig. 8.16.

A second 3D Multiproject program that includes 20 participants has begun. Additional topics to be studied are optimum 3D layout to control heat dissipation, heat extraction from 3D circuits, performance enhancements with scaled 3D vias, and the utility of a interconnect tier to provide signal distribution and redundancy as shown in Fig. 8.17. Test data are not available as of this writing but the results are sure to lead to further understanding of the promise of 3D design.

8.5 Three-Dimensional Circuit and Device Results

The "low-hanging fruit" of 3D technology is focal plane design and fabrication because the imaging tier has a 100% fill factor, and the analog and digital processing tiers are below the imaging tier. Using Lincoln Laboratory's 3D technology discussed above, we were successful in the design, fabrication, and operation of avalanche photodiode (APD) imagers and 3D visible imagers.

8.5.1 Three-Dimensional-LADAR Chip

The 3D-LADAR chip [27] is a laser radar imager based on Geiger-mode APDs integrated with high-speed all-digital CMOS timing circuits. A target is illuminated with a laser, and the chip measures the arrival time of a single reflected photon that is incident on an APD such that the diode avalanches and produces a digital pulse that functions as a stop signal to a fast digital counter in the pixel circuit. This "photon-to-digital conversion" yields quantum-limited sensitivity and noiseless readout, enabling high-performance LADAR systems. Previous visible LADAR systems [32] used 32×32 silicon APD arrays epoxy-bonded to bulk 0.35-μm CMOS foundry chips. The pixel spacing was 100 μm and was limited primarily by the area required for the pixel circuitry. The circuits achieved a 0.5-ns timing quantization. The 3D-LADAR chip is a 64×64 imager composed of three tiers with 50-μm pixel spacing. Tier 1 is the 30-V Geiger-mode APD array and is the imaging tier. It is fabricated in a bulk wafer with an epitaxial layer for improved quantum efficiency. Tier 2 is fabricated in a 3.3-V, 0.35-μm SOI process. It contains circuitry that arms and disarms an APD and generates a stop signal when the APD has avalanched. Tier 3 is fabricated in a 1.5-V, 0.18-μm SOI process. It includes a 9-bit pseudorandom counter that is connected to a 3-bit register in tier 2 to improve the timing resolution of the pixel. The isometric drawing of a single pixel in Fig. 8.18 shows the five 3D via connections between tiers and the APD arming, pixel timing,

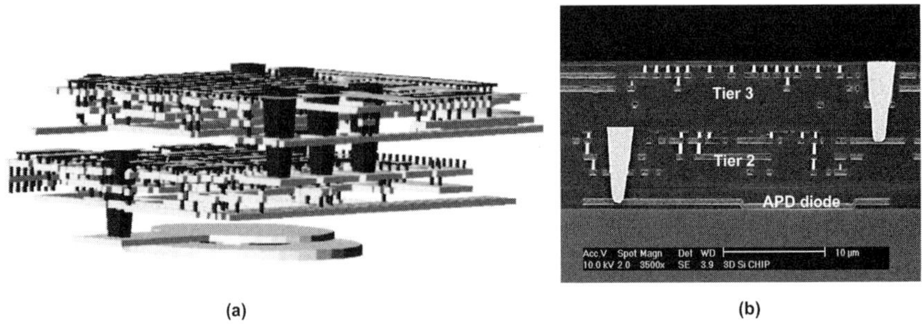

Fig. 8.18 (**a**) Isometric drawing of a 3D-LADAR pixel showing a single 3D via connecting the APD diode to tier 2 and five 3D vias between tiers 2 and 3; there are 88 3.3-V FDSOI transistors in tier 2 and 138 1.5-V transistors in tier 3. (**b**) Cross-sectional SEM of a pixel intersecting two 3D vias. From [27] © 2006 IEEE

and control circuitry that are beneath the image plane. Tier 1 would be thinned after 3D integration in order that the LADAR chip performs as a backside imager similar to backside-illuminated charge-coupled devices (CCDs). Figure 8.19 shows an imaging target, which is a cone, and a range image of the cone produced by the 64 × 64 imager after processing 210 frames to eliminate spurious counts. These results indicate that with a redesign to improve power distribution, the combined effect of the array size, the reduced pixel size, and the faster timing circuitry will result in a ∼80× reduction in voxel volume compared to previous [32] designs. This 3D-LADAR chip represents the first functional three-tier circuit with active circuits

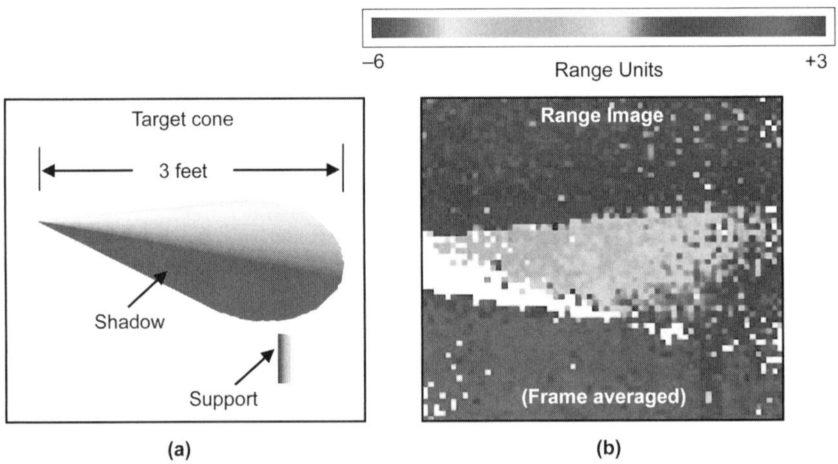

Fig. 8.19 (**a**) A cone was illuminated by a 4-kHz pulsed laser, λ = 532 nm. The reflected pulses were incident on the 3D 64 × 64 LADAR imager located 5 m from the cone where the pulses were detected, multiplexed, and sent to an off-chip processor to create the image shown in (**b**). The range units of the 9-bit counter of tier 3 correspond to ∼0.1 meter per count

and devices on all tiers and is the first demonstration of three different process technologies integrated into one 3D chip.

8.5.2 1024 x 1024 3D Visible Imager

A two-tier 1024 × 1024 visible imager with 8-μm pixels was demonstrated [33]. Tier 1 is a pn photodiode, and tier 2 is a FDSOI tier operated at 3.3 V. This is the most dense 3D imager circuit developed using this 3D circuit technology and illustrates the realization of a 100% imager fill factor by 3D technology. Each pixel of the 1024 × 1024 array includes a reverse-biased p^+n diode (in tier 1), a reset transistor, a source follower transistor, and a select transistor (in tier 2), as shown in Fig. 8.20b. Measured responsivity from the PMOS reset imager with added in-pixel capacitance was ~2.7 μV/e^- and from the NMOS reset imager was ~9.4 μV/e^-, corresponding to a charge-handling capacity of 350 000 e^- and 85 000 e^-, respectively. Measured pixel operability is in excess of 99.9%; the principal yield detractor was column or row dropouts and was not due to defective 3D vias. The high degree of pixel functionality is seen in Fig. 8.20a, an image acquired by projecting a 35-mm slide onto the CMOS circuit side of the 3D integrated imager.

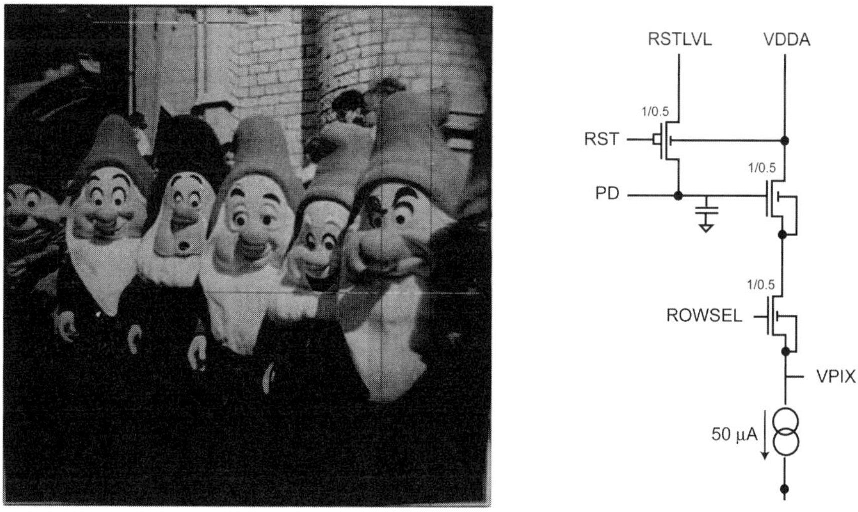

(a) (b)

Fig. 8.20 (**a**) Output from a 1024 × 1024 3D imager with 8-μm pixel and (**b**) the pixel circuit in tier 2, the 3.3-V FDSOI tier. The photodiode (PD) is in tier 1. The imager has a 100% fill factor. From [1] © 2006 IEEE

8.5.3 Heterogeneous Integration

The feasibility of heterogeneous integration of dissimilar materials while building 3D-IR focal planes was recently demonstrated [34]. Six-inch indium-phosphide wafers were bonded to oxidized silicon substrates, as illustrated in Fig. 8.21, and high-yield 3D via chains fabricated to show the applicability of the 3D technology to materials other than silicon. This is a major step toward the integration of mixed materials in 3D ICs for digital and imaging applications.

8.6 Summary

Functional 3D circuits of two and three tiers were fabricated with the wafer-scale 3D circuit integration technology. The technology permits the unrestricted placement of dense-vertical interconnections between tiers and the fabrication of 3D circuits with different technologies and eventually different materials. The advantages of the technology are obvious for imaging applications where a 100% fill factor can be obtained with a chip that includes complex pixel circuits. Current work is directed toward the application of 3D technology to mixed-signal applications as part of the DARPA-funded multichip program as well as the development of techniques to control and remove heat from the 3D stack. Results from the first DARPA Multiproject program suggest that (1) 3D SRAM memories will have higher bandwidth performance due to reduced interconnect delay, (2) 3D FPGAs will offer greater design and programmability flexibility, (3) 3D interconnect tiers

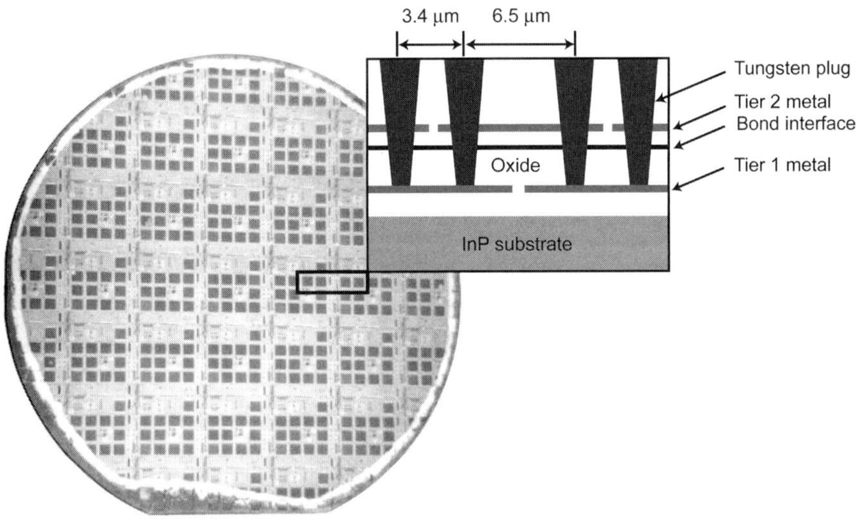

Fig. 8.21 150-mm InP wafer is shown bonded to an oxidized silicon wafer with 3D via chains; 10000 vias were fabricated per die. The chain yield was 100% and the average 3D via resistance was less than 1 Ω. From [34] © 2006 IEEE

will provide reduced signal delay and latency and better RF isolation between and within tiers, and (4) we can expect microelectromechanical (MEMS) to be integrated into 3D structures.

Acknowledgments This authors wish to acknowledge the dedication and persistence of the Microelectronics Laboratory staff and the editorial assistance of Karen Challberg.

The work was sponsored by the Defense Advanced Research Projects Agency under Air Force contract #FA8721-05-C-0002. Opinions, interpretations, conclusions, and recommendations are those of the authors and are not necessarily endorsed by the United States Government.

References

1. Burns JA, Aull BF, Chen CK, Chen CL, Keast CL, Knecht JM, Suntharalingam V, Warner K, Wyatt PW, Yost DRW (2006) A wafer-scale 3-D circuit integration technology. IEEE Trans Electron Devices 53(10):2507–2516.
2. McClelland RW, Bozler CO, Fan JCC (1980) TP-A2 the cleft process: a technique for producing many epitaxial single-crystal GaAs films by employing one reusable substrate. IEEE Trans Electron Devices 27(11):2188.
3. Sailer PM, Singhal P, Hopwood J, Kaeli DR, Zavracky PM, K. Warner K, Vu DP (1997) Creating 3D circuits using transferred films. IEEE Circuits Devices Mag 13(6):27–30.
4. Burns J, McIlrath L, Hopwood J, Keast C, Vu DP, Warner K, Wyatt P (2000) An SOI-based three-dimensional integrated circuit technology. In: IEEE international SOI conference proceedings, pp 20–21.
5. Burns J, McIlrath L, Keast C, Lewis C, Loomis A, Warner K, Wyatt P (2001) Three-dimensional integrated circuits for low-power, high-bandwidth systems on a chip. In: Digest of technical papers of IEEE international solid-state circuits conference, pp 268–269.
6. Reif R, Fan A, Chen KN, Das S (2002) Fabrication technologies for three-dimensional integrated circuits. In: Proceedings of the IEEE international symposium quality electronic design, pp 33–37.
7. Chan VWC, Chan PCH, Chan M (2000) Three dimensional CMOS integrated circuits on large grain polysilicon films. In: Technical digest of IEEE international electron devices meeting, pp 161–164.
8. Fukushima T, Yamada Y, Kikuchi H, Koyanagi M (2005) New three-dimensional integration technology using self-assembly technique. In: Technical digest of IEEE international electron devices meeting, pp 359–362.
9. Lea R, Jalowiecki I, Boughton D, Yamaguchi J, Pepe A, Ozguz V, Carson J (1999) A 3-D stacked chip packaging solution for miniaturized massively parallel processing. IEEE Trans Adv Packaging 22(6):424–432.
10. Fukushima T, Yamada Y, Kikuchi H, Koyanagi M (2005) New three-dimensional integration technology using self-assembly technique. In: Technical digest of IEEE international electron devices meeting, pp 359–362.
11. Topol A, Tulipe D, Shi S, Alam S, Frank D, Steen S, Vichiconti J, Posillico D, Cobb M, Medd S, Patel J, Goma S, DiMilia D, Farinelli M, Wang C, Conti R, Canaperi D, Deligianni L, Kumar A, Kwietniak T, D'Emic C, Ott J, Young A, Ieong M (2005) Enabling SOI-based assembly technology for three-dimensional (3D) integrated circuits (ICs). In: Technical digest of IEEE international electron devices meeting, pp 363–366.
12. MITLL Low-Power FDSOI CMOS Process Design Guide, Revision (2006:7), Advanced Silicon Technology Group, MIT Lincoln Laboratory, 244 Wood St., Lexington, MA 02420, USA.
13. Warner K, Chen C, D'Onofrio R, Keast C, Poesse S (2004) An investigation of wafer-to-wafer alignment tolerances for three-dimensional integrated circuit fabrication. In: IEEE international SOI conference proceedings, pp 71–72.
14. Metra 2100 Process Engineer's Manual, Optical Specialties Inc., 1993.

15. Chen CK, Warner K, Yost DRW, Knecht JM, Suntharalingam V, Chen CL, Burns JA, Keast CL (2007) Scaling three-dimensional SOI integrated-circuit technology. In: IEEE international SOI conference proceedings, pp 87–88.
16. Warner K, Burns J, Keast C, Kunz R, Lennon D, Loomis A, Mowers W, Yost D (2002) Low-temperature oxide-bonded three-dimensional integrated circuits. In: IEEE international SOI conference proceedings, pp 123–124.
17. Maszara WP, Goetz G, Caviglia A, McKitterick JB (1988) Bonding of silicon wafers for silicon-on-insulator. J Appl Phys 64(10):4943–4950.
18. Knecht J, Yost D, Burns J, Chen C, Keast C, Warner K (2005) 3D via etch development for 3D circuit integration in FDSOI. In: IEEE international SOI conference proceedings, pp 104–105.
19. Burns J, Warner K, Gouker P (2001) Characterization of fully depleted SOI transistors after removal of the silicon substrate. In: IEEE international SOI conference proceedings, pp 113–114.
20. Gouker P, Burns J, Wyatt P, Warner K, Austin E, Milanowski R (2003) Substrate removal and BOX thinning effects on total dose response of FDSOI NMOSFET. IEEE Trans Nucl Sci 50(6):1776–1783.
21. Chen CL, Chen CK, Burns JA, Yost DR, Warner K, Knecht JM, Wyatt PW, Shibles DA, Keast CL (2007) Thermal effects of three dimensional integrated circuits stacks. In: IEEE international SOI conference proceedings, pp 91–92.
22. Mentor Graphics, IC Nanometer Design Tool Suite.
23. Cadence Virtuoso Design Tool.
24. Courtesy of Dr. Rhett Davis, North Carolina State University.
25. Reber M, Tielert R (1996) Benefits of vertically stacked integrated circuits for sequential logic. In: Proceedings IEEE international symposium circuits systems, vol. 4, pp 121–124.
26. Tyrrell B (2004) Development of an architectural design tool for 3-D VLSI sensors. MS Thesis, Massachusetts Institute of Technology, Department of Electrical Engineering.
27. Aull B, Burns J, Chen C, Felton B, Hanson H, Keast C, Knecht J, Loomis A, Renzi M, Soares A, Suntharalingam V, Warner K, Wolfson D, Yost D, Young D (2006) Laser radar imager based on three-dimensional integration of Geiger-mode avalanche photodiodes with two SOI timing-circuit layers. In: Dig. tech. papers IEEE international solid-state circuits conference, pp 304–305.
28. Courtesy of Dr. J. McDonald, Rensselaer Polytechnic Institute.
29. Lee SM, Park H, Wooley BA (2006) Per-pixel floating-point ADCs with electronic shutters for a high dynamic range, high frame rate infrared focal plane array. In: Proc. IEEE Custom Integrated Circuits Conf., pp. 647–650.
30. Gu Q, Xu Z, Ko J, Chang MCF (2007) Two 10 Gb/s/pin low power interconnect methods for 3D ICs. In: Dig. Tech. Papers IEEE Int. Solid-State Circuits Conf., pp. 364–365.
31. Culurciello E, Weerakoon P (2007) Three-dimensional photodetectors in 3D silicon-on-insulator technology. IEEE Electron Device Lett 28(2):117–119.
32. Aull BF, Loomis AH, Gregory J, Young D (1998) Geiger-mode avalanche photodiode arrays integrated with CMOS timing circuits. In: IEEE annual device research conference Dig., pp 58–59.
33. Suntharalingam V, Berger R, Burns JA, Chen CK, Keast CL, Knecht JM, Lambert RD, Newcomb KL, O'Mara DM, Rathman DD, Shaver DC, Soares AM, Stevenson CN, Tyrrell BM, Warner K, Wheeler BD, Yost DRW, Young DJ (2005) Megapixel CMOS image sensor fabricated in three-dimensional integrated circuit technology. In: Dig. tech. papers IEEE international solid-state circuits conference, pp 356–357.
34. Warner K, Oakley DC, Donnelly JP, Keast CL, Shaver DC (2006) Layer transfer of FDSOI CMOS to 150 mm InP substrates for mixed-material integration. In: International conference indium phosphide related materials, pp 226–228.

Chapter 9
3D Fabrication Options for High-Performance CMOS Technology

Anna W. Topol, Steven J. Koester, Douglas C. La Tulipe, and Albert M. Young

9.1 3-D Technology

9.1.1 Introduction

The last several decades have seen an incredible increase in the functionality of computational systems. At its core, this capability has been driven by the scaling of semiconductor devices, from fractions of millimeters in the 1960s to tens of nanometers in today's technologies. The scaling has enabled the number of transistors on a single chip to correspondingly grow at a geometric rate, roughly doubling every 18 months; a trend is that now referred to as Moore's law [1]. The impact of this trend cannot be underestimated and the resulting increase in computational capacity has had major impacts on almost every facet of society.

For this reason, there is a tremendous push to continue along these same trends. However, several serious roadblocks exist. The first is the limits to lithographic scaling. The second is that power densities will not allow reliable systems to be fabricated even if lithographic scaling could continue. Therefore, it becomes a big challenge to increase system performance.

Three dimensional (3D) integration technologies offer the promise of increasing system performance even in the absence of scaling. However, the main advantages of 3D integration can be summed up as follows: (1) 3D decreases the interconnect distance between regions of a chip, decreasing wiring parasitics and interconnect delay times, (2) 3D can dramatically increase the number of interconnects and therefore increase the aggregate communication bandwidth between chips, and (3) 3D can allow dissimilar functions, technologies, and materials to be integrated.

In the next part of this chapter we will talk about the 3D technology landscape and focus on IBM's wafer-level integration research and development (R&D) efforts. We will describe layer transfer processing and wafer-level thinning, discuss bonding and interconnection processes, and show successful implementation of the

A.W. Topol
IBM T. J. Watson Research Center, Yorktown Heights, NY, USA
e-mail: atopol@us.ibm.com

C.S. Tan et al. (eds.), *Wafer-Level 3D ICs Process Technology*,
DOI: 10.1007/978-0-387-76534-1_9, © Springer Science+Business Media, LLC 2008

through-silicon via (TSV) process and Cu bonding for advanced microprocessor applications.

The remainder of this chapter will describe our progress on one specific 3D integration scheme based upon oxide fusion bonding. This technology is by no means the preferred-by-all 3D technology, but is perhaps ultimately the most capable technology to achieve the highest density of 3D interconnects, and thus enables the most aggressive 3D architectures. The scheme could also be readily applicable to multiple stack architectures, and it uses a conventional complementary metal-oxide semiconductor (CMOS) processing approach. We describe the latest progress on the development of this technology, and also point out some of the remaining challenges needed to bring it into manufacturing for mainstream CMOS applications.

9.1.2 Three-Dimensional Technology Landscape

At IBM, 3D integration is being pursued for various applications that have very different technology requirements. As depicted in Fig. 9.1 these applications span from 3D packaging to high-density 3D integrated circuit (IC) implementations. More specifically, for wireless communications, where TSVs are needed for making low-inductance contacts to the backside ground plane, a very low density of TSVs are required. More recently IBM has investigated 3D technology to integrate decoupling capacitors with microprocessor to stabilize power delivery to the chips [2]. In these applications, TSV densities on the order of the standard C4 (controlled collapse chip connection flip-chip package) pitch are required. In the most extreme cases, for highly integrated microprocessor configurations, a much higher density of 3D vias are needed (10^5–10^6 pins cm^{-2} range) [3]. These requirements drive the highest-density 3D processes that will be the main topic of the remainder of this

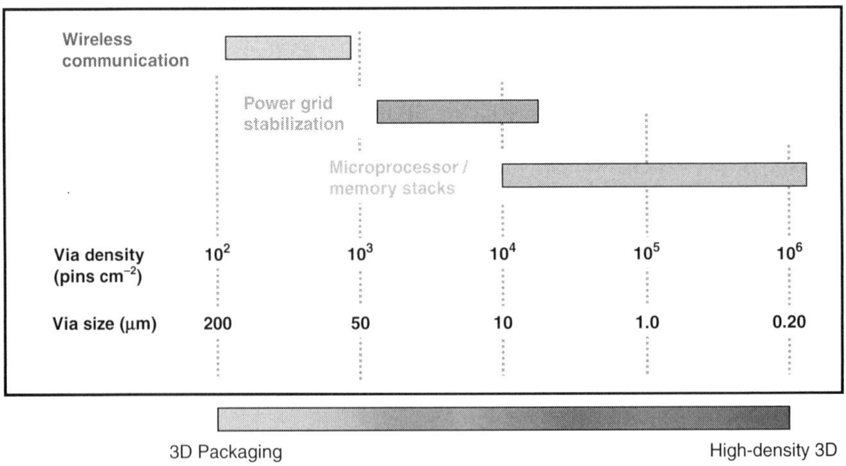

Fig. 9.1 Chart showing range of 3D technology applications being implemented at IBM

chapter. Data points in Fig. 9.1 are approximate but are representative of implementations of 3D technologies that have been demonstrated by IBM or others, and hence the via size and density values are not directly correlated.

9.1.3 Wafer-Level 3D Integration

Since many 3D integration research programs have been underway for years, several of the critical technology components have been demonstrated. As discussed in previous chapters, these include low-cost fine via hole formation and highly reliable via-filling technologies, bonding methodology, as well as layer transfer techniques. In addition, 3D-specific process equipment (such as bonding and alignment tools) and materials (such as good quality bonding mediums) are just now becoming available. For some applications industry is beginning to move past the feasibility (R&D) phase into the commercialization phase. A TechSearch International report highlights the major processes and materials used by various companies, and forecasts the market for 3D integration [4]. Even though much work still remains to be accomplished related to 3D design, in the areas of thermal and test issues for example, much progress has been made. In the end economic realities will determine if and in which markets the 3D technologies will be adopted.

Various 3D integration schemes have been developed using similar technology elements. The main difference between these schemes relates to the choice of the so-called "face-to-back" or "face-to-face" options. The "face-to-back" method is based on bonding the front side of the bottom wafer with the backside (usually thinned) of the top wafer and correspondingly is often referred to as a "front side-to-backside" approach. The height of the structure and therefore the height of the interconnecting via depends on the thickness of the thinned top wafer. Figure 9.2a–c illustrates various options for this technology based on bonding and via formation process assuming that silicon-on-insulator (SOI) wafers are used and hence all of the Si can be removed to provide for the shortest interwafer via connections.

Similarly the "face-to-face" approach focuses on joining the front sides of two wafers and hence is also referred to as "front side-to-front side" wafer-level transfer. Figure 9.2d–h depicts potential integration options assuming use of SOI wafers. If bulk wafers are used, the final structures would look similar except that they would contain a thin layer of Si. As a consequence the minimum height of the interconnecting via represented numerically on the right-hand side of all the structures in Fig. 9.2 would have to be increased by at least a few microns (minimum reliable Si thinning process result). Due to processing limitations, as the height of the via increases the minimum width of the via has to increase as well to enable the subsequent metallization process.

Assessment of various integration options is ongoing (e.g., Sematech in 2005 initiated such an evaluation), as tool and process benchmarking develops and 3D infrastructure, materials, and unit processes mature [5]. Frequent technology questions include: "Is there a one common approach to 3D integration and if not how to determine which one should be used?" "What is driving the choice for particular 3D

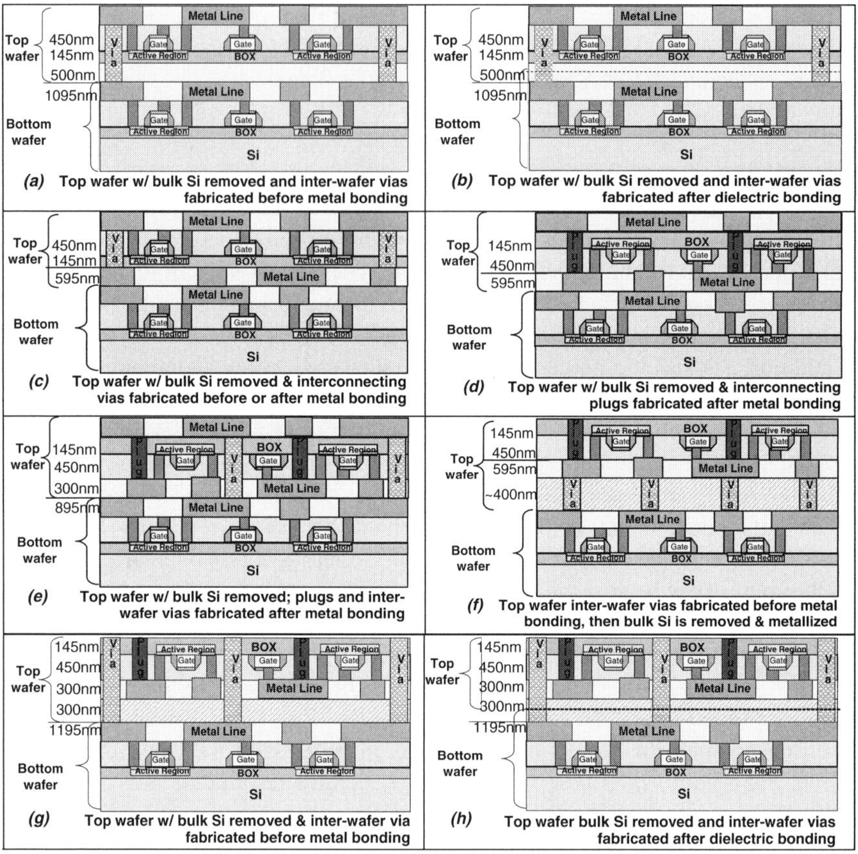

Fig. 9.2 Cross-sectional diagrams (**a**)–(**c**) show a final structure of potential "face-to-back" or "front side-to-backside" integration schemes; (**d**)–(**h**) depict "face-to-face" or "front side-to-front side" 3D integration schemes. Numbers on the right side of the images highlight the minimum height of interconnecting vias needed to provide electrical signal between the two wafers assuming 90-nm SOI-based technology (buried oxide layer (BOX) of 1450 nm). *Dashed lines* in (**b**) and (**h**) indicate bonding interfaces

assembly technology?" "What materials and processes should be used?" "Should vias be formed during the wafer-stacking process or implemented later in the process? "What are the limiting factors to the adoption of various 3D technologies?"

As discussed in previous paragraphs, 3D integration process flow depends largely on the application requirements. More specifically, based on the application specifications – which may include input/output (I/O) number and interconnect density, number of layers to be stacked, expected performance and bandwidth requirements, cost, design limitations (including thermal and routing aspects), technology choices (bulk vs. SOI etc.), or architectural level (device, macro, circuit, system, chip, package) – different 3D technology elements can be chosen. Specifically the choice of bonding medium, use of carrier wafer, order of processing steps, incorporation of additional thermal cycles (such as curing, bond-strengthening, and outgassing), choice of metallization techniques, implementation of seals, or any other yield/reliability enhancers (such as via redundancy) are all dictated by application needs.

However, in general all 3D integration schemes which utilized the handle or carrier wafer have six common technology elements. These are (1) a handle wafer methodology (e.g., a temporary bonding and release); (2) wafer thinning; (3) vertical interconnects; (4) alignment; (5) bonding; and (6) final metallization and testing. In sections below we describe these elements, and give an update on the maturity of their implementation.

9.1.4 IBM 3D Integration Approaches

In the next two sections we will present two very specific applications of 3D technology and discuss the rationale behind process options utilized to build these structures. Both of these 3D constructions, namely system-level 3D implementations and 3D circuits have been chosen as they represent two distinct cases of different 3D options for high-performance CMOS technology. The first application falls under the microprocessor/memory stacking regime (see Fig. 9.1), while the second one is limited to high-density IC implementations. Both of the examples were fabricated using a face-to-back layer transfer process. Each of them benefits from 3D implementation in a unique way, but both of them are being developed by IBM as potential candidates for replacement of traditional planar circuit layout to enable future advanced CMOS technologies.

9.1.4.1 Copper Bonding with TSVs for Processor/Memory Stacking

As depicted in Fig. 9.3, the "face-to-back" approach can be used to connect functional blocks of CMOS circuitry. More specifically, this method uses logic and memory components, that traditionally in 2D chip layouts reside side by side, and

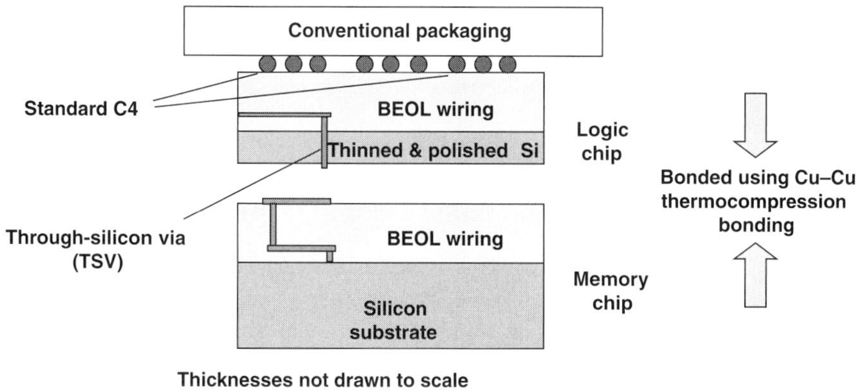

Fig. 9.3 Graphical representations of stacked memory/logic components integrated in a 3D chip. The structure is based on face-to-back stacking technology and TSV interconnection. Bonding of functional levels is accomplished using a Cu-to-Cu thermocompression bonding process

stacks them on top of each other providing a 3D IC solution. The resulting "sandwich" components enable reduced overall chip/package size and increased speed of data flow among the various functional blocks of the chip.

In the next subsections, we describe the process components needed to fabricate the functional-block-level 3D integration scheme described above.

Temporary Bonding and Release

In the "front-to-back" layer transfer approach, the use of carrier wafers is often required to protect the circuit layers and to provide mechanical stability during the layering process. IBM has historically used a glass handle wafer which has a thermal coefficient of expansion (TCE) matched to silicon. In this process after the top wafer is fabricated and metallized a protective coating (usually nitride layer) is deposited, followed by the application of polymer-based layers that act as a bonding medium [6]. Once a glass carrier is attached to the top wafer the critical circuit layers are protected and removal of the bulk Si can take place. IBM uses this method as we have a long and successful history related to packaging of implementing glass and polyimide and then detaching glass by means of laser ablation. During this process the polyimide is ablated and after the glass is removed an ash process is implemented to clean residuals. The protective nitride layer is critical for success of this process as it shields Cu-based metal layers from corrosive oxide chemistry of the ash process.

Other than glass a Si wafer can be used as a carrier. Also, beyond the use of polymeric materials other temporary attachment methods can be implemented (glues, adhesives, porous layers, etc.). The critical aspects of the temporarily attachment is that it has to be strong enough to withstand a mechanical and grinding process used during Si bulk removal. Also, once the layer transfer is completed the detachment of the carrier has to be easy to provide for reliable, fast, and cost-efficient methodology. If the top wafer layer is not thinned below $\sim$100 μm it is possible to omit the handle wafer process all together, but protective thick layers have to be deposited to shield the circuitry during the thinning process (when wafers are attached to the grinding or polishing wheels and exposed to strong mechanical forces).

Wafer Thinning

Wafer thinning is a necessary component of 3D integration as it provides capability of bringing layers closely together. However, its biggest challenge is the necessity of thinning a full Si wafer down to $\sim$5%–10% of its original thickness, with a required uniformity of $\sim$1–2 μm. This is because, when bulk Si wafers are used in the creation of the integrated CMOS circuitry, there is no natural etch stop like in the SOI-wafer case, and hence some of the original Si remains after thinning of the wafer is complete. The final thickness depends on thinning process control capabilities and is limited by the thickness-uniformity specifications of the silicon removal process (that being mechanical grinding and polishing, wet or dry etching). Successful thinning to a uniform Si thickness of few microns has been demonstrated,

Table 9.1 List of common steps used in the thinning process. Steps are usually combined to achieve better thickness uniformity

Process step	Initial thickness	Final thickness	Additional comments
Coarse grinding	~725 μm	~150–125 μm	~400 mesh; process tool/vendor dependent
Fine grinding	~150–125 μm	~125–100 μm	1800–2000 mesh; process tool/application specific
Mechanical polishing	~125–100 μm	~60–30 μm	Tool/vendor-specific process
Wet etch	~125–100 μm	Down to BOX – for SOI wafer	Mainly used for SOI wafers; or in combination with dry etch (TMAH removal rate for Si is ~0.5 μm min^{-1})
Dry etch	~725 μm or other initial thickness	Down to SOI box or any final thickness	SF_6 chemistry with O_2 (oxygen added to increase the etch rate); ~10% thickness uniformity

but usually ~30–60 μm of Si remains. Standard process steps for Si thinning are listed in Table 9.1.

Vertical Interconnect

Perhaps the most important technology element for 3D integration is the 3D interconnect. This is sometimes referred to as the TSV or the through-silicon interconnect, though in the case of our SOI 3D scheme, the via does not need to go through silicon since the substrate is removed. A vertical interconnect is necessary for 3D integration to truly take advantage of 3D for system-level performance, since without it interconnects would be limited to the periphery of the chip, and in this case, the interconnect density would be no greater than in conventional 2D technology. This interconnection method is essentially the same as a contact hole (or back-end-of-the-line (BEOL-like) process, with the difference that a much deeper hole has to be created vertically through the silicon material using a special etch process. IBM, Samsung, Tessera, Intel, Elpida, IMEC, and others are developing their TSV methodology optimizing the patterning and metallization process for their applications [7, 8].

A variety of vertical through-silicon interconnect technologies has been developed by IBM and has been described in the literature [9, 10, 11]. Via-last approaches have been outlined that have the advantage of being compatible with fully processed silicon circuits. In this case, silicon is removed from the via region by backside deep reactive-ion etching. Insulation is subsequently applied to the interior of the via and selectively removed from the base to allow electrical contact. Metallization of large vias prepared in this manner has been demonstrated by using an initial partial fill with plated copper followed by evaporation of Cr/Cu ball-limiting metallurgy(BLM) and Pb/Sn solder. For via-first process flows, metallization of large vias has also been demonstrated using a hybrid approach that consists of an initial partial fill with plated copper followed by an additional composite-paste fill. Use of this hybrid approach reduces TCE mismatch for larger via sizes; this is important because large metal-filled features are often not thermally compatible with silicon.

Via-first approaches enabled by using annular-via geometries have also been described. The use of relatively narrow shapes in an annular geometry allows complete fill of the etched structure with a variety of materials. For annuli with small-enough central cores (the region defined by the inner diameter of the annulus), the annular region can be filled with isolation dielectrics, and the central core subsequently etched and metallized. For annuli of larger dimensions, the central core can be left untouched, and the region in the annulus itself can be filled with a variety of conductors, which could include doped polysilicon, electroplated copper, or chemical vapor deposition (CVD) tungsten. Figure 9.4 shows a scanning electron micrograph (SEM) cross-section of one such via. This via has been fabricated with tungsten deposited in the annular-shaped via region. The process flow for fabricating this type of via has been described elsewhere; in short, it involves deep reactive-ion etching of the annular-via pattern, thermal oxidation of the etched sidewalls, tungsten metallization of the annular region, and additional BEOL processing. After fabrication of BEOL-wiring levels, the wafer can be attached to a handle wafer, thinned, and backside processed to expose the vias from the backside.

Alignment

In the TSV approach to 3D integration, the alignment requirement is not as stringent as in the 3D IC case (which will be discussed later). An alignment tolerance of ∼3 μm is often sufficient, using Cu–Cu bonding pads of ∼10 μm, a tolerance which is within the capabilities of a standard alignment tool for bonding applications. However since this is a wafer-to-wafer 3D integration scheme, alignment and bonding are still key fabrication challenges. The key to good process control is the ability to separate the alignment and pre-bonding steps from the actual bonding process. Such integration design allows for better understanding of the final alignment

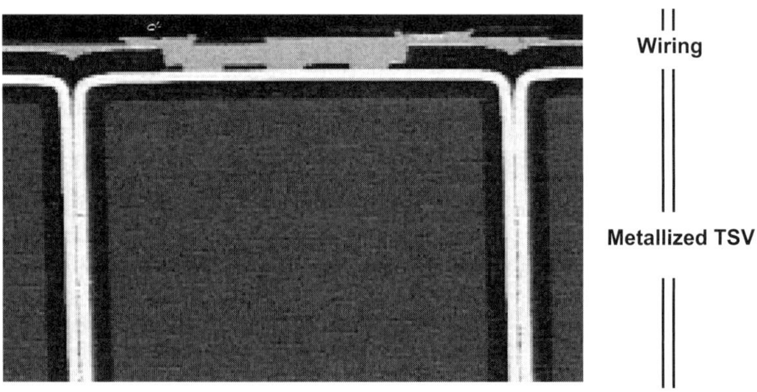

Fig. 9.4 Cross-sectional SEM of Si region with metallized TSV and connecting wiring level

error contributions. Especially in applications where thermal-compression bonding is utilized in addition to the standard pre-bonding alignment error problems with good temperature uniformity or uniformity of pressure across the wafer can greatly influence the final alignment accuracy [12]. Hence the accepted procedure is to use aligner tools (which have high throughput) and then move wafers to specialized bonding tools with double-sided heating chucks for good control of temperature and pressure of the wafer and the gradient across the stack.

Bonding

Attachment of two of the functional levels of this 3D structure is completed by using a metal bonding process. This process is chosen as it not only secures two parts together but at the same time it provides for electrical connection between them. More specifically copper-to-copper (Cu–Cu) thermocompression bonding is used. Optimization of the quality of this bonding process is a key issue being addressed and includes provision of various surface preparation techniques, post-bonding straightening thermal annealing cycles, as well as use of optimized pattern geometry [13, 14, 15]. These solutions combined with use of various seal designs address a key challenge of 3D technology, namely reliability and manufacturability of 3D ICs and packages. Figure 9.5 shows a close-up of the cross-sectional SEM image of a Cu via successfully bonded to a Cu pad.

Final Metallization and Testing

In the case of stacking memory and logic, or any other application in which more complicated circuits/systems are involved, one has to consider an appropriate design for 3D interconnects (including choice of geometry and metallization) and optimized layout for metal levels of the various components. The critical challenge is to build 3D elements in parallel to a point where they can be pre-tested before layer transfer so that their functionality and fabrication process is verified. Then one has to be able to re-test circuit components right after their vertical stacking so that any problems may be corrected before 3D parts are processed through any additional steps. To implement this reliability procedure the use of additional verification test

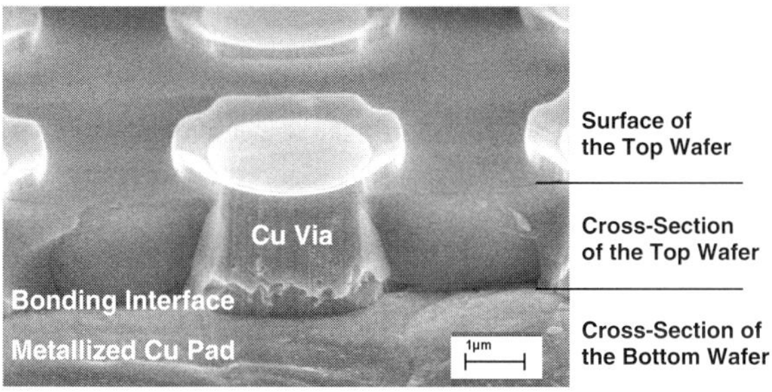

Fig. 9.5 A close-up of the cross-sectional SEM image of a Cu via successfully bonded to a Cu pad

circuits are often implemented. This may cost an increased design cycle for 3D structures but immensely helps in debugging process and layout problems. The 3D structures have nonstandard requirements; for example, the interconnecting vias are no longer the representation of the traditional BEOL contact level as they now often go through the front-end-of-line (FEOL) layers. Hence, special care to protect this FEOL environment from potential exposure to BEOL materials (like Cu contamination/diffusion in the active regions) is required, but also new rules (like exclusion of dummy fills in active regions) have to be implemented to enable fabrication of these new structures to FEOL components.

Hence new computer-aided design (CAD)-based layout programs have to be developed to help designers address the nonstandard needs of 3D circuits/systems. A key question is how many metal layers one needs to build before the layer transfer and how many of them are to be processed after the stacking of 3D elements is completed. To answer this question a good understanding of available 3D processing technology and available test circuits at every metal level is a must. For example, a simple ring oscillator structures can be tested at first metal level, but functional static random access memory (SRAM) structures may need to be built to at least 4th metal level.

9.1.4.2 SOI-Based 3D Integration for Ultrahigh-Density and Device-Level Stacking

Aggressive scaling of CMOS device structures coupled with new materials systems and integration techniques have been primary enablers of increased performance and density-reduction trends in the industry. However, to keep up with semiconductor technology roadmap projections, innovation in transistor structures and integration of novel materials may no longer be adequate [16]. Three-dimensional integration at the device level is an attractive option as it provides a venue not only to potentially improve device performance as it could enable heterogeneous technologies each with their own unique performance enhancement strategies to be stacked, as well as allowing the development of new integration options [17]. For example, it provides for separate but parallel processing of diverse p- and n-type transistors enabling optimization of each of these devices independently prior to attaching them together in a common circuit [18].

Temporary Bonding and Release

The majority of our SOI-based 3D bonding experience has been focused on the "face-to-back" stacking of device layers as described earlier. We have utilized transparent Borofloat glass substrates and a temporary bonding process to transfer device layers from one silicon wafer, remove the silicon from those devices and bond them to devices fabricated on another silicon wafer [19]. This approach was chosen primarily to address deficiencies in 3D-specific alignment tooling which preclude an automated blind alignment of two patterned surfaces to within several hundred nanometers of each other. It has been necessary to make use of a transparent substrate to allow critical pre-bonding inspection, and manual alignment adjustment on even the most advanced automated alignment tooling [20].

Several factors must be considered when choosing a temporary bonding process for such an application. Of primary importance is the chemical resistance of the thin, primarily organic, bonding films used for such an application. Once attached to a glass substrate these materials must be impervious to highly corrosive wet clean and etch chemistries; for example, several hours of exposure to hot TMAH, as well as highly alkaline/oxidizing wet cleans required to ensure defect-free, reactive bonding surfaces. Thermal stability is of equal concern as substrates can be cycled through several hundred degrees centigrade during the deposition and preparation of oxide films for thermal fusion bonding. Any thermal or stress-induced flow of polymer films could cause localized pattern shifts. We have shown that wafer-scale deformation of the glass substrates, due in part to unoptimized TCEs of the polymers, are directly responsible for pattern overlay errors which cannot be easily compensated for in subsequent lithographic steps [20]. Finally, by definition, this bonding process is meant to be temporary and therefore a mechanism for release of the glass substrate from the bonded silicon wafer must be considered. We have described a laser-ablation process in which a polymer layer in our temporary bonding stack is highly absorbing and instantly carbonized when exposed to high-energy mid-UV laser pulses [21]. Others are actively considering thermal and flood UV release mechanisms for other polymer systems [22].

Although it is clear that such a layer transfer technique can bring immediate advantages to the bonding alignment process in the form of improved alignment accuracy and the potential for pre-bonding rework, it is difficult to visualize such an approach being used to produce high-density, wafer-scale bonding with $<500\,\text{nm}$ 3σ pattern overlay error. It is very possible that once 3D bonding alignment tools are capable of routinely achieving automated wafer-scale alignment accuracy in this range, there will be a shift to "face-to-face" integration schemes and temporary layer transfer processes will no longer be necessary.

Wafer Thinning

Whether a "face-to-back" or "face-to-face" 3D integration scheme is being followed, the use of SOI substrates is critical to realizing a high-yield oxide fusion bonding process. It is well known that high-quality thermal oxide can be used as an extremely effective etch stop when etching silicon in aqueous TMAH or SF_6-based RIE processes. We have found that mechanical grinding and polishing processing can be used with great efficiency to remove the bulk of the sacrificial silicon wafer, but limitations in these processes preclude stopping at a specified point within a silicon substrate, or to remove an entire substrate without detrimentally affecting the device layers. It is possible to remove up to 600 μm of the silicon wafer by these mechanical techniques with up to 10-μm uniformity across a wafer and without inducing "grinding" damage to the attached device layers. We have employed 25% TMAH at 80°C to etch (40 μm h^{-1} etch rate) the last 100 μm of silicon down to the buried oxide layer which provides a natural etch stop (>300:1 etch selectivity) during the removal process. The overwhelming advantage of such an approach is that all of the Si can be uniformly removed, leaving a very smooth (<10 nm) surface for the application of bonding oxide films, and the layer of transferred circuits is automatically of minimal thickness across the wafer facilitating the fabrication of high-density interlevel wiring later in the process flow.

Bonding

Dielectric fusion bonding techniques have been applied to many semiconductor-related technologies and mechanisms for many material systems have been described in detail elsewhere. Independent of material system, the technique requires absolute <10 nm RMS control of surface smoothness, the ability to precisely control surface chemistry and cleanliness, and a thermal activation energy for the fusion bonding process which is compatible with the structures being bonded. The technique is particularly well suited for application to 3D device layer stacking since materials which are common to the intrinsic dielectric films used for CMOS device fabrication can be employed. Figure 9.2b shows a previously documented [23] approach in which an oxide-to-oxide bonding process has been used to attach two parts of the 3D structure. Unlike many applications where thermal fusion bonding processes are employed, application of the technique in 3D integration is not straightforward. Compatibility with precise device doping profiles, exotic BEOL low-κ dielectrics and Cu metallurgy, and processing-induced surface topography impose strict limitations on a fusion bonding process. Thermal cycles of less than $<450^\circ$C must be maintained for these applications and the impact on dielectric film purity and final bond strength can be dramatic. To enable successful low-temperature ($<450^\circ$C) oxide-to-oxide bonding a detailed study of this process has been published [24].

We have found low-temperature, $<300^\circ$C, plasma-enhanced chemical vapor deposition (PECVD) oxide (LTO) films to contain significant concentrations of H_2 as deposited and tend to absorb large amounts of water vapor without subsequent thermal processing. Post-deposition annealing of the films, at or above the oxide deposition temperatures, is used to drive out contaminants and densify the oxide films. We have found that high-energy, broad-spectrum UV-assisted anneals can produce high-quality bond interfaces and reduce overall cost of ownership as manufacturable solutions. Figure 9.6 shows a series of TEMs taken of the oxide fusion bond interface after various post-oxide-deposition anneal conditions. Gas bubbles, 10–20 nm in diameter, which form in the films during post-bond oven anneals tend to coalesce at the bond interface and negatively impact bond energies in films which see high-temperature UV-assisted post-deposition anneal processes. Films processed at UV cure temperature conditions closer to the oxide deposition temperature show a more even distribution of voids throughout the film and result in higher bond energies. We propose that high-temperature post-deposition cures, while effective at driving impurities from the oxide films, provide densification of the films but are not capable of absorbing impurities anywhere but at the bond interface.

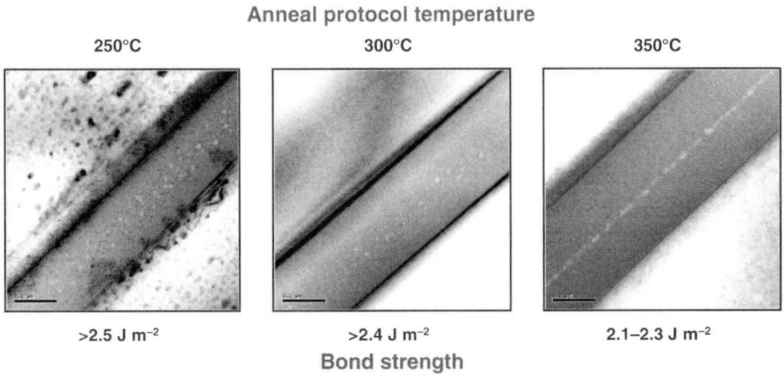

Fig. 9.6 Plot showing oxide–oxide bonding interfaces after various degrees of UV cure thermal treatment prior to bonding

Chemical–mechanical polishing (CMP) is capable of smoothing a dielectric surface to <10 nm RMS per 30 μm^2 roughness, with film thickness uniformity <5% 3σ, across a 200-mm Si wafer. Unfortunately, actual device wafers can pose unique topography issues if one measures over longer ranges (>30 μm). Topography introduced by CMP "dishing" effects introduced during BEOL damascene metal polishes can produce pattern-dependent voiding at the oxide fusion bond interface. Figure 9.7 shows bonding voids produced between device-level probe pads when a blanket low-temperature oxide (LTO) film is bonded over them. While it is possible to minimize the effect by careful consideration of pattern loading factors during the mask design process, we have also found that a process-driven solution is also necessary. In particular, as shown in Fig. 9.8 some of the pattern-dependent topography can be minimized by implementation of sequential PECVD oxide deposition and multiple CMP processing steps. Figure 9.8 shows surface topography range for two examples in which 2*x* and 4*x* oxide was deposited over target "short" (<100-μm, probe pads) and "long" (>800-μm, full scan) pads with single or double CMP steps,

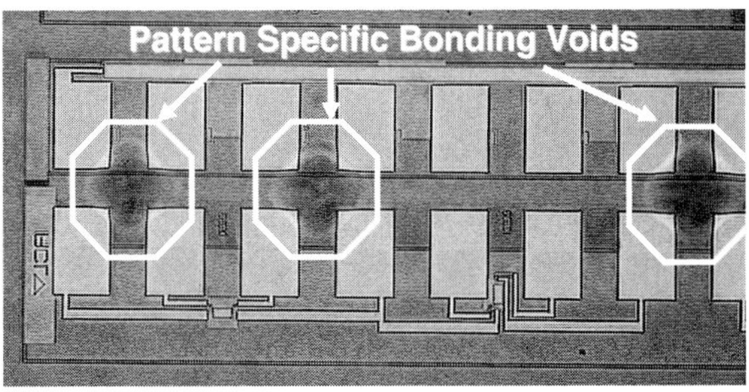

Fig. 9.7 Example of pattern-dependent voiding at the oxide fusion bonding interface possible caused by the CMP "dishing" effects introduced during BEOL damascene metal polishes

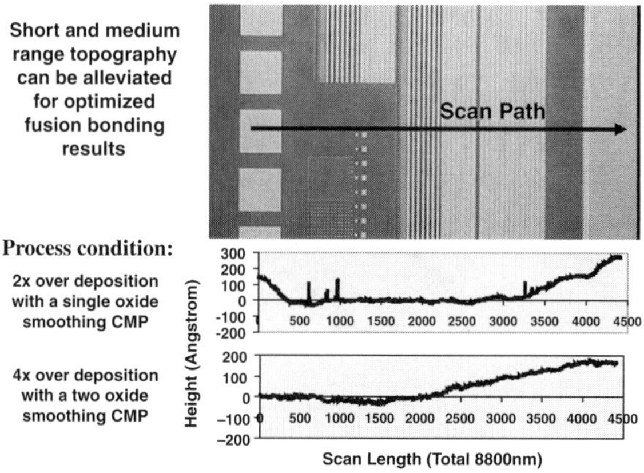

Fig. 9.8 Surface optimization focused on minimization of the pattern-dependent voiding can be accomplished by sequential deposition and CMP processes

respectively. As seen the topography is decreased by ~50% and much improved smoothing is achieved.

After bonding oxide films are annealed and planarized, they must be cleaned to remove particulates, particularly CMP slurry residues, and activated to ensure that the highest bond strengths can be achieved. It is critical that these processes be carried out in a very clean environment in order to achieve the highest chip yield per bonded substrate. We have found that the smallest particles ($\leq 1\ \mu m$) can result in millimeter-size bonding defects having a dramatic impact on wafer-scale chip yields. Many types of surface activation techniques have been reported in the literature and many others are undoubtedly kept proprietary. Activation techniques ranging from aqueous (hydrogen peroxide) [25] to dry plasma activation have been used to produce surface chemistries which are conducive to producing high-strength, low-temperature fusion bonds. The two surfaces must be modified to have high densities of hydroxyl groups (OH) so that upon contact they form weak hydrogen bonds between them [26].We have found that a combination of hydrogen bonding (Si–OH:HO–Si) and van der Waals attraction is enough to secure the surfaces well enough, about $0.7\,J\ m^{-2}$, that pattern shifts due to thermal expansion differences between the glass and Si substrates during subsequent thermal anneal cycles do not cause measurable pattern shift [20].

Once bonded, substrates must be annealed to convert the weaker hydrogen bonds to the desired Si–O–Si, high-strength configuration. We have found that oven annealing with slow ramp rates is required since differences in thermal expansion characteristics of the two substrate materials can produce cracks and subsequent yield loss. As we have discussed earlier, the evolution of water vapor is a natural consequence of the bond-strengthening process and careful control of the post-bonding anneal cycle as well as bonding film porosity can help minimize defects generated due to outgassing of the films.

Alignment

Since stacking of the devices has to occur on the device level the alignment tolerance and interconnection geometry requirements are much more aggressive when compared to any other 3D application. In particular since interconnecting vias with bottom diameter of $\sim$0.25 μm are typically expected to land on the pads designed with about two times the tolerance; the tolerance for misalignment needs to be controlled to $<$0.5 μm 3σ. As discussed earlier, wafer-scale bonding alignments can be significantly affected by the physical characteristics of the substrates themselves. For instance, we have shown that variations in substrate flatness, or bow, drive significant wafer-scale expansion errors [20]. We have also seen that localized imperfections in bonding films or insufficient bonding oxide surface preparation can drive lower (localized) bond energies and result in pattern overlay errors in those areas. Another concern for wafer-scale bonding alignment is the effect of bonding substrates which have been patterned independently on a variety of lithography steppers. Fortunately we have found that a variety of new process enhancements including wafer bow reduction, lithographic compensations and marking, use of glass carrier wafers, as well as new alignment test structure design have been demonstrated to enable this tight alignment requirement [20, 27, 28].

Vertical Interconnect

The vertical interconnections in aggressive 3D IC applications need to meet the density requirement of the FEOL technology. To achieve this goal IBM has developed 3D via technology. This via technology allows substantially denser 3D interconnect pitch than is possible with TSVs, due to the limited aspect ratio of the TSV etch-and-fill technology. This 3D via approach builds upon BEOL metallization technology; the aspect ratio of vias (ratio of the height of via to its width) is enhanced, but typically not higher than 8:1. This rigorous specification can be achieved with a SOI-based layer transfer process which minimizes the 3D stack thickness. Figure 9.9 shows the structural comparison of TSVs [29] and 3D via technology [23], highlighting the difference in via geometry and pitches.

Final Metallization and Testing

Once interconnecting vias are processed the final metallization is based on the standard BEOL technology. Depending on the application the number of metal levels may vary. Also additional metallization levels can be built before the stacking process. One has to only take into consideration the fact that interwafer vias have limited aspect ratio and hence they can "go through" only so many layers. Best designs are based on connecting the uppermost layer of the bottom wafer with the lowest metallization layers of the top wafer.

As always, testing of the stacked components is critical. Specialized test structures can be designed not only to provide information regarding performance and reliability of interwafer connecting vias and bonding interface [24] but also to enable

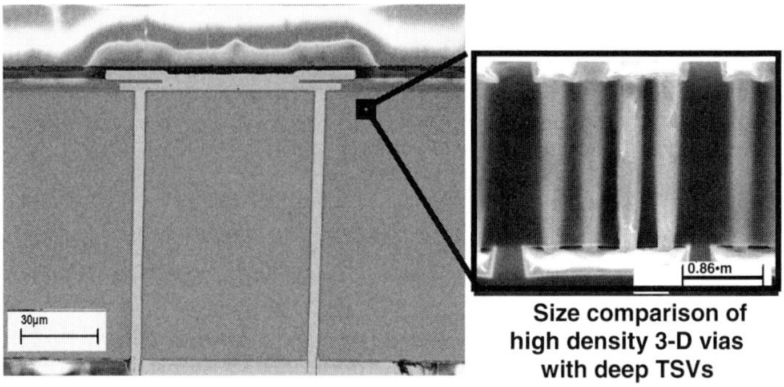

Fig. 9.9 Comparison of TSVs from Tsang et al. (*left*) fabricated on bulk Si wafers to high-density 3D vias from Topol et al. (*right*) fabricated using SOI-based 3D process

process optimization. For example, we have fabricated electrical and mechanical structures which were sensitive to via size (indicated minimum via size required) and provided layer-to-layer registration (alignment) information [28].

As we are progressing in qualifying 3D technology, verification test structures for circuit power and thermal management are becoming more critical. More specifically, to evaluate the thermal aspects of the 3D versus 2D ICs test structures based on subthreshold slope and polysilicon resistance measurements have been implemented. In our work, we have designed and built structures that address both the spread-heating and self-heating issues of 3D stacking [18]. We have also designed and implemented, circuit performance integrity test structures, like ring oscillator (RO) structures, single transistors (FETs), and inverter circuit structures. These structures help verify if thermal cycling and mechanical stresses, brought on by the layer transfer processes during 3D IC fabrication, have an effect on overall performance of the 3D structures. They also show if precise alignment and low parasitic connection requirements have been met for stacking and interconnecting the multiple layers [28].

9.2 Future Development Activities for 3D Integration

Clearly, in spite of great effort and work scope already covered in the area of 3D integration, several development activities must still be carried out in the fields of system and device design and modeling, frontend processes, backend processes, manufacturing issues, electrical testing strategies, yield, and reliability.

9.2.1 Thermal Dissipation in Bonded Structures

Initial assessment of the impact of 3D layer transfer on circuit performance has been reviewed [30]. However, additional evaluation of thermal dissipation in bonded structures is critical. One aspect of the new 3D technologies that has to be addressed is the thermal impact of the frontend and backend technology. Some of the issues that need to be investigated are self-heating effect on SOI wafers, Joule heating in the backend (electrical resistance increases with T – nonlinear); low-κ materials (thermal conductivity is a strong function of κ); heat flow through the interconnect (complex geometry); and use of thermal vias and design rules to minimize temperature rises (optimization). Joshi et al. [31] showed that simulation results clearly indicate the importance of the thermal effects in 3D integration structures. For example, the inclusion of Joule heating in the backend was found to be important. Therefore, detailed thermal analysis needs to be performed in early design stages, and it has to include specific choice of materials used in 3D stacking. Also, layout issues and design rules need to be analyzed in the light of thermal impact.

9.2.2 Noise in 3D Integrated Structures

Newest efforts related to 3D modeling focus on better understanding of power supply noise in 3D integrated systems [32, 33]. Simplified circuit models for power supply delivery in 3D systems have been implemented analyzing the worst-case peak noise for various layers. These calculations showed that in some cases a prohibitively large power supply noise can be reached and by adding P/G IOs, it can be reduced to acceptable levels. The 3D power delivery problem has to be further evaluated to provide a high I/O design solution.

9.2.3 Bandwidth Utilization and Smart Power-Efficient Designs (Lower-Power Voltage Operations, Energy Efficiency)

9.2.3.1 Bandwidth Advantages of 3D

Perhaps the most compelling advantage of 3D integration for system applications is the massive communication bandwidth that can be achieved between the chips in the 3D stack. Figure 9.10 shows graphical representation of the 3D system performance over the 2D solutions. The reason why this represents such an advantage for chip performance can be understood by realizing that, in the last few years, the growth in microprocessor performance has been driven more by the utilization of multithreaded, multicore architectures than by increasing clock frequencies. Such chip architectures, combined with specialized accelerator functions designed to further enhance chip performance, significantly increase the need for cache availability to support the many different threads running on a chip. In typical 2D chips, the fast cache memory is composed of SRAM or embedded DRAM (e-DRAM) [34] that is integrated directly onto the chip. Three-dimensional integration enables greater total cache bandwidth in two ways. First, it allows the cache to be much larger, since in 3D, the amount of total chip area that can fit in a given footprint is greater that of a 2D chip. Secondly, and perhaps more importantly, 3D allows a much larger number of interconnects to be made to the cache, since the cache can be integrated

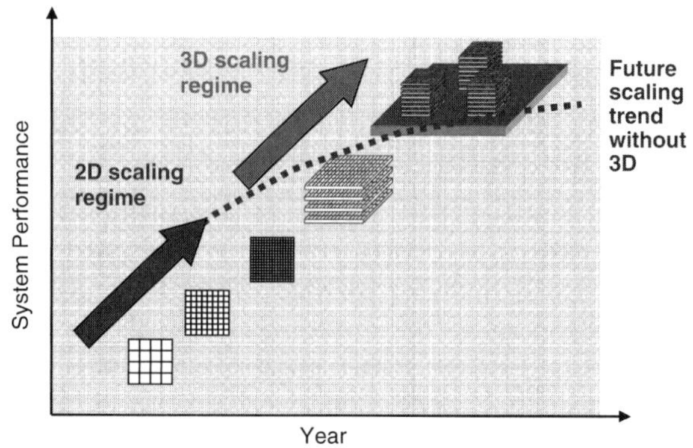

Fig. 9.10 Comparison of system performance scaling trends for 2D and 3D solutions

vertically with respect to the logic portion of the chip, thereby eliminating the wiring blockages that limit that 2D intrachip interconnect densities. Due to the small length of these vertical interconnects (a few microns to a few tens of microns), much higher data rates can be transmitted, at much lower power, compared to if they had to be transmitted across the chip, or off-chip.

9.2.4 Power-Efficient 3D Possibilities

Ultimately, the performance of many high-performance processors is limited by the total power that can be supplied to the chip, as well as the ability to dissipate the resulting heat. Given this constraint, one apparent way of improving the chip performance would be to increase the power efficiency of the logic function. For instance, greatly improved efficiency can be accomplished by reducing the supply voltage. For instance, reducing the supply voltage from 1 V to 0.5 V could increase the logic efficiency by a factor of close to 4*x*. Of course, this efficiency comes at the cost of reduced processor speed, and must be made up for with parallelism if overall system performance is to be maintained. Such parallelism may be difficult to implement in 2D systems, since the increase in silicon area would require multiple chips, and the resulting power needed to drive the communications between the chips would negate the advantages of the reduced supply voltages. However, in 3D architectures, the increased silicon area could be accommodated by stacking, while still maintaining low-power interconnection pathways, preserving the power advantages. In addition, novel power-saving architecture strategies could be employed, such as utilizing SRAM or e-DRAM caches stacked in multiple layers above a dedicated logic core. Three-dimensional integration could provide elegant power management solutions, such as the utilization of voltage-regulation modules that could be implemented to provide fine adjustment of the supply voltage to regions of the logic chip in order to counterbalance local process variability, or provide fine-grain power management.

9.3 Summary

Motivations for 3D die stacking include reduction in system size, interconnect delay, and power dissipation, and enabling hyperintegration of chips fabricated using disparate process technologies. To enable 3D opportunities, IBM has developed fabrication technology based on both oxide- and Cu-bonding and has proven the viability of layer transfer processing to accomplish stacked 3D structures. We have fabricated, tested, and proven reliability of various metallization schemes enabling interconnection between various stacked layers, which included TSV technology and Cu-based IC interconnection technology. We have also pushed the alignments limits and shown that with an optimized process, alignment accuracy below 0.5 μm can be achieved.

Going beyond the optimization of the fabrication process, IBM is currently focused on remaining 3D technology challenges, which include heat analysis and removal, understanding of modeling and design requirements for device and system level, utilization of bandwidth provided by 3D with connection with smart low-power designs. We are designing and creating new test and reliability structures to measure noise immunity, heat dissipation, and system/circuit performance integrity. We are engaged in further modeling and design activities to take full advantage of the bandwidth benefits offered by 3D solutions and to create the future advanced CMOS-based 3D systems.

Acknowledgments The authors wish to thank for their contributions (1) the following IBM personnel: Steven Steen, Kuan-neng Chen, Leathen Shi, Cornelia Tsang, Paul Andry, David Frank, Jyotica Patel, James Vichiconti, Deborah Neumayer, Narender Rana, Robert Trzcinski, Latha Ramakrishnan, Roy Yu, James Tornello, Michael Lofaro, Gill Singco, John Ott, David DiMilia, William Price, Jesus Acevedo, and (2) the following RPI personnel: Dr. James Lu, Dr. Sang Hwui Lee, Dr. Ravi Kumaroh. The authors also acknowledge the support of IBM's MRL and CSS as well as staff at EV Group and Suss MicroTec. This project was partially funded by DARPA under contract numbers N66001-00-C-8003 and N66001-04-C-8032.

References

1. Moore G (1965) Cramming more components onto integrated circuits. Electronics 38(8): 114–117
2. IBM press release (2007) IBM moves Moore's law into the third-dimension http://www-03.ibm.com/press/us/en/pressrelease/21350.wss
3. Guarini K, Wong, H-SP (2004) Wafer bonding for high performance logic applications. In: Alexe M, Goesele U (eds) Wafer bonding: applications and technology. Springer-Verlag, Berlin, pp 157–160
4. http://www.techsearchinc.com/ (2007) In: Through Silicon Via Technology: The Ultimate Market for 3D Interconnect. TechSearch International Report
5. http://www.sematech.org/research/3D/
6. Topol AW, Guarini KG, Yu R, Shi L, Newport MR, J. Tornello J, Melick D, O'Neil PA, Colburn M, Singh DV, Cohen GM, Krishna M, Ruiz N, Pogge HB, Ieong M, Purushothaman S, Haensch WE (2003) Demonstration of wafer-level layer transfer of high performance devices and circuits for three-dimensional integrated circuit fabrication. AVS 4th International Conference on Microelectronics and Interfaces, Santa Clara, CA, USA, pp 5–7

7. Jonson SC (2007) 3-D TSV cips take off. Semiconductor International, July, p 40
8. Vardaman J (2007) 3-D Through-silicon vias become a reality. Semiconductor International, June, pp 40–45
9. Andry PS, Tsang C, Sprogis E, Patel C, Wright SL, Webb BC (2006) A CMOS-compatible process for fabricating electrical through-vias in silicon. Proceedings of the 56th Electronic Components and Technology Conference, San Diego, CA, pp 831–837
10. Tsang CK, Topol AW (2006) 3D integrated circuits and silicon carrier packaging realization. Proceedings of 23rd VLSI VMIC Conference, Fremont, CA, September 25–28, 2006, VMIC no 06 IMIC-050 pp 61–69
11. Patel CS, Tsang CK, Schuster C, Doany FE, Nyikal H, Baks CW, Budd R,Buchwalter LP, Andry PS, Canaperi DF, Edelstein DC, Horton R, Knickerbocker JU, Krywanczyk T, Kwark YH, Kwietniak KT, Magerlein JH, Rosner J, Sprogis E (2005) Silicon carrier with deep through vias, fine pitch wiring and through cavity for parallel optical transceiver. Proceedings of the 55th Electronic Components and Technology Conf, Lake Buena Vista, FL, pp 1318–1324
12. Topol AW, Furman BK, Guarini KW, Leathen S, Cohen GM, Walker GF (2004) Enabling technologies for wafer-level bonding of 3D MEMS and integrated circuit structures. 54th Electronic Components and Technology Conference (ECTC), Las Vegas, NV; June 1–4, 2004; vol 1, no s21p2, pp 931–938
13. Chen KN, Tan CS, Fan A, Reif R (2004) Morphology and bond strength of copper wafer bonding. Electrochem Solid-State Lett 7(1):G14–G16,
14. Chen KN, Tsang CK, Topol AW, Lee SH, Furman BK, Rath DL, Lu J-Q, Young AM, Purushothaman S, Haensch W (2006) Improved manufacturability of Cu bond pads and implementation of seal design in 3D integrated circuits and packages. 23rd International VLSI Multilevel Interconnection (VMIC) Conference, Fremont CA, September 25–28, 2006, VMIC Catalog No 06 IMIC-050, pp 195–202,
15. Chen KN, Lee SH, Andry PS, Tsang CK, Topol AW, Lin YM, Lu JQ, Young AM, Ieong M, Haensch W (2006) Structure design and process control for Cu bonded interconnects in 3D integrated circuits. International Electron Devices Meeting (IEDM) Technical Digest, Session 13.5, pp 20–22
16. Ieong M, Narayana V, Singh D, Topol A, Chan V, Ren Z (2006) Transistor scaling with novel materials and integration. Mater Today 9(6):26–31
17. Ieong M, Guarini KW, Chan V, Bernstein K, Joshi R, Kedzierski J, Haensch W (2003) Three dimensional CMOS devices and integrated circuits. Proceedings of the IEEE Custom Integrated Circuits Conference, pp 207–213
18. Topol AW, La Tulipe DC, Shi L, Frank DJ, Bernstein K, Steen SE, Kumar A, Singco GU, Young AM, Guarini KW, Ieong M (2006) Three-dimensional integrated circuits. IBM J Res Dev 50(4/5):491–506
19. Guarini KW, Topol KW, Ieong M, Yu R, Shi L, Newport MR, Frank DJ, Singh DV, Cohen GM, Nitta SV, Boyd DC, O'Neil PA, Tempest SL, Pogge HB, Purushothaman S, Haensch WE (2002) Electrical integrity of state-of-the-art 0.13 lm SOI CMOS devices and circuits transferred for three-dimensional (3D) integrated circuit (IC) fabrication. IEDM Tech. Digest, pp 943–945
20. La Tulipe D, Shi L, Topol A, Steen S, Pfeiffer D, Posillico D, Neumayer D, Goma D, Vichiconti V, Rubino J, Young A, Ieong M (2006) Critical aspects of layer transfer and alignment tolerances for 3D integration processes. International Conference and Exhibition on Device Packaging (IMAPS), Scottsdale, AZ, March 20–24, 2006, Journal of Microelectronics and Electronic Packaging vol 3, no 1
21. Narayan C, Purushothaman S, Doany F, Deutsch A (1995) Thin film transfer process for low cost MCM-fabrication. IEEE Transactions on Components, Packaging, and Manufacturing Technology (Trans CPMT) Part B, vol 18, no 1, pp 42–46
22. Ultra Thin Wafer Processing Solutions. June 20, 2007, (http://www.semiconductor.net/webcastsDetail/2140049068.html). Semiconductor International Webcast

23. Topol AW, La Tulipe DC, Shi L, Alam SM, Frank DJ, Steen SE, Vichiconti J, Posillico D, Cobb M, Medd S, Patel J, Goma S, DiMilia D, Robson TM, Duch E, Farinelli M, Wang C, Conti RA, Canaperi DM, Deligianni L, Kumar A, Kwietniak KT, D'Emic C, Ott J, Young AM, Guarini KW, Ieong M (2005) Enabling SOI based assembly technology for three-dimensional (3D) integrated circuits (ICs). IEDM Tech. Digest, pp 363–366
24. Guarini KW, Topol TA, Singh DV, La Tulipe DC, Shi L, Young AM, Alam A, Frank DJ, Neumayer DA, Vichiconti J, Sicina RM, Conti RA, Wang C, Canaperi DM, Deligianni L, Kwietniak KT, Steen SE, Robson M, Gibson GW, Posillico D, Ieong M (2005) Process technologies for three dimensional integration. Proceedings of the 6th Annual International Conference on Microelectronics and Interfaces, American Vacuum Society, pp 212–214
25. Warner K, Burns J, Keast C, Kunz R, Lennon D, Loomis A, Mowers W, Yost D (2002) Low-temperature oxide-bonded three-dimensional integrated circuits. Proceedings of the IEEE International SOI Conference, Oct 7–10 2002, pp 123–125
26. Publications inclue: Tong QY et al (1996) Proceddings of the IEEE International SOI Conference, pp 36–37; Stengl R et al (1989) Jpn J Appl Phys 2810:1735–1741; Tong QY et al (1995) Electrochem Soc Proc 95(7):78–95
27. Steen SE, LaTulipe D, Topol A, Frank D, Belote K, Posillico D (2007) Wafer scale 3D integration: overlay as key to drive potential. Microelec Eng 84(5–8):1412–1415, May–August 2007
28. Topol AW, La Tulipe DC, Shi L, Alam SM, Young AM, Frank DJ, Steen SE, Vichiconti J, Posillico D, Canaperi DM, Medd S, Conti RA, Goma S, Dimilia D, Wang C, Deligianni L, Cobb MA, Jenkins K, Kumar A, Kwietniak KT, Robson M, Gibson GW, D'Emic C, Nowak E, Joshi R, Guarini KW, Ieong M (2005) Assembly technology for tree dimensional integrated circuits. Proceedings of 22 nd International VLSI Multilevel Interconnection Conference (VMIC), a special conference of 2005 Institute for Microelectronics on Chip Interconnection (IMIC), Fremont CA, Oct 4–6 2005, pp 83–88
29. Tsang CK, Andry PS, Sprogis EJ, Patel CS, Webb BC, Manzer DG, Knickerbocker JU (2006) CMOS-compatible through sillicon vias for 3D process integration. Pannel Discussion on Enabling Technologies for 3D Integration; MRS Symposium Proceedings November 27–29, 2006 Boston, MA, vol 970, pp 145–153
30. Guarini KW, Topol AT, Ieong M, Bernstein K, Xiu K, Joshi RV, Yu R, Shi L, Newport MR, Singh DV, Cohen GM, Pogge HB, Purushothaman S, Haensch WE (2003) The impact of wafer-level layer transfer on high performance devices and circuits for 3D fabrication. Proceedings of Electrochemical Society (ECS) Annual Meeting, Paris, France, Abstract #431, April 27–May 2, 2003, PV 2003-20, ISBN 1-56677-403-9, In: New Trends in Intercalation Compounds for Energy Storage Conversion (eds) Zaghib K, Julien CM, Prakash J, pp 390–404
31. Joshi RV, Smy T, Banerjee K, Topol A (2007) Thermal dissipation in bonded structures. In: Thermal and Design Issues in 3D ICs, SEMATECH Workshop, October 9–11, 2007, in Albany, NY
32. Huang G, Bakir M, Naeemi A, Chen H, Meindl JD (2007) Power delivery for 3D chip stacks: physical modeling and design implication. IEEE 16th Conference on Electrical Performance and Electronic Packaging, Atlanta, October 28–31 2007, Session IX, No 5
33. Huang G, Sekar D, Naeemi A, Shajeri K, Meindl JD (2007) Physical model for power supply noise and chip/package co-design in Gigascale Systems with consideration of hot spots. IEEE Custom Integrated Circuits Conference (CICC), San Jose, September 16–19 2007, no 27.1
34. Iyer S, Barth JE, Parries PC, Norum IP, Rice JP, Logan LR, Hoyniak D (2005) Embedded DRAM: technology platform for the blue gene/L chip. IBM J Res Dev 49(2–3):333–350

Chapter 10
3D Integration Based upon Dielectric Adhesive Bonding

Jian-Qiang Lu, Timothy S. Cale, and Ronald J. Gutmann

10.1 Introduction

Wafer bonding with intermediate polymer adhesives is one of the viable bonding approaches for three-dimensional (3D) integration. This chapter begins with an overview of the fundamentals of adhesive wafer bonding, and the desired properties of bonding adhesives. This is followed by a description of two wafer-level 3D integration technology platforms developed at Rensselaer. One of these platforms utilizes a "via-last" process flow in which wafers are bonded using a blanket (i.e., unpatterned) adhesive layer. The second platform discussed is based upon recent research with a patterned metal/adhesive layer compatible with the desirable "via-first" process flow. As with other 3D platforms, there are four key processing challenges for 3D platforms using blanket adhesive or patterned metal/adhesive bonding, namely wafer-scale alignment accuracy, bonding integrity, wafer thinning and leveling control, and interwafer (or interstrata) interconnection. Each of these processes must be compatible with semiconductor-processing constraints such as pressure and temperature. The development and evaluation of these key processes are summarized in this chapter. The emphasis is on back-end-of-line (BEOL)-compatible process with micron-sized vias for high interstrata interconnection density.

Most of the chapter focuses on two Rensselaer technology platforms that use dry-etch divinylsiloxane bis-benzocyclobutene (BCB) as a bonding adhesive and copper as interstrata interconnects. Benzocyclobutene is a thermosetting polymer; the extent of BCB curing impacts on 3D processing, particularly alignment accuracy. Integrity characterization of both blanket adhesive films for a via-last 3D process flow and patterned metal/adhesive films for a via-first 3D process flow are discussed. The process flows and interstrata via resistance obtained using 200-mm diameter wafers are presented. Thermomechanical modeling results that indicate the need for interstrata via design rules in order to prevent plastic deformation of interstrata copper vias are summarized.

J.-Q. Lu
Rensselaer Polytechnic Institute, Troy, New York, USA
e-mail: luj@rpi.edu

C.S. Tan et al. (eds.), *Wafer-Level 3D ICs Process Technology*,
DOI: 10.1007/978-0-387-76534-1_10, © Springer Science+Business Media, LLC 2008

While applications of 3D integration are covered elsewhere in this book, some possible applications enabled by 3D integration pursued at Rensselaer are briefly discussed with references for details at the end of this chapter.

10.2 Adhesive Bonding Mechanisms and Dielectric Adhesives

Bonding two substrates together has been an important process in the fabrication of both microelectronics systems and microelectromechanical systems (MEMSs) for years. This includes wafer-bonding techniques such as direct bonding, anodic bonding, solder bonding, eutectic bonding, ultrasonic bonding, metal fusion bonding, thermocompression bonding, low-temperature melting glass bonding, and adhesive bonding. In adhesive bonding, an intermediate adhesive layer is used to create a bond between two wafer surfaces to hold them together. Recently developed reliable and high-yield adhesive bonding processes have made adhesive wafer bonding a generic, and in some cases enabling, wafer-bonding technique for a variety of applications [1–10]. The desired properties of adhesive materials to achieve strong and defect-free bonding interfaces for large substrates are summarized in Section 10.2.1.

In most commonly used adhesive wafer-bonding processes, a polymer adhesive is applied to one or both of the wafer surfaces to be bonded, as shown in Fig. 10.1. After joining the wafer surfaces that are covered with the polymer adhesive, pressure is applied to force the wafer surfaces into intimate contact. The polymer adhesive is then converted from a liquid or viscoelastic state into a solid state, usually by exposing the polymer adhesive to heat or UV light. During this conversion, the adhesive film may reflow to accommodate surface topography or particles, and the adhesive is cured to form cross-link networks (indicated schematically in Fig. 10.1), providing a robust bonding interface with high thermal and mechanical stability over the ranges of BEOL and packaging processing conditions. Section 10.2.2 summarizes the adhesive wafer-bonding process flow.

The main advantages of adhesive wafer bonding, relative to other approaches, include

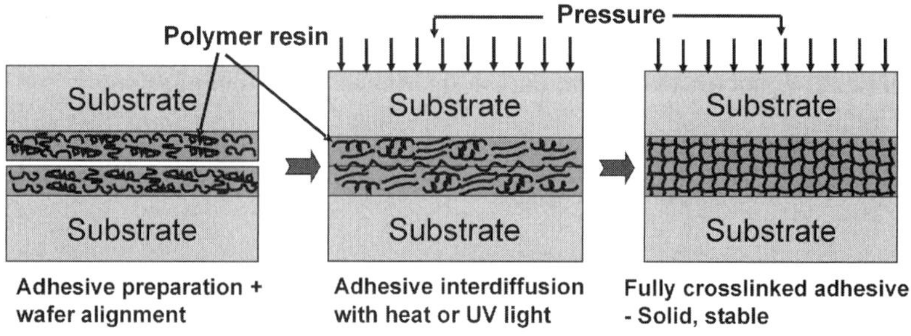

Fig. 10.1 Bonding mechanism for representative polymer adhesive used in 3D integration

- relatively low bonding temperatures (between room temperature and 450°C, depending on the polymer material),
- insensitivity to the topography of the wafer surfaces,
- compatibility with standard complementary metal-oxide semiconductor (CMOS) wafers,
- ability to join practically any materials.

Adhesive wafer bonding does not require special wafer surface treatments such as planarization and extensive cleaning. Structures and particles at the wafer surfaces can be tolerated and accommodated to some extent by the polymer adhesive. While adhesive wafer bonding is a comparably simple, robust, and low-cost process, and desirable properties of adhesive materials can be listed (next section), there is limited information about long-term stability of many polymer adhesives in the demanding environments seen in 3D integration.

10.2.1 Desired Properties of Polymer Adhesive Materials for Wafer Bonding

The polymeric adhesive layer for wafer bonding in 3D integration must provide a seamless interface and strong adhesion to prevent delamination, be sufficiently thin to minimize the via aspect ratio, be processed at modest temperatures to avoid any effects on device and circuit performance and reliability, and be thermally and mechanically stable after bonding. Desirable properties of a bonding adhesive include:

- good adhesion (adhesive to wafer) and cohesion (adhesive to adhesive) to prevent delamination,
- no outgassing during bonding to avoid void formation,
- high thermal and mechanical stability over the ranges of BEOL and packaging processing conditions; that is, a high glass transition temperature and rigid structure after bonding,
- low stress relaxation and creep,
- low moisture uptake,
- ability to form uniform and micron-thick films over entire wafers.

A variety of polymers have been explored for wafer bonding; reference [1] provides a review. Based on the desired properties listed above, Rensselaer evaluated a number of polymers, including spin-on polymers Flare (polyaryl ether), methylsilsesquioxane (MSSQ), BCB, and hydrogen silsesquioxane (HSQ), as well as the vapor-deposited polymer Parylene-N. These polymers can be deposited as thin films, have relatively well-known chemical and physical properties, and importantly, are compatible with integrated circuit (IC) processing to some extent. Figure 10.2 shows some challenges of adhesive wafer bonding [11]: voids due to outgassing with MSSQ, low cohesive bond strength with Parylene, and adhesive films that

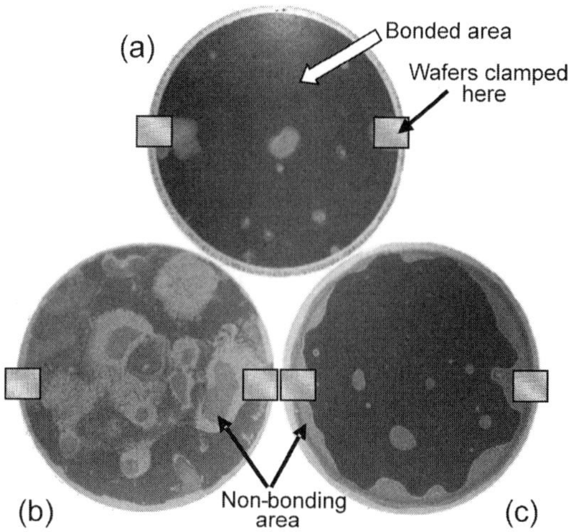

Fig. 10.2 Bonded wafers (200-mm Corning 7740 glass wafer to silicon wafer) using (**a**) Flare, (**b**) MSSQ, and (**c**) Parylene-N. Highly contrasted (*light gray*) areas show voids or nonbonding areas. From [11] © 2001 IEEE

are too thin (<0.3 μm) to accommodate large particles with Flare (although the bond strength with Flare is very high). Figure 10.3 shows several other examples of adhesive bonding used for 3D integration [3, 8, 12, 13].

Among the adhesives evaluated at Rensselaer, BCB was selected as a model adhesive for wafer bonding and 3D integration research because of its properties and availability. Benzocyclobutene has been used in various semiconductor applications, is compatible with most CMOS and packaging processes, and there is more information about its relevant use and performance than other adhesive polymers.

10.2.2 Adhesive Wafer-Bonding Technology

Most 3D IC applications of adhesive wafer bonding require high yield, strong bond interfaces, and often precise alignment of the bonded wafers. In order to repeatably achieve high-quality bonding results, the bonding process and parameters, as well as wafer-to-wafer alignment where required, must be precisely controlled. This is often accomplished with a cluster tool consisting of a wafer aligner and bonder. A wafer bonder typically consists of a vacuum chamber, a mechanism for joining the wafers inside the vacuum chamber, a wafer chuck, and a bond tool, as shown in Fig. 10.4. Bonding parameters such as bonding pressure, bonding temperature, chamber pressure, and temperature ramping profile can significantly impact the resulting bonding quality and defect density. The wafer stack is placed between the bottom wafer chuck and the bond tool. Thus, the wafer stack can be pressed together with the bond tool using a controlled pressure (force per wafer or bond area). The wafer stack can be heated through the bottom wafer chuck and the top chuck. The top chuck can be a stiff flat plate or stiff flat plate with a soft plate or sheet in between the top chuck and the wafer stack. Soft plates or sheets typically

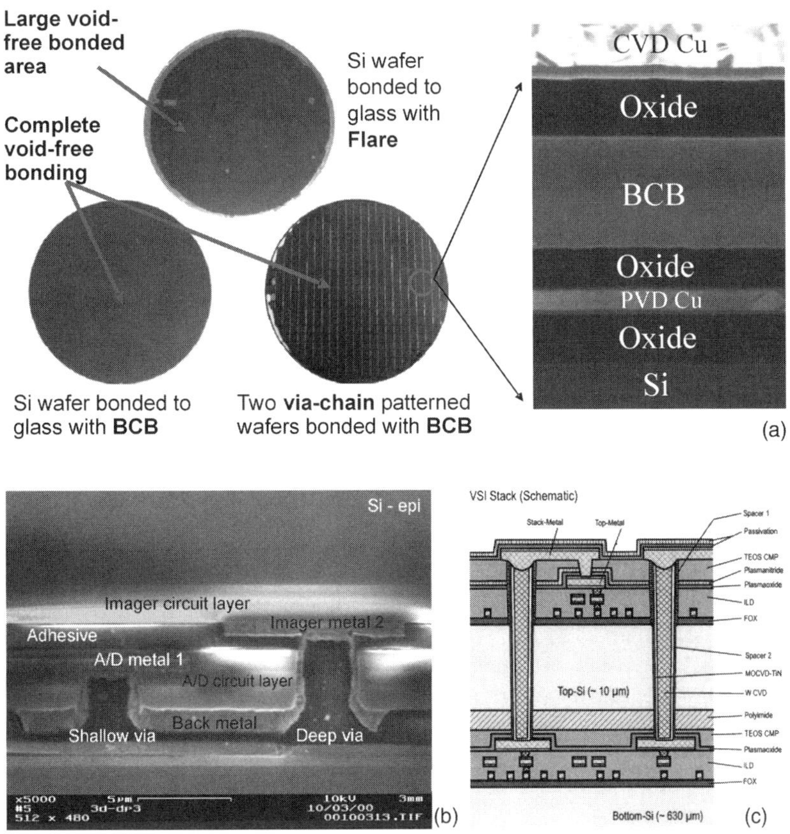

Fig. 10.3 (**a**) Wafer bonding results: Corning glass 7740 bonded to prime Si with Flare (*top*), PG&O glass 1737 bonded to prime Si with BCB (*left*), two via-chain-patterned wafers bonded with BCB (*right*), with an SEM image showing seamless bonding interface. From [8] Copyright 2003 GaAs MANTecH and [12]. (**b**) An SEM of a 3D ring oscillator section showing the adhesive bond between the imager and A/D wafers. From [3] © 2001 IEEE. (**c**) Polyimide as a bonding adhesive. From [13] © 2002 MRS

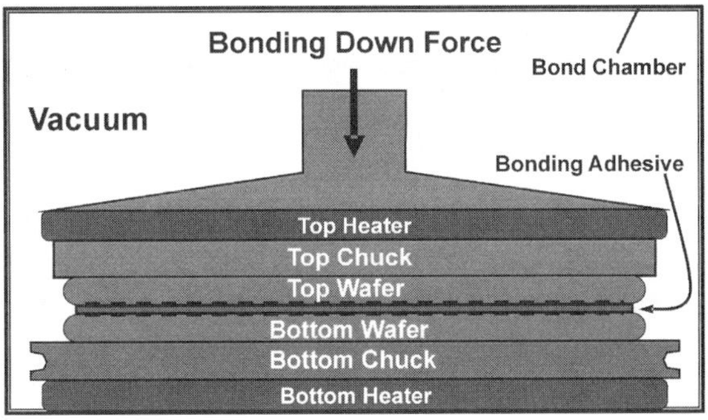

Fig. 10.4 A representative commercial wafer bonder

adapt better to nonuniformities of the wafer stack and thus distribute the pressure more evenly over the wafer stack.

Table 10.1 describes a typical process scheme for wafer bonding with an intermediate polymer adhesive. The process flow involves the use of a wafer bonder, and applies generally to bonding with thermoplastic polymer adhesives or thermosetting polymer adhesives.

Table 10.1 Typical process steps for adhesive wafer bonding [1, 2, 11, 12]

No.	Process step	Purpose of the process step
1	Cleaning and drying the wafers	Remove particles, contaminations, and moisture from the wafer surfaces
2	Treating the wafer surfaces with an adhesion promoter (optional)	Adhesion promoters can enhance the adhesion between the wafer surfaces and the polymer adhesive
3	Applying the polymer adhesive to the surface of one or both wafers; patterning the polymer adhesive (optional)	The most commonly used application method is spin coating, with spray coating and vapor deposition as alternatives
4	Soft-bake or partial cure of the polymer	Solvents and volatile substances are removed from the polymer coating. Thermosetting adhesives may only be partially polymerized. Thermoplastic adhesives may be completely polymerized
5	Aligning wafers (optional)*	Two wafers are aligned using alignment marks on both wafers in a wafer aligner, and then clamped on the bottom bond chuck.
6	Placing the aligned wafers (on the bottom chuck) in the bond chamber, establishing a vacuum atmosphere.	If wafers are not aligned, the wafers can be placed on a bottom chuck in the bonder. *Caution*: To prevent voids, any trapped gasses at the bond interface should be pumped away before the bond is initiated.
7	Applying pressure to the wafer stack with the bond tool	The wafer and polymer adhesive surfaces are forced into intimate contact over the entire wafer. The bonding pressure may be applied before or after the bonding temperature is reached
8	Ramping up to bonding temperature to re-melt or cure the polymer adhesive while applying pressure.	The hardening procedure depends on the curing mechanism of the used polymer adhesive and may take few minutes to hours. The reflow of the polymer adhesive is typically triggered through elevated temperature
9	Chamber cool-down, purge, and bond pressure release.	To solidify the polymer adhesive, the bond pressure is not released until cooling down to certain temperature.

*Wafer-to-wafer alignment prior to wafer bonding is a critical step in most of the 3D integration process flow. The alignment marks are usually metal patterns produced along with the top metal interconnects of the processed wafers that are to be aligned and bonded.

After wafer alignment and bonding, the 3D integration is completed by wafer thinning and through-silicon via (TSV) formation (also called, interwafer interconnect or interstrata interconnect formation). Details of wafer thinning and TSV formation are described in Sections 10.4 and 10.7.

10.3 Wafer-Level 3D Integration Platforms Based upon Adhesive Bonding

With the adhesive bonding technology described in previous section, two technology platforms for wafer-level BEOL-compatible hyper-integration (the term "hyper-integration" denotes the integration of various materials, processing technologies, and functions beyond conventional ultra large-scale integration (ULSI) or gigascale integration (GSI) of planar ICs) are discussed in this section, that is, via-first and via-last 3D platforms, as illustrated in Figs. 10.5 and 10.6, respectively. After a general introduction of these two platforms, some details specific to adhesive wafer bonding are presented in Sections 10.3.1 and 10.3.2. These platform details highlight key 3D processes, and serve as a basis for further discussions in later sections of this chapter.

For the via-last 3D platform, as shown in Fig. 10.5, the interstrata vias are formed after the wafers are aligned and bonded, and the top wafer is thinned by removing most or all of its silicon. Since these vias are usually formed through a silicon layer, they are also called TSVs. A TSV usually consists of an electrical isolation layer (e.g., silicon dioxide or other dielectrics), a liner or barrier layer (e.g., titanium, tantalum, TiN or TaN), and a via metal (e.g., copper, tungsten, or highly doped polysilicon). The wafers can be attached by either adhesive-to-adhesive bonding [2, 14, 15] or oxide-to-oxide bonding [16, 17].

For the via-first 3D platform as shown in Fig. 10.6, the majority of interstrata vias are formed during bonding of the aligned wafer; by contacting exposed metal on both wafers. These vias are called "bond vias" to differentiate them from TSVs. Similar to TSV, the bond via also consists of an electrical isolation layer, a liner or barrier layer, and a via metal. Candidates for the via metal might be elemental metals (e.g, copper, gold), or eutectic conductors or solders (e.g., InAu, CuSn). One approach to exposing the via metal on both wafers to be bonded is to recess the dielectric – that is, have the metal stick out of the dielectric. Wafer bonding results in an air gap around metal via [18, 19]. A more exacting but attractive approach is to bond wafers using both metal-to-metal and adhesive-to-adhesive bonding [10, 20, 21] (or even more exacting, using oxide-to-oxide bonding [22]). After this bonding step, the "top" wafer is thinned to form TSVs that are needed to bring the electrical input and output connections (I/Os) to and from the stack, and/or to interconnect the two bonded strata to upper-level strata as shown in Fig. 10.6.

Note that the interstrata vias that pass through the CMOS device layer in these two platforms are really TSVs only if they are formed through a thin silicon layer

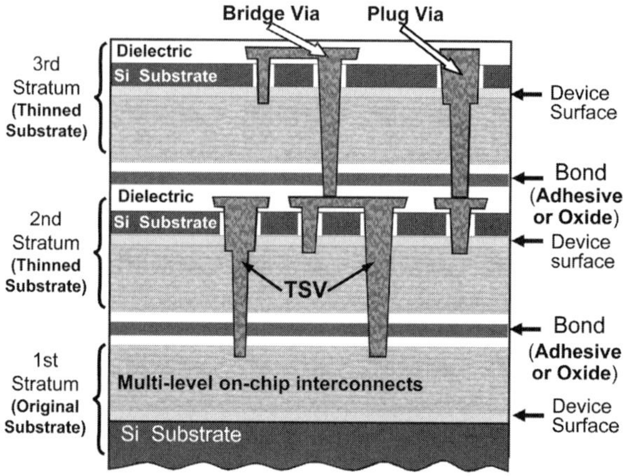

Fig. 10.5 Schematic cross-section of a three-stratum stack of via-last 3D platform, showing bonding interface and vertical interstrata vias (TSVs). The bonding interface can be either adhesive-to-adhesive bonding [2, 14, 15] or oxide-to-oxide bonding [16, 17]. From [15] © 2003 IEEE

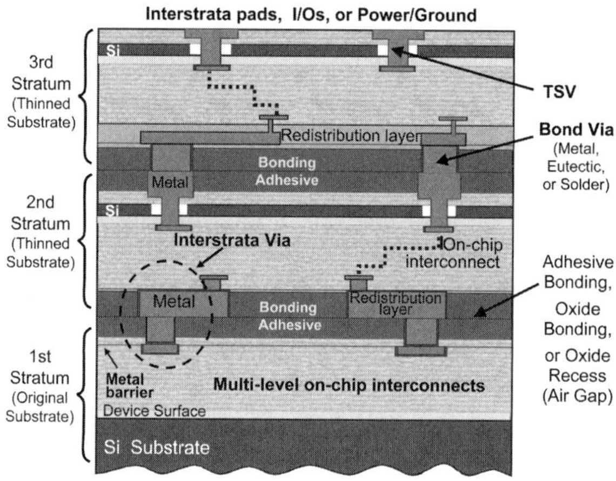

Fig. 10.6 Schematic cross-section of a three-stratum stack of via-first 3D platform, showing bonding interface and vertical interstrata vias (bond vias and TSVs). Besides the metal bonding that forms the interstrata interconnects, the bonding interface can include an adhesive-to-adhesive bonding [10, 20, 21], or oxide-to-oxide bonding [22], or leave an air gap by recessing the dielectrics surrounding the metal bonding posts [18, 19]. From [10] © 2005 IEEE

that remains after thinning a bulk silicon substrate. If a silicon-on-insulator (SOI) wafer is used as the top wafer, there is no silicon layer for these interstrata vias to go through because all the silicon substrate of the SOI wafer is removed and these interstrata vias are formed through the SiO_2 layer surrounding the CMOS devices. Therefore, the term "TSVs" could be used more generically for "through-strata vias".

10.3.1 Via-Last 3D Platform Using Blanket Adhesive Wafer Bonding and Cu Damascene Interstrata Interconnects

Figure 10.5 shows a representation of a structure that might result from a via-last 3D platform for monolithic 3D hyper-integration using adhesive polymer [2, 14, 15, 23]. A 3D IC using this platform starts with two processed IC wafers, each containing an active device layer and multilevel on-chip interconnects. These wafers are coated with an adhesive layer (on each wafer or one of the two wafers); the adhesive layer is soft-baked with time and temperature-dependent on the adhesive (see Table 10.1). The two wafers are then aligned to tolerances within 1 μm using a wafer aligner. These two aligned wafers are loaded into a wafer bonder and bonded by the adhesive under conditions compatible with CMOS processing and packaging. After bonding, the top wafer in the two-wafer stack is thinned to $\leq$10 μm by a three-step thinning process: (1) backside grinding, (2) chemical–mechanical polishing (CMP), and (3) wet etching to an etch stop, for example, an implanted layer, epitaxial layer, oxide layer, or a buried oxide (BOX) layer in SOI technology. The bottom wafer retains its full thickness as a "base wafer" of the stack. Subsequently, "bridge-type" and/or "plug-type" interstrata vias that electrically connect the circuits in these two strata are formed using a Cu damascene patterning process on top of the stack. This damascene patterning process, developed jointly by Rensselaer and the University at Albany [23, 24], involves high-aspect-ratio (HAR) via etching and cleaning, barrier and copper deposition, and CMP. Repeating this process flow, the third wafer (or more) can then be aligned, bonded, thinned, and interconnected. The final "wafer" with multiple strata of circuits is singulated into individual 3D chips. Thus, a long-distance interconnect that might run a centimeter across a conventional 2D chip may be replaced by a 2–10-μm (vertical length) HAR via between strata in the 3D chip.

Three key advantages of this approach compared to other 3D approaches are [24]:

- the ability of the dielectric adhesive to accommodate wafer-level nonplanarity (e.g., surface structures and wafer bow) and particulates at the bonding interfaces,
- no handling wafers are required, that is, thinned silicon is not transferred,
- stacks of three or more wafers can be fabricated without changing the processing approach.

10.3.2 Via-First 3D Platform with Wafer-Bonding of Damascene-Patterned Metal/Adhesive Redistribution Layers

This via-first 3D technology platform developed at Rensselaer employs wafer bonding of "damascene-patterned metal/adhesive redistribution layers" on two wafers, as shown in Fig. 10.6. This approach provides interstrata electrical interconnects (bond vias) and adhesive bonding of two wafers in one unit processing step [10, 20, 21]. The feasibility of this via-first 3D approach was demonstrated using copper/tantalum (Cu/Ta) bond vias and BCB adhesive bonding. Processing of a 3D IC using this platform begins with a damascene-patterned Cu/BCB layer that is formed over the uppermost metal layer of each of two processed wafers, each of which has an active device layer and multilevel on-chip interconnects. Similar to the via-last 3D approach, one of the two wafers is then flipped, aligned, and bonded to the other wafer (i.e., the "base wafer" of the stack with full thickness) through the patterned Cu/BCB layers. A key difference from the via-last 3D approach is that wafer bonding results in interstrata electrical interconnects and physical wafer bonding by Cu-to-Cu thermocompression bonding and adhesive bonding. The substrate of the face-down-bonded second wafer is then thinned, and TSVs are formed on top of the stack to bring the I/Os to and from the stack. The process can be extended to multiple wafer stacks by creating another damascene-patterned Cu/BCB layer with TSVs on top of the stack, providing a Cu/BCB structure that mates with the bond vias on a third wafer.

The patterned Cu/BCB layer on each wafer can be a Cu/BCB "redistribution layer" if needed, or just a layer with Cu posts/pads inlaid in the BCB for wafer bonding. Moreover, an extra Cu/oxide (or Cu/BCB, or other metal/dielectric) redistribution layer, for example, the uppermost metal layer of the third wafer as shown in Fig. 10.6, can be added prior to patterning process of any Cu/BCB bonding layer. This extra redistribution layer simplifies the patterning of the Cu/BCB bonding layer because only Cu-bonding posts (vias) are needed, and offers a simple bonding scheme; that is, assuming minimal misalignment, one always bonds Cu posts to Cu pads and BCB field to BCB field. This minimizes undesirable contact (i.e., bonding) of long Cu lines with BCB; for example, see the bonding layers between second and third wafers shown Fig. 10.6. This extra redistribution layer also provides additional redistribution capability compared with the approach of combining Cu-bonding vias and the redistribution layer (e.g., the Cu/BCB redistribution layer on front side of the second wafer as shown in Fig. 10.6).

This via-first 3D approach using wafer bonding of damascene-patterned metal/adhesive redistribution layers provides [20, 21]:

- both electrical and mechanical interstrata connections/bonds (combining advantages of both BCB/BCB and Cu/Cu bonding),
- high interstrata interconnectivity while allowing larger alignment tolerance by eliminating deep interstrata vias,
- redistribution layers for interstrata interconnect routing for wafers, on which the interstrata interconnect pads are not matched, which further reduces the process flow complexity and is compatible with wafer-level packaging (WLP) technologies,
- *Thermal management options*: Cu/BCB redistribution layers can serve as a thermal conductor and/or spreader (with large percentage of Cu area), or as a thermal insulator (with large percentage of BCB area), or as a thermal conductor for some selected areas and a thermal insulator for other areas.

The bonding process for such a technology platform is challenging, as a variety of surfaces are exposed, including the adhesives, diffusion barriers, and electrical conductors. Ideally, all should be capable of being bonded to one another without interfering with the electrical characteristics of the Cu-to-Cu interconnection. Surface preparation techniques for improving adhesion of BCB to silicon and silicon nitride [25] as well as copper [26] have been discussed in the literature, but not with respect to wafer bonding. Although damascene patterning of copper in fully cured BCB has been demonstrated [27], soft-baked BCB cannot withstand the Cu/BCB CMP process and fully cured BCB does not form further cross-link networks at the bonding interface to enhance the wafer bond. Considering these factors, a partially cured BCB layer was selected as it would offer the best compromise between patterning capability for Cu/BCB redistribution layer and bond quality of BCB-to-BCB bonding [10].

10.4 Impacts of Soft-Baked BCB and Partially Cured BCB

A degree of freedom when using dry-etch BCB is the extent of curing before aligning and bonding (via-last and via-first 3D) and/or damascene processing (via-first 3D). This section discusses some aspects of BCB curing and its impacts.

Adhesive wafer bonding, like most bonding techniques, is based on the fact that atoms and molecules fuse and adhere to each other when they are brought in sufficiently close contact. When an intermediate polymer adhesive is used to join two solid-state surfaces, the adhesive deforms to fit the surfaces to be bonded. Dry-etch BCB is a thermosetting polymer that undergoes cross-linking during curing to form a stable polymeric network. Benzocyclobutene flows for a short time during the cure process to achieve cross-linking and cannot be remelted or reshaped after curing (unlike thermoplastics). Dry-etch BCB is supplied by the Dow Chemical Company as a liquid polymer precursor that consists of the BCB polymer, which is approximately 35% cross-linked and dissolved in the solvent mesitylene. After applying the adhesion promoter to the wafer surfaces, the BCB precursor is spin-coated on the wafer surfaces and soft-baked at below 170°C for a few minutes in order to evaporate the solvent. This soft-bake process does not significantly increase the level of BCB cross-linking. The level of BCB cross-linking is increased using a pre-curing step, and is a function of the pre-curing time and the pre-curing temperature as shown in Fig. 10.7. For example, a pre-curing temperature of 190°C together with a pre-curing time of 30 min results in an increase of the BCB cross-linking level from 35% to about 45%. In the interest of 3D integration, we use the term "partially cured-BCB" for cure level at 45% or higher, that is, BCB is rigid enough (behaving similar to a rubber or gel glass as shown in Fig. 10.7) to be further processed, but not fully cured in order to facilitate further wafer bonding.

A number of wafer-bonding experiments have been performed to investigate the influence of soft-baked BCB and partially cured BCB on 3D processing, particularly on the post-bond alignment accuracy and the uniformity of the post-bond dielectric layer thickness [30]. The percentage level of cross-linking of the BCB

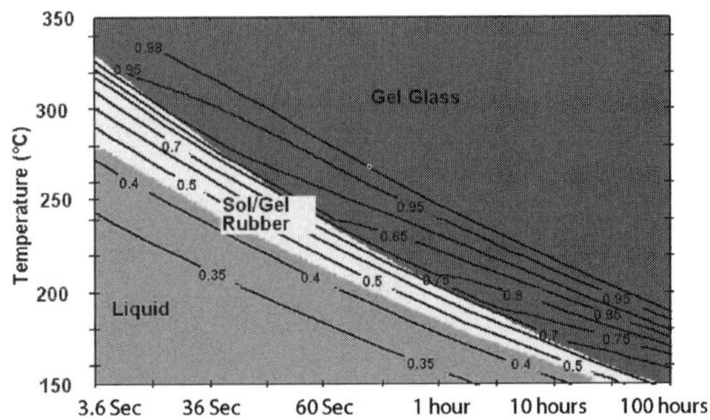

Fig. 10.7 Curing percentage of BCB as a function of time and temperature, adapted from [28, 29]. Partially cured BCB is defined as the BCB cure level reaches 45% or higher, i.e., BCB is rigid and behaves like a rubber or gel glass. From [36] © 1997 Elsevier

prior to bonding, the BCB coating thickness, and the temperature ramping during the wafer bonding process were varied in these experiments. For all experiments, 200-mm-diameter silicon wafers with a nominal thickness of 725 μm were used. The alignment structures on the wafers consist of copper damascene patterns. A wafer bonder from EVGroup was used for the wafer-bonding experiments and a SmartView™ system from EVGroup was used for wafer-to-wafer alignment with the patterned structures. The processing procedure for all these experiments consisted of the following steps:

1. wafer clean and spin–rinse dry,
2. spin-on of the adhesion promoter (AP3000 from the Dow Chemical Company) on both wafers,
3. spin-on of a BCB layer on both wafers,
4. soft-bake (and partial-cure) of the BCB layers,
5. wafer alignment with the SmartView™ system followed by clamping the aligned wafers on the bond fixture (for the alignment experiments),
6. load the bond fixture with the wafers into the bond chamber, followed by pump-down to a vacuum environment of 4×10^{-2} Pa,
7. apply a force of 10 kN with the bond tool to the wafer stack,
8. temperature ramp-up of the bond tool to 250°C for a duration time of 1 h at 250°C while applying a force of 10 kN to the wafer stack,
9. bond force release and cool-down.

10.4.1 Thickness Uniformity of BCB Layers

When using adhesive wafer bonding for wafer-level fabrication of 3D ICs, the uniformity of the dielectric polymer layer at the bond interface is important. A total of four bonding experiments have been performed to investigate the thickness uniformity of the dielectric BCB adhesive after bonding [30]. The thickness nonuniformity

Table 10.2 Process parameters of bonding experiments for monitoring BCB thickness nonuniformity after bonding (data from [30])

BCB total thickness (μm)	BCB treatment prior to bonding	Cross-linking to bonding (%)	BCB thickness nonuniformity after bonding
2.6	Soft-bake	35	~10% of nominal BCB thickness
2.6	Soft-bake + partially cure	43	~0.4% of nominal BCB thickness
0.6	Soft-bake	35	~15% of nominal BCB thickness
0.6	Soft-bake + partially cure	43	~0.4% of nominal BCB thickness

of the BCB coatings prior to bonding, as determined by optical refraction, was quite small (less than 0.3%); for example, about 0.2% for 0.3-μm-thick BCB coatings. The process parameters of the experiments and the resulting thickness nonuniformity of the BCB adhesive after the wafer bonding are listed in Table 10.2.

The thickness uniformity of the BCB layer after wafer bonding was also measured using refraction, after one of bonded silicon wafers was completely removed by mechanical grinding to 50 μm, followed by wet etching using tetramethylammonium hydroxide (TMAH). Since the BCB is quite inert to TMAH, the processing is assumed not to impact BCB thickness uniformity.

It is believed that in the experiments with the soft-baked, low-viscous BCB, the BCB flows and redistributes during the bonding process, thereby potentially creating large thickness nonuniformities. The BCB layer thickness in the experiments with the partially cured BCB does not reflow or redistribute; the low pre-bond layer thickness nonuniformity of about 0.4% of the nominal BCB layer thickness is maintained. This small thickness nonuniformity is desirable if a thick BCB bonding layer and short interstrata vias are used for a specific 3D integration application.

10.4.2 BCB Bond Strength and Impact on Bonding Void and Defects

Bond strengths are determined using the four-point bending technique discussed in Section 10.5.2. The critical adhesion energy of the wafer bonds with 2.6-μm-thick soft-baked BCB as the bonding adhesive were measured to be approximately $31\,\mathrm{J\,m^{-2}}$ for silicon-to-silicon wafer bonding and $9\,\mathrm{J\,m^{-2}}$ for silicon-to-glass wafer bonding [31]. In the four-point bending experiments, the bond failure occurred at the Si–BCB interface in the former and at the glass–BCB interface in the latter case; not at the BCB–BCB interface. The bond strength using partially cured BCB as the bonding adhesive is similar to that using soft-baked BCB (see also Section 10.6). The bond failure during the four-point bending experiments occurred at the interface between substrate and partially cured BCB, indicating that the critical adhesion energy at the interface between the partially cured BCB layers is higher than that at the interface between substrate and partially cured BCB. Therefore, the bond strengths resulting from partially cured BCB bonding are sufficient to withstand the polishing and grinding steps required for 3D IC process integration.

Though the bonding strength for wafer pairs bonded by partially cured BCB is similar to that by soft-baked BCB, it is clearly observed that partially cured BCB is more sensitive to defects or particles at the bonding interface. Partially cured BCB accommodates well the surface topography and provides a uniform bonding intermediate layer; however, voids and defects are often observed at partially cured BCB bonding interfaces. This is because soft-baked BCB reflows during wafer bonding, and is more accommodating to particles or defects at the BCB surface, while partially cured BCB does not reflow. The particles or defects at the BCB surface may be introduced during the necessary wafer-alignment or other wafer-handling processes. Therefore, bonding using soft-baked BCB is more robust for 3D integration than that using partially cured BCB.

10.4.3 Impact on Wafer-to-Wafer Alignment Accuracy

Although bonding with soft-baked BCB is simpler and more robust than bonding with partially cured BCB, and often results in a void-free bonding interface, a large, nonreproducible misalignment ($\geq 10\,\mu$m) is sometimes observed after alignment and bonding of processed wafers [30]. This large misalignment is believed to be induced by wafer bonding, that is, wafers shift relative to each other during bonding. Misalignments of wafers after the SmartView^{TM} aligner are on the order of 1 μm, depending upon details [11, 12, 23, 32]. The shifting of the wafers may be attributed to inhomogeneities in BCB reflow process during the bonding/curing process under compression and/or slight shear forces at the bonding interface.

Without any reflow, minor shear forces at the bonding interface would not induce such large, nonreproducible misalignment. Since partially cured BCB does not reflow, wafer alignment and bonding experiments were performed to evaluate the bonding-induced wafer-to-wafer misalignment using partially cured BCB as the bonding adhesive [30]. The wafer-to-wafer alignment accuracy is measured before and after wafer bonding with infrared microscopy, which has a resolution in the range of $\pm 0.5\,\mu$m. The difference between pre-bond and post-bond alignment accuracy represents the shift of the wafers relative to each other during the wafer-bonding process. The results are plotted in Fig. 10.8 for wafer pairs bonded using a 43% partially cured BCB. The partial cure was done on a hotplate for 30 min at 190°C, with a nitrogen atmosphere to prevent BCB oxidation. In these experiments the mean value of the wafer shift relative to each other during the wafer-bonding process was 0.8 μm at the right and left alignment keys. The dramatic improvement of the bonding-induced misalignment is attributed to the different BCB viscosity during bonding. The partially cured BCB maintains a high viscosity during the adhesive wafer-bonding process as compared to soft-baked BCB; thus, the BCB does not flow much during the wafer-bonding process.

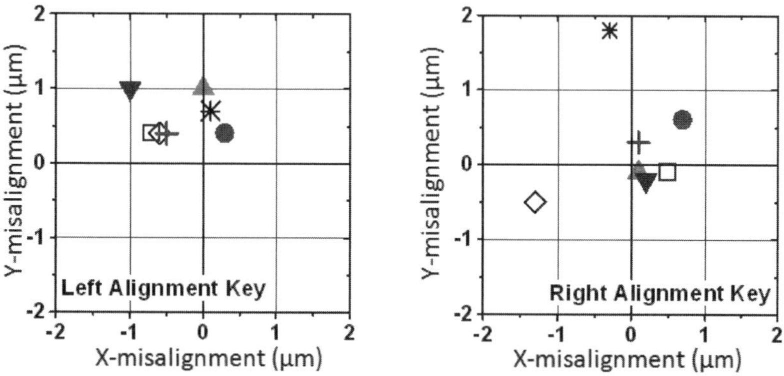

Fig. 10.8 Relative wafer-to-wafer shift during adhesive wafer bonding using partially cured BCB. The left graph represents the left alignment keys and the right graph represents the right alignment keys, adapted from [30]. Reproduced by permission of ECS – The Electrochemical Society 2006

10.5 Integrity Characterizations of Blanket Adhesive Wafer Bonding

It is critical to achieve a robust bond between wafers with minimal bonding voids or defects for any 3D integration platforms. Though such a wafer bond is easier to achieve using blanket adhesive wafer bonding than using other bonding approaches, Fig. 10.2 indicates that challenges still exist. The bond strength, as well as the void and defect densities at the bond interface in adhesive wafer bonding are influenced by the polymer adhesive, wafer materials, diameter, thicknesses and surface topographies, particle densities and their size distributions at the wafer surfaces, polymer thickness, degree of pre-curing (level of polymerization) of the polymer adhesive, as well as the time trajectories of adhesive curing/bonding conditions, that is, pressure, temperature, and bond chamber ambient. Bonding characterizations/evaluations are critical in identifying the issues before steps can be taken to increase the bond strength and avoid voids and defects in the bonds. Several bonding evaluation methods presented below demonstrate the bonding integrity and robustness of wafers bonded using soft-baked BCB [2, 11, 15, 30–38].

10.5.1 Optical Inspection Using Glass Wafer

Bonding defects and voids can be quickly checked by optical inspection if one of the wafers is glass, as shown in Figs. 10.2 and 10.3a. To translate the results reasonably to silicon wafers, for example, the coefficient of thermal expansion (CTE) of the glass wafer used matches that of silicon fairly well. The defect-free bonding of 200-mm damascene-patterned wafers, as shown in Fig. 10.3a, indicates that the nonplanarity of damascene-patterned Cu/oxide via-chain structures

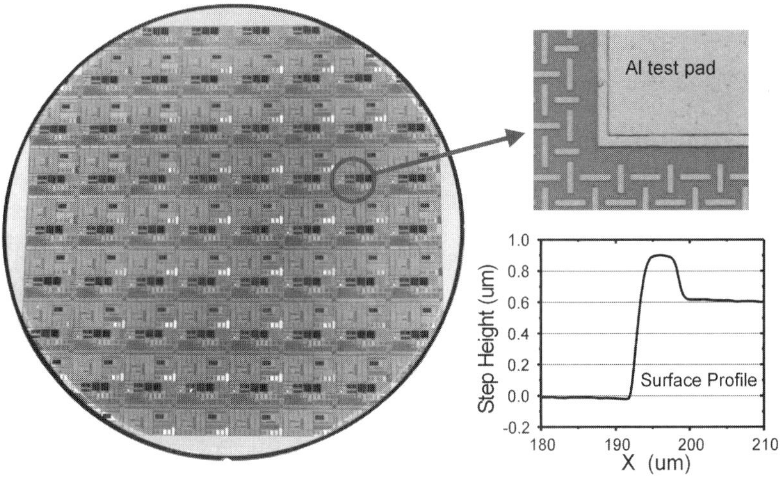

Fig. 10.9 Optical image of a semitransparent wafer with two-level Cu/oxide interconnect structures, bonded to a glass wafer using BCB, taken after the Si substrate was removed from the copper interconnect wafer (processed wafer provided by SEMATECH) [15, 38]. Void-free bonding is obtained and maintained after wafer thinning, although the surface profile (right) shows a step height of ~900 nm across the Al pads. Reproduced by permission of ECS – The Electrochemical Society 2003

is accommodated by a 2.6-μm-thick BCB adhesive layer. In fact, wafers with relatively large topological features are routinely bonded without noticeable void or defect problems. Figure 10.9 shows an optical image of a wafer containing die with two-level copper interconnect test structures, bonded to a glass wafer using BCB, after the silicon substrate was removed from the copper interconnect wafer that was provided by SEMATECH. Though the surface profile shows a step height of ~900 nm across the Al pads on the processed wafers, void-free bonding is obtained and maintained after the silicon substrate was removed from the damascene wafer with the two-level copper interconnect test structures. It is noteworthy that the backside of the damascene wafer has been ground and polished completely away (to the silicon-oxide interface) without adhesion failure of the BCB [11]. The interconnect structures on such wafers are damage-free after wafer bonding and thinning processes.

10.5.2 Bonding Strength Characterization Using Four-Point Bending

In addition to having defect/void-free bonding interfaces, it is critical to obtain high bond strengths (critical adhesion energies). Four-point bending tests can be used to quantify bond strength and to identify the weak bond interface [31, 33, 34]. This technique was developed to measure the critical adhesion energy at the interface in a beam specimen by analyzing a load-displacement curve. The beam specimen is

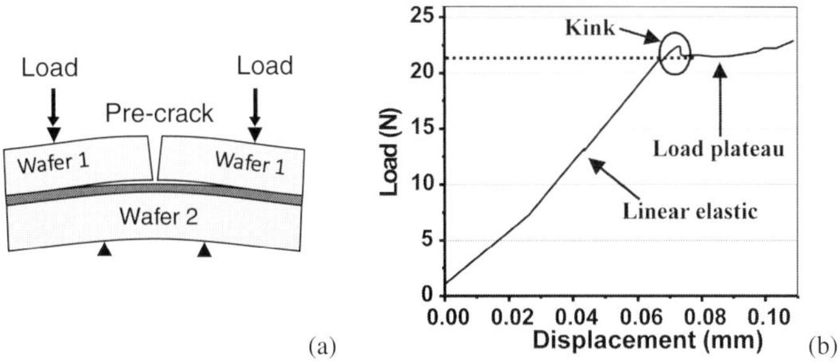

Fig. 10.10 (**a**) Beam specimen geometry for four- point bending experiments with four pins and pre-crack shown. (**b**) Representative four-point bending result: a load-displacement curve [30]. Reproduced by permission of ECS – The Electrochemical Society 2006

mounted in a special fixture (see Fig. 10.10a). In this fixture, a load cell measures applied load, and an actuator measures displacement of the "pins" that hold the specimen. The experiment essentially consists of measuring the load required to maintain a constant displacement rate. During displacement, the pre-crack propagates vertically to the weak interface, after which the crack proceeds along that interface. A typical load versus displacement curve is shown in Fig. 10.10b [34].

Figure 10.11 shows the critical adhesion energy determined using four-point bending for different wafers bonded using BCB [15]. The critical adhesion energy of 32 J m^{-2} measured for BCB bonding of oxidized silicon wafers is well above the value of 22 J m^{-2} determined for BCB bonding of an oxidized silicon wafer to a wafer with Cu interconnect structures with SiO_2 dielectric, and much higher than the value of 6 J m^{-2} for BCB bonding of an oxidized silicon wafer to a wafer with Cu interconnect structures with JSR porous low-κ dielectric [15]. Direct observations indicate that failure occurs within the interconnect test structures with either oxide or JSR porous low-κ dielectrics, and not at either bonding interface. Even with a

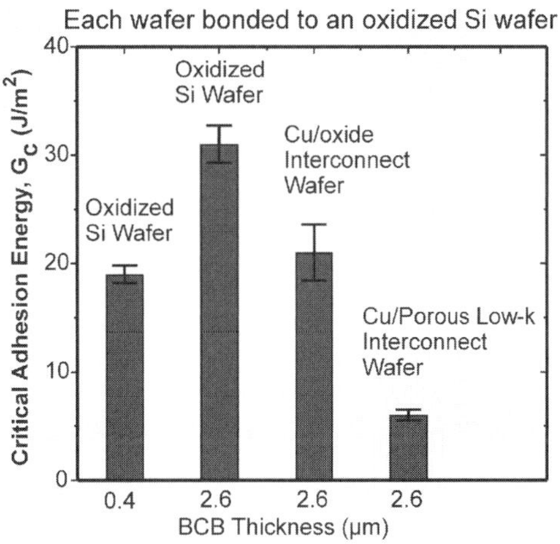

Fig. 10.11 Bond strengths (critical adhesion energies) for BCB-bonded wafers: wafers with oxidized surfaces bonded together, wafers with oxidized surfaces bonded to Cu/oxide damascene wafers, and wafers with oxidized surfaces bonded to Cu/low-κ interconnect damascene wafers (data from [15])

BCB bond layer thickness as thin as 0.4 μm, the measured critical adhesion energy of 19 J m^{-2} is sufficiently strong, that is, much stronger than that of Cu/porous low-κ interconnect structures.

10.5.3 Thermal Cycling Test

Since wafer bonding and post-bonding processing, such as via metallization, involve temperature changes, any bonding interface has to withstand thermal cycling tests at BEOL-processing-compatible temperatures. The effects of thermal cycling on bond strength at the interface between BCB and SiO_2-coated silicon wafers are evaluated by four-point bending and wafer-curvature techniques [34, 35]. Figure 10.12 shows an example of such a thermal cycling test. The SiO_2 films are deposited by plasma-enhanced chemical vapor deposition (PECVD), and the wafers are bonded using the established (baseline) process described in Section 10.2. Thermal cycling is done between room temperature and a maximum or peak temperature. In thermal cycling experiments performed with 350°C and 400°C peak temperatures, the critical adhesion energy increases significantly during the first thermal cycle. The increase in critical adhesion energy is attributed to relaxation of residual stress in the PECVD SiO_2. This relaxation leads to increases in deformation energy due to residual stress and critical adhesion energy [34]. Thermal cycling also cures the BCB beyond the ~88% achieved in the baseline process, and the residual stress in the BCB is reset at the glass transition temperature corresponding to the increased BCB cure conversion. Although BCB bond cannot withstand a thermal cycling at peak cycle temperature of 450°C, it is clear that the BCB bond passes the thermal cycling tests at BEOL-compatible temperatures, which are lower than 400°C.

10.5.4 Packaging Reliability Tests

The wafer bond in any 3D platform has to pass the standard IC package reliability tests. These tests include autoclave, thermal shock, and die sawing. In order to

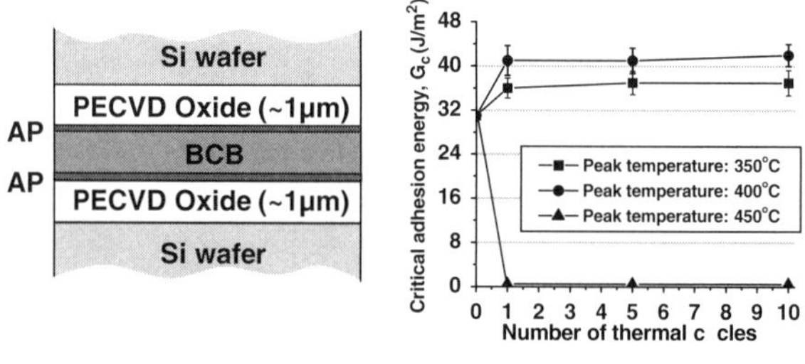

Fig. 10.12 Beam specimen configurations (*left*) and critical adhesion energy (*right*) for 1, 5, and 10 thermal cycles at peak cycle temperature of 350°C, 400°C, and 450°C. BCB layer is sandwiched by PECVD oxide-coated silicon wafers [34]. AP is adhesion promoter. Reproduced by permission of ECS – The Electrochemical Society 2005

perform the packaging reliability tests, two sets of samples are bonded using our baseline bonding process with a nominal 2.6-μm BCB as the bonding adhesive, and sawn into chips [36]. One set of samples consists of two thermally oxidized prime silicon wafers, while another set consists of one thermally oxidized wafer and one SOI wafer with shallow-trench-isolation (STI) patterns. The samples were either put in the autoclave chamber or directly in a carrier in the liquid-to-liquid thermal-shock (LLTS) system. The die-level autoclave tests are conducted at conditions of 100% humidity, 2 atmospheres, and 120°C for 48 h or 144 h, as usually done for packaging reliability tests. The LLTS tests were conducted between –50°C and 125°C for 1000 cycles. The four-point bending technique was used to measure the bond strengths before and after these tests. The high critical adhesion energies of wafers bonded using BCB, as shown in Fig. 10.11, are not degraded after these standard IC package reliability tests. This result indicates that BCB-bonded wafers pass these standard packaging reliability tests. Although BCB is known to absorb moisture (~0.12% uptake by weight at a relative humidity level of 81% [29]), moisture reversibly leaves the BCB when subsequently exposed to a dry environment.

Die edge integrity testing was also performed on bonded wafers using a double-bonding/thinning and BCB ashing procedure as described in Section 10.5.6 and Fig. 10.16 [2,15]. Doubly bonded and thinned wafers with four-level Cu/low-κ CMOS SOI test structures were randomly sawn to evaluate die chipping and cracking. In the post-saw inspections, the die edge roughness is comparable to that obtained with conventional silicon and SOI wafers with varying metal density in the saw path as shown in Fig. 10.13. A focused ion beam (FIB) image near the saw path of a CMOS SOI wafer after double-bonding/thinning and sawing is shown in Fig. 10.14 [37]. No evidence of BCB delamination was observed, as anticipated considering the high critical adhesion energy.

10.5.5 Wafer Thinning

Current wafer thinning of bonded silicon wafers involves mechanical backside grinding, then polishing (CMP), and then either wet etching or reactive-ion etching (RIE). The mechanical grinding and CMP can be a harsh mechanical test of the wafer-bonding integrity, because severe shear forces are applied. If the bond strength of the bonded wafer is low, or there are voids/gaps or defects at the bonding

Fig. 10.13 Top-down view (× 50) of a sawn section of a wafer with four-level Cu/low-k CMOS SOI test structures BCB-bonded to an oxidized prime silicon wafer. From [36] © 2004 IEEE

Fig. 10.14 FIB cross-section near the saw path of a CMOS SOI wafer that is BCB-bonded to an oxidized prime silicon wafer. Note that (1) the silicon substrate was completely removed from the CMOS SOI wafer, (2) a second prime silicon wafer that was BCB-bonded to the top of the CMOS SOI wafer was also completely removed, and (3) the BCB was ashed away. The CMOS devices and circuits on this processed wafer were electrically tested. From [37] © 2003 MRS

interface, the top wafer of the bonded wafer pair will break or shatter during grinding and/or CMP. Moreover, if the wafer edge is not well bonded, backside grinding and CMP can result in a formation of sharp wafer edges. Such edges, combined with damage induced during the grinding, lead to edge chipping and/or wafer breakage. Edge chipping has a deleterious effect on the useful wafer real estate and post-thinning processing.

Figure 10.15 shows a typical pre-etch wafer thinning result for a silicon wafer from a BCB-bonded 200-mm silicon wafer pair [38]. The silicon substrate of the top wafer is thinned from ~725 μm to a mean thickness of 45 μm with a 4-μm standard deviation by mechanical grinding and polishing, as shown in Fig. 10.15a. Similar results are routinely obtained without wafer breakage for wafers bonded by BCB, such as the results shown in Figs. 10.3a and 10.9. A "horseshoe pattern" in the profile is observed in all the thinned wafers, which is attributed to grinding and polishing process nonuniformity. Post-grinding techniques such as polishing and wet or dry chemical etching have been shown to significantly improve the edge quality and increase the wafer survival probability after thinning. Minimal wafer edge chipping (<1 mm), as shown in Fig. 10.15b, has been observed. This excellent edge quality is attributed to the full edge coverage of the spin-on BCB.

10.5.6 Electrical Characterization of Adhesive Wafer-Bonding Integrity

While BCB provides a defect-free bonding interface, sufficient bond strength with high temperature stability, IC packaging compatibility, and insensitivity to surface

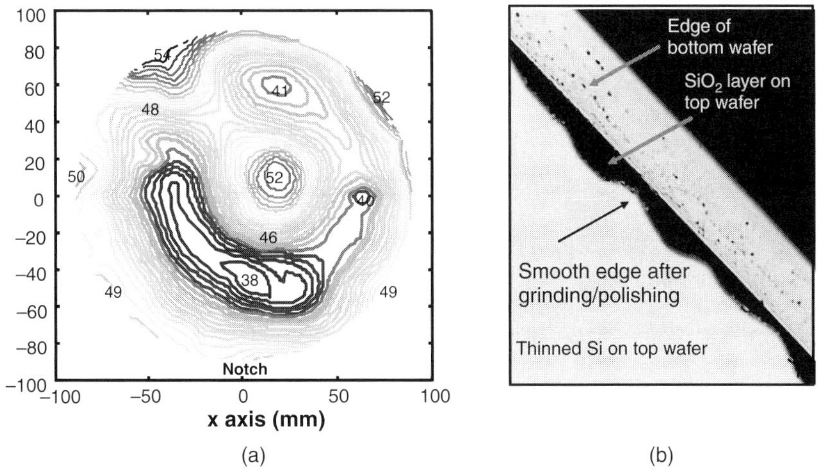

Fig. 10.15 Result of thinning BCB-bonded 200- mm wafers, adapted from [38]. (**a**) Remaining Si thickness contour profile after grinding and polishing to an average thickness of 45 μm (with a 4-μm standard deviation). (**b**) Top view of the edge of two SiO_2/Si wafers bonded using BCB after grinding and polishing the top wafer to ~50 μm. Note the smooth edge of the thinned top silicon wafer over its own oxide layer, which is still bonded by BCB to the bottom wafer. Reproduced by permission of ECS – The Electrochemical Society 2003

conditions (particles, roughness, and planarity), it is also critical to evaluate bonding and thinning impacts on the electrical performance of processed wafers. The impact of these processes was evaluated with a double bonding and thinning procedure [2, 15, 37], as shown in Fig. 10.16. Electrical characteristics before and after this procedure are almost identical for wafers with two-level copper interconnect structures provided by SEMATECH [37] and for wafers with four-level Cu/low-κ CMOS SOI test structures provided by Freescale Semiconductor [15]. An FIB image of a CMOS SOI wafer after this procedure is shown in Fig. 10.14.

Figure 10.17 shows a typical ring oscillator delays before and after double bonding/thinning and BCB ashing for wafers that have state-of-the-art 130-nm technology CMOS SOI test structures with four-level Cu/low-κ (organosilicate glass) interconnects and aluminum bond pads provided by Freescale Semiconductor. Note that the silicon substrate of the CMOS SOI wafer was completely removed during the double bonding/thinning process; only the transistors, circuits, and interconnects on the SOI layer with the BOX layer are BCB-bonded on another silicon wafer (see Fig. 10.14). All changes in electrical parameter values, including other tested parameters not shown here, are less than one-third of the 10%–90% spread in the distribution of original parameter values, indicating that the double bonding and thinning process flow has not significantly affected 130-nm technology CMOS device parameters [37].

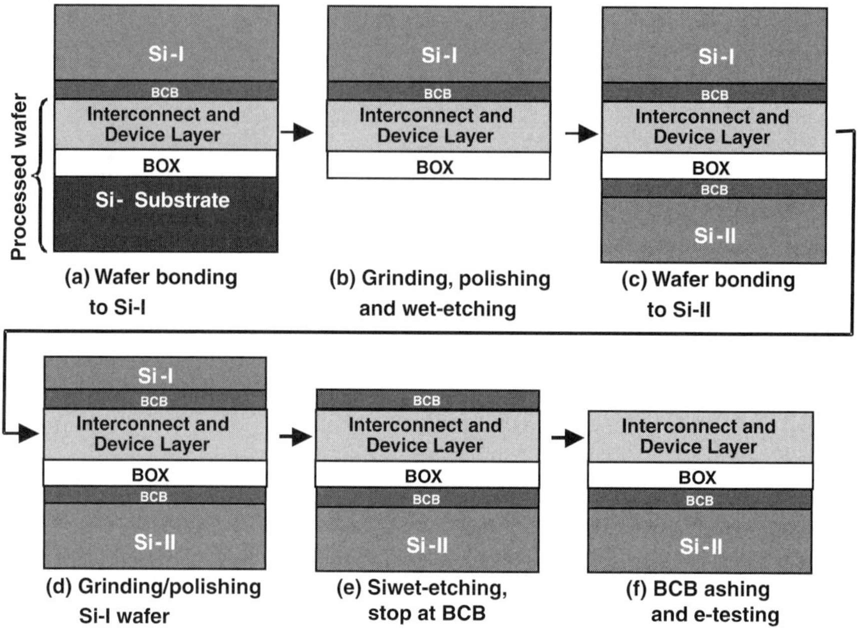

Fig. 10.16 Process flow of a typical double bonding/thinning and BCB ashing procedure, permitting bonding and thinning integrity evaluation without interstrata interconnect processing. From [2] © Springer 2005 and [15, 37]

Fig. 10.17 Ring oscillator delay before and after double bonding/thinning. From [37] © 2003 MRS

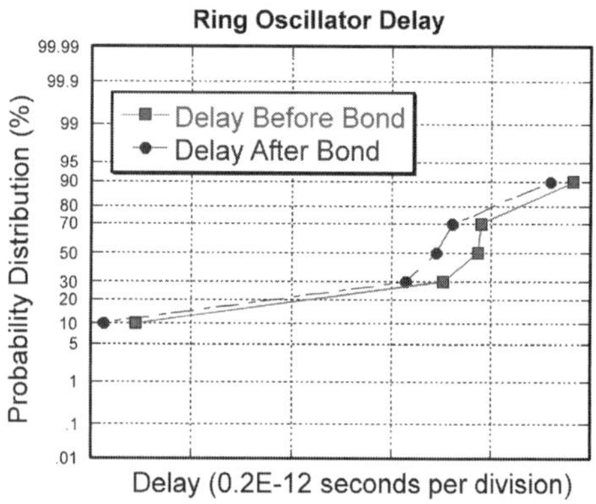

10.6 Integrity Characterizations of Patterned/Processed Adhesive Wafer Bonding

The previous section summarizes characterization methods and results of blanket adhesive wafer bonding, where the soft-baked blanket BCB is used as the bonding adhesive for the via-last 3D platform. The patterned and/or processed adhesive in this section refers to the partially cured BCB that is mainly used for the via-first

3D platform; while the partially cured BCB can also be used for via-last 3D platform if a uniform BCB thickness and precise wafer alignment are required. All the characterization methods used for blanket adhesive wafer bonding, discussed in Section 10.5, can be applied to patterned/processed adhesive wafer bonding. This section summarizes only the characterizations unique to patterned/processed adhesive wafer bonding.

10.6.1 Partially Cured BCB for Copper Damascene Patterning

As mentioned in Section 10.3.2, partially cured BCB is selected for the via-first 3D platform using wafer bonding of patterned Cu/BCB redistribution layers. To do so, the feasibility of using partially cured BCB for copper damascene patterning with commercially available CMP slurries was evaluated. For this purpose, BCB was spun onto 200-mm wafers and cured at temperatures ranging from 190°C to 250°C, providing a wide range of cross-link percentage by changing the cure time. Commercially available copper CMP slurries were used for application to these films; responses considered were removal rate, surface damage (surface scratching and embedded abrasives), and planarity [39].

For a partially cured BCB with a cross-link percentage of about 45±5%, cured at 190°C for 30 min, 220°C for 10 min, and 250°C for 1 min, optical inspection revealed that nearly all of the BCB cured at 190°C was removed after 1 min CMP, while the films cured at 220°C and 250°C were still intact. Further inspection showed very little scratching of the high-temperature cured films, with only 3 of 18 wafers polished having macroscopically visible scratches. Subsequent tests were conducted on films cured at a temperature of 250°C for various times. Figure 10.18 shows film thickness data versus time at nine points within 180 mm of 200-mm wafers, after polishing a 85% cross-linked film [39]. Using the central nine points of the map, the removal rates of 95%, 85%, and 50% cross-linked BCB are 11.1±6.6, 13.7±8.0, and 112±33 nm/min, respectively. The CMP removal process is edge-fast, meaning that removal rates are higher at the edge of the wafer. It is possible

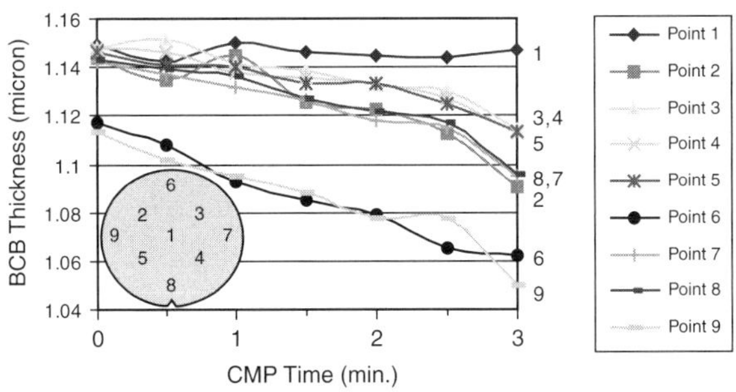

Fig. 10.18 BCB thickness data over CMP time of 85% cross-linked BCB undergoing CMP in a mixture of EKC Microplanar 9003 and 9011, a barrier slurry formulated to be selective to the barrier relative to copper. Nonuniform removal rate across the wafer is observed [39]

that nonuniformities in the bake temperature (and thus final cross-link percentage), and/or the CMP process, contribute to this nonuniform removal rate. These results indicate that a BCB partially cured at 250°C with a cure percentage of higher than 50%, withstands the CMP process. Though the process needs to be optimized for practical applications, CMP can be used for Cu/BCB damascene patterning, which is a critical process step for via-first 3D integration.

10.6.2 Bond Strength and Voids/Defects

As mentioned in Section 10.4.2, the bond strength for wafer pairs bonded with partially cured BCB is similar to that with soft baked BCB, at least for low extents of pre-curing. Bond strengths and voids/defects at the bonding interface were evaluated for various cross-link percentages, surface treatments such as CMP, brush cleaning, and nitrogen plasma exposure, as well as different bonding conditions [10]. Except for nitrogen plasma treatment, the bond strengths for all cases are similar to that using soft-baked BCB. However, voids/defects are observed. Careful measures such as cleanness control to avoid particles introduced prior to wafer bonding, BCB curing in a controlled nitrogen environment to prevent BCB oxidation during partial cure process, post-CMP vacuum bake, and topography control of damascene-patterned Cu/BCB films, have to be taken to prevent voids and defects at the bonding interface. The issue with topography control for damascene-patterned Cu/BCB film is discussed in next section.

10.6.3 Surface Topography of Damascene-Patterned Cu/BCB Layer

Surface topography of damascene-patterned copper conductors with tantalum liners and partially cured BCB plays a role in the bondability of patterned wafers. Surface profilometry has been used to evaluate the step heights of various features on patterned wafers, and its impact on bonding uniformity [10, 21, 39, 40]. Step heights over patterned features, going from copper to partially cured BCB, varied from 60 nm to 120 nm over the radii of 200-mm wafers in a bonded pair. Bonding results indicate that good bonding can be obtained when step heights are smaller than 60 nm. Further work evaluated the effect of wafer-scale topography by employing a nonideal die layout as shown in Fig. 10.19 [21, 40]. Die with Cu/BCB test structures were placed with closely spaced rows, but widely spaced columns. Results show that the BCB field was high relative to the BCB/Cu-patterned regions. This non planarity was approximately 500 nm over a distance of several die, and as much as 2.5 μm over the entire mapped area of $100 \times 100\,\text{mm}^2$, centered on the wafer. To facilitate wafer bonding using damascene-patterned Cu/BCB layer, this large nonplanarity must be avoided by tightly packing the Cu/BCB patterns with a high copper density; perhaps using dummy copper vias.

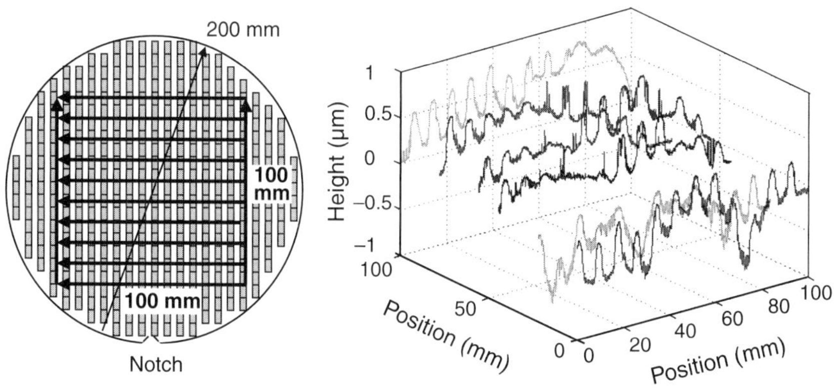

Fig. 10.19 (*Left*) Die layout and profilometry scans over a 200-mm wafer for wafer-level nonplanarity study. (*Right*) Wafer-level nonplanarity for CMP over a nonideal die layout. Adapted from [21] © 2006 MRS and [40] © 2005 SMTA

10.7 Feasibility Demonstrations of Wafer-Level 3D Integration

With the integrity characterizations of wafer bonding using soft-baked BCB and partially cured BCB as discussed in previous sections, this section summarizes the feasibility demonstrations of the two wafer-level 3D integration platforms using 3D via-chain structures.

10.7.1 Via-Last 3D Platform Feasibility Demonstration

An interstrata interconnect via-chain structure was designed in order to develop 3D unit processes and demonstrate the feasibility of our via-last approach to 3D processing with 200-mm wafers [2, 12, 14, 24]. Figure 10.20 shows the process flow for the via-chain structure fabrication. Conventional Cu/oxide damascene BEOL processes are used for Cu metallization on the bottom wafer (M1) and the top wafer (M2), while bridge metal (M3) and interstrata plug and bridge vias are fabricated using an interstrata damascene process. Based on screening tests of various dielectric bonding adhesives, soft-baked BCB was selected as the baseline wafer bonding. The three-step thinning process was used to thin the backside of top silicon wafer, that is, (1) silicon substrate mechanical grinding, (2) polishing to remove damage and stresses caused by grinding and reduce top wafer thickness to ~35 μm, and (3) wet chemical etch to remove the remaining silicon. TMAH is used to etch silicon due to its compatibility with CMOS processing and excellent etch selectivity for silicon relative to SiO_2. The interstrata interconnects are formed by deep via etching and cleaning, CVD of TaN liner, Cu CVD fill, and CMP. Additional process information on the 3D via-chain fabrication is presented elsewhere [24].

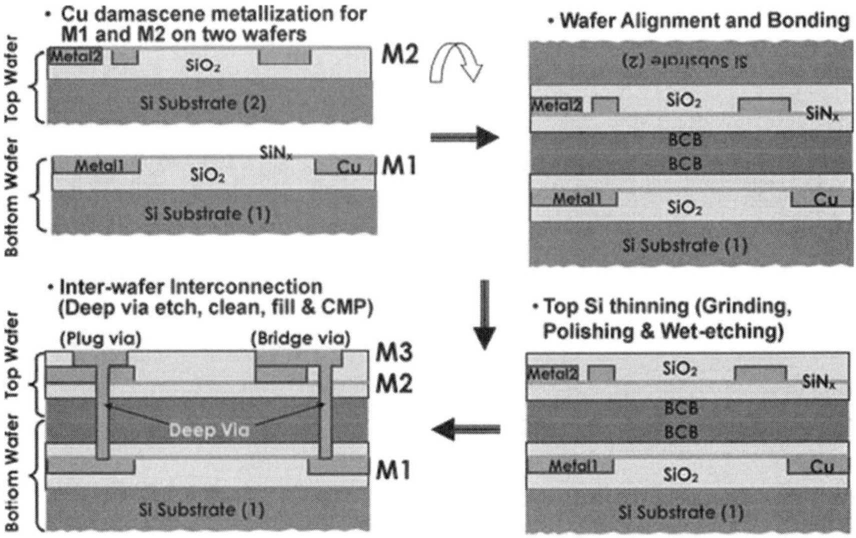

Fig. 10.20 Via-chain process flow for demonstration of via-last 3D integration using adhesive wafer bonding and copper damascene patterning [23, 24]. Reproduced by permission of ECS – The Electrochemical Society 2003

As shown in Fig. 10.21, two Cu damascene-patterned wafers with soft-baked BCB after face-to-face aligning, bonding, and thinning of the top wafer reveal wafer-scale and die-scale alignment within 1 μm [11, 24, 42]. These promising results indicate that a very high density of interstrata interconnects (via diameter ~2 μm) is feasible with this via-last 3D platform. There was no difference in the patterned

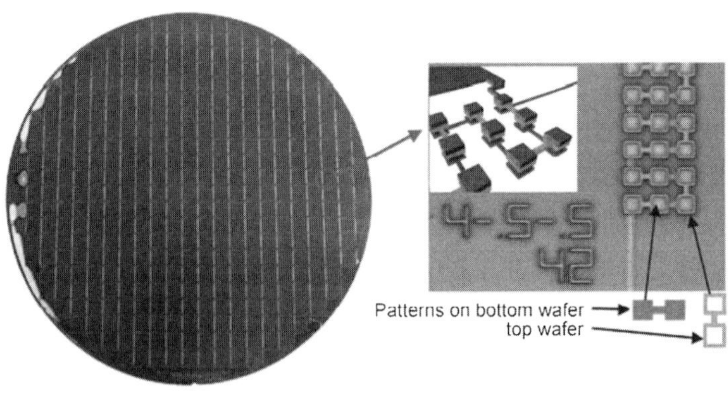

Fig. 10.21 Image of 3D via-chain test structures on 200-mm wafer after Cu-damascene patterning on two wafers, wafer alignment, bonding using soft-baked BCB, and thinning of the top wafer using backside grinding/polishing and wet etching. The irregular patterns near the wafer edge were induced by wafer-scale nonuniformities of the processing tools used for oxide etch and Cu sputter deposition. From [41] © 2004 MRS and [42]

wafer surface quality before and after top silicon wafer thinning to ~35 μm by mechanical grinding and CMP [42]. The wafer pair showed good silicon thickness uniformity without edge cracks, revealing good bonding and thinning integrity of the stacked wafers.

An FIB cross-section and via-chain resistances from a functional wafer pair are shown in Fig. 10.22 [2, 24]. Though the via-chain-specific contact resistance of ~5 μΩ-cm^2 is much larger than expected, the linear relationship between the chain resistance and chain length indicates that continuous and uniform 3D via-chains were fabricated for nominal interstrata via diameters of 2, 3, 4, and 8 μm. The high contact resistance is attributed to inadequate cleaning of via-etch residue prior to metallization.

10.7.2 Via-First 3D Platform Feasibility Demonstration

A 3D via-chain structure similar to that used for the via-last 3D platform demonstration is used to develop the unit processes and demonstrate the feasibility of the via-first 3D integration with wafer bonding of Cu/BCB redistribution layers [10, 20, 21]. The process flow for the via-first 3D via-chain fabrication is developed based on that for the via-last 3D via-chain fabrication. First, a redistribution layer of damascene-patterned Cu in partially cured BCB was fabricated on each of two 200-mm silicon wafers as follows. After a 2-μm-thick thermal oxide was grown, BCB was spun onto these substrates to a nominal thickness of 1.2 μm. These films were then partially cured at 250°C for 60 s in a coat/bake track modified to include a nitrogen purge, resulting in ~55% BCB cross-linking. The partially cured BCB was photolithographically patterned using an i-line stepper, and then etched in an inductively coupled plasma (ICP) etcher using C_4F_8 and oxygen as reactive species. The Ta liner and Cu were sputtered over the patterned BCB at a low power levels suitable

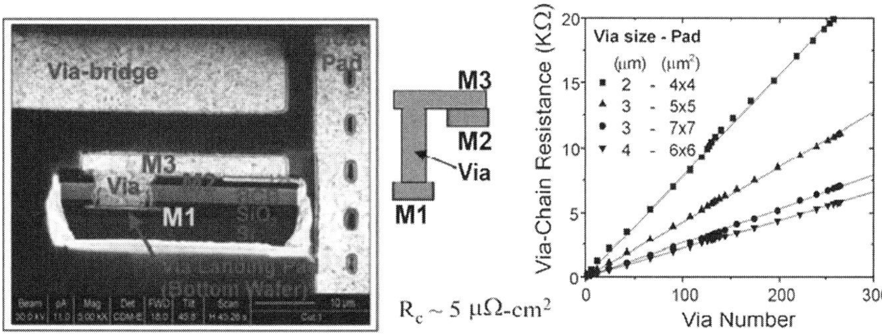

Fig. 10.22 Feasibility demonstration of via-last 3D platform with a via-chain structure: (*left*) a FIB cross-section of the via-chain structure showing Cu metallization: bottom wafer metal (M1), top wafer metal (M2), bridge metal (M3), and the interstrata vias; (*right*) via-chain resistance vs. via-chain length (via number) for nominal via diameters of 2, 3, and 4 μm. From [2, 24] © 2005 Springer and [24]

for deposition onto a polymer dielectric. CMP was carried out on these films using conventional rotary CMP equipment, with commercially available slurries and pads, until the patterns were well-defined. Post-CMP cleaning was done with deionized water and PVA brushes. These wafers were then aligned and subsequently bonded under vacuum. The bond process used a mechanical downforce of 10 000 N and a specially designed temperature profile to bond both BCB and Cu. This temperature profile includes a temperature ramp to 250°C and "soak" for 60 min to complete the BCB–BCB bonding, followed by a ramp to 350°C and soak for 60 min to produce the Cu–Cu bond, followed by cooling to room temperature. One of the bonded wafers was thinned to a nominal thickness of 50 μm using grinding and polishing, followed by a wet chemical etch in TMAH, which has high Si-to-SiO_2 selectivity, to completely remove the remaining silicon. Figure 10.23 shows four plan-view optical photographs of the fabricated via-chain test structures, showing a wafer-to-wafer alignment accuracy of better than 2 μm [21].

After the above fabrication procedure, the bonded wafer pair was sectioned, allowing a lot split for characterization of structural and electrical properties. The structures were observed using cross-sectional FIB/SEM. One resulting cross-section of the bonded area near the wafer center is shown in Fig. 10.24 [10, 20]. This figure shows a bonded Cu-to-Cu interface, a bonded BCB-to-BCB interface, a Cu-to-BCB interface that appears to be in intimate contact but may not be well-bonded across the wafer, and voids at the copper-to-copper interface and at the Cu-Ta-BCB interface. A twin boundary is seen penetrating the original Cu–Cu bonding interface, indicating that intimate contact between the two copper structures has been achieved.

For electrical characterization, the surface oxide (originally the isolation layer under the Cu/BCB redistribution layer) is removed to allow access to the via-chains.

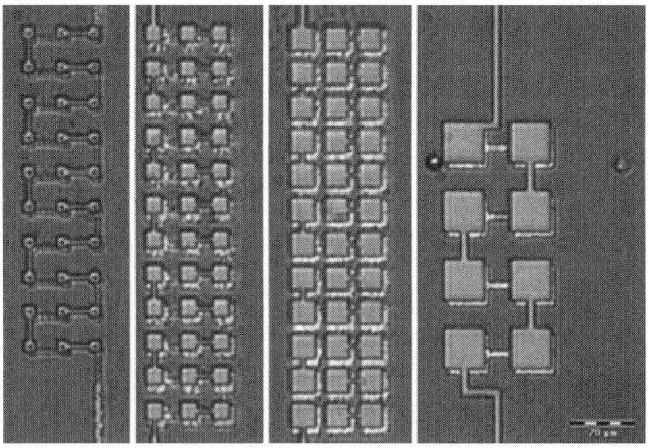

Fig. 10.23 Four plan-view photographs of via-chain test structures after complete removal of the top silicon substrate, showing an alignment accuracy of better than 2 μm. Via pad sizes vary from 3 μm to 16 μm. Tantalum remains on via pads from top wafer, and copper interconnect on bottom wafer is visible. Views are from different areas of the wafer with same scale bar of 20 μm. Adapted from [21] © 2006 MRS

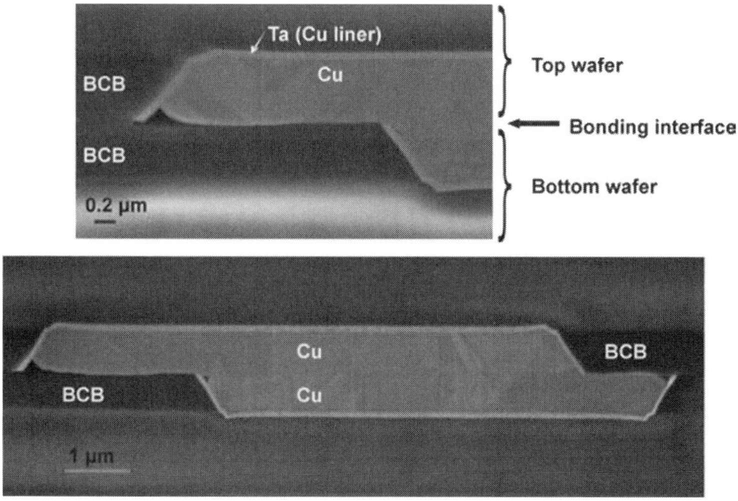

Fig. 10.24 Cross-sectional FIB/SEM image of bonded damascene-patterned Cu/BCB wafers, showing well-bonded Cu-to-Cu and BCB-to-BCB interfaces. Adapted from [10] © 2005 IEEE and [20]

A two-point probe setup was used to measure electrical resistance on several inter-strata vias, and optical microscopy was used to measure the overlap area. The measured results are shown in Fig. 10.25, demonstrating that continuous via-chains were fabricated. Specific contact resistance was measured on the order of 10^{-7} Ω-cm^2, which is higher than previously reported values for bonded evaporated copper [21].

These results are quite promising, and demonstrate the feasibility of this via-first 3D approach. However, key challenges need to be overcome and are currently under investigation. These challenges include (1) optimizing the BCB partial-curing process, to provide sufficient bond strength and serve as a damascene-patterning-compatible dielectric material for Cu/BCB redistribution layer fabrication, (2) achieving wafer-level, feature-scale planarity of the Cu/BCB redistribution layer fabricated by single-level CMP, (3) developing post-CMP surface treatment and wafer-bonding schemes to facilitate bonding of Cu-to-Cu, BCB-to-BCB, and

Chain Size (µm)	Number of Contacts	Resistance (Ω)	Contact Area (cm^2 × 10^{-8})	Specific Contact Resistance (10^{-7}Ω-cm^2)
4	6	20.5	5	1.7
8	2	0.35	63	1.1
8	8	1.18	63	0.9
8	32	14.0	63	2.7

Fig. 10.25 (*Left*) Electrical testing of 12 µm pads (8-µm via interconnects). (*Right*) Summary of several tested structures bonded in a vacuum environment at 10 000 N and a temperature profile of 250°C for 60 min followed by 350°C for 60 min. From [21] © 2006 MRS

Cu-to-BCB over the entire wafer in one unit process step, (4) lowering the electrical resistivity of the interstrata Cu interconnection, and (5) developing reliable redistribution layer design and fabrication protocols.

10.8 Thermomechanical Modeling

A finite element (FE)-based analysis was used to determine if there are potential reliability concerns due to thermally induced stresses in interstrata copper via structures in 3D ICs when BCB is used as the dielectric adhesive to bond wafers [43]. The FE model was first partially validated by comparing computed results against two types of experimental data from planar ICs:

1. *Volume-averaged thermal stresses measured by X-ray diffraction in an array of parallel Cu lines passivated with TEOS*: This study showed the quantitative predictability of the modeling approach used.
2. Studies of failures induced by thermal cycling via chain structures embedded in SiLK or SiCOH. This work showed that SiLK-based via chains would fail, while SiCOH via chains would not fail; both results agree with IBM data [44].

The modeling approach is then employed to study thermal stresses in interstrata Cu vias in 3D IC structures bonded with BCB.

The FE simulations were performed to calculate stresses in a representative "unit cell". The model structure consists of seven layers with embedded, regularly spaced, Cu interconnects (Fig. 10.26). In order to derive design-related parameters and make the simulation tractable, a repeatable unit cell in the 3D IC structures was selected. Moreover, due to symmetry in the x–z and y–z planes (Fig. 10.26), only one-quarter of a unit cell is used in the simulations. The structure is assumed stress-free at 250°C (wafer-bonding temperature using BCB). Circular Cu vias, with an assumed 23.6-μm height, connect 10-μm-thick multilevel metal (MLM) on-chip interconnects (L4, L7) in the two wafers. Benzocyclobutene Young's modulus is found to decrease with temperature from 2.5 GPa at 25°C to 0.3 GPa at 180°C [28].

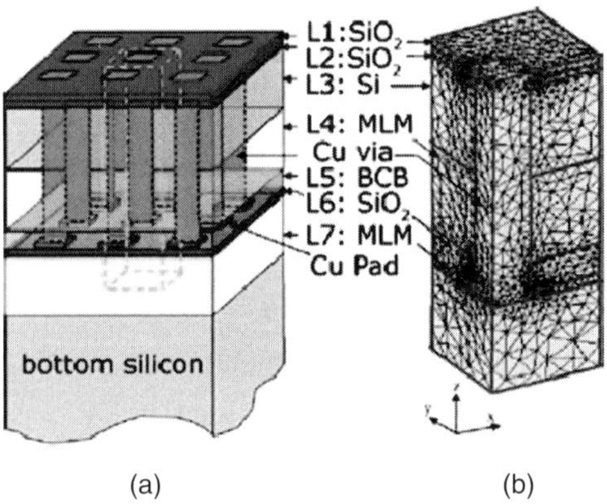

Fig. 10.26 Model 3D structure. (**a**) Several cylindrical vias connecting square landing pads through a layered structure (not to scale). Dashed block is a unit cell. (**b**) Representative FE mesh. Taking advantage of symmetry, only one quarter of a via and pad has been included in geometric model. From [43] © 2006 IEEE

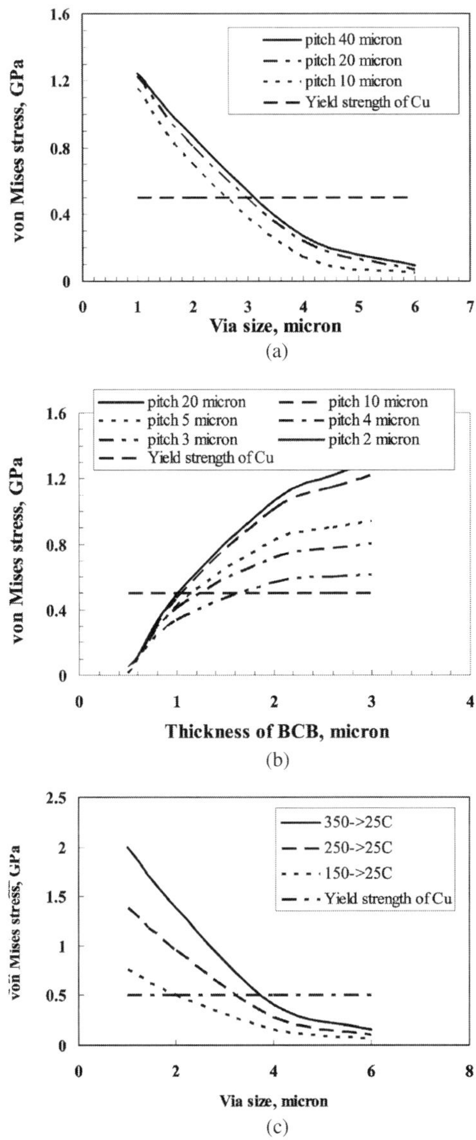

Fig. 10.27 Computed maximum von Mises stress in Cu via [43]: (**a**) different via size and pitches for BCB thickness of 2.6 μm; (**b**) 1-μm Cu vias with different BCB thicknesses and via pitches; and (**c**) different via size cooling from 150°C, 250°C, and 350°C to 25°C. Note that the horizontal dashed lines refer to the lower limit of yield strength (500 MPa) for a conservative design. From [43] © 2006 IEEE

To be conservative, we use the value at 25°C, which will tend to overpredict the stresses in the Cu; these stresses are induced. Benzocyclobutene has a much higher CTE than Cu (the problem with SiLK relative to SiCOH in the IBM studies). Other parameters, material properties, and mechanical boundary conditions are described elsewhere [43]. The FE mesh is refined in areas where large stress gradients are expected.

Figure 10.27 shows the maximum von Mises stress computed in Cu vias 23.6 μm in height for different via sizes (diameters), via pitches, BCB thickness and temperature changes. The calculated von Mises stress decreases with increasing via size and decreasing via pitch, and increases with BCB thickness. A Cu yield strength of 500 MPa is used to be reasonably conservative. As shown in Fig. 10.27a, for a BCB thickness of 2.5 μm, the via diameter should be larger than 3 μm at a pitch of 10 μm, and 3.5 μm at a pitch of 40 μm.

For a via diameter of 1 μm, as shown in Fig. 10.27b, the simulations show that the BCB thickness should be less than 1 μm to avoid plastic yield of Cu vias for pitches larger than 3 μm. This is due to the relatively high CTE of BCB compared to Cu. By reducing the BCB thickness and via pitch, plastic yield of Cu vias can be avoided.

Our simulations also show that the magnitude of temperature change is a major factor determining the stress level in the interstrata Cu vias. Fig. 10.27c shows computed stresses assuming different initial stress-free temperatures, after cooling down to room temperature. As expected, the stress increases as the initial temperature increases if the structures cool down to room temperature. Larger via size is desired to prevent the plastic yield of Cu vias.

In summary, simulations indicate that there is a concern regarding the stability of interstrata Cu vias. Guidelines for design rules using these results are estimated, for example, interstrata via size, pitch, and BCB thickness. For 2.6-μm-thick BCB and 23.6-μm-long Cu vias, computations indicate that via size (diameter) should be larger than 3 μm at a pitch of less than 10 μm to avoid plastic yield of Cu vias. The results also support the goal of using thin BCB. In general, thin BCB has been targeted in order to keep via aspect ratios down; however, it also reduces the via stresses due to BCB contractions and expansions. Further simulations with more accurate physical parameters, and perhaps improved models for interfaces, are needed to better understand failure of the interstrata vias. Simulation work also indicates that stresses in the thinned Si layers are an issue. Though initial simulation work makes dielectric buffer films look promising, this topic deserves more attention. More generally, the impacts of stresses on interconnect electromigration and on the performance of devices surrounding the vias would be of interest for both via-first and via-last 3D platforms.

10.9 General Discussions on 3D Platforms and Applications

Based on 3D technology developments to date, it is clear that the key advantage of the via-last 3D approach is that the wafer-bonding process is separated from interstrata via formation. Therefore, a robust mechanical/chemical wafer bond, such as using BCB as shown in Fig. 10.5, can be easily achieved compared to the via-first 3D approach. The key concern for the via-last 3D approach is that the TSV density could be low because of the large via diameter required for long TSVs to pass through the thinned silicon substrate, particularly if a bulk silicon wafer is used. If an SOI wafer is used for the top wafer, along with a thin adhesive bonding layer, a high density of short TSVs can also be achieved using BEOL damascene patterning through the BOX/SOI layer. This is because the silicon substrate of the SOI wafer can be completely removed, leaving only the BOX/SOI layer that has a thickness of perhaps a few hundred nanometers [36, 45].

Moreover, this simpler via-last 3D platform is particularly attractive for applications where a limited density of TSVs are needed, such as a 3D memory stack, mixed signal/RF circuitry with separate passive stratum stacked on CMOS circuitry stratum, and other heterogeneous sensor systems; for example, MEMSs, optical MEMS, bio-MEMS, and bio/chemical/mechanical sensors.

The key advantages of the via-first 3D approach are (1) a high density of "bond vias" can be formed at the bonding interface, (2) the formation of bond vias and wafer bonding are done in one unit process, and (3) redistribution layers and/or thermal management layers can be introduced at each and every interstrata interface. Particularly, as shown in Fig. 10.6, the metal/adhesive via-first 3D platform offers combined process integration advantages of metal-to-metal electrical bonding (e.g., Cu-to-Cu) with the increased adhesion strength and robustness of dielectric adhesive bonding (e.g., using partially cured BCB). A redistribution layer formed by damascene-patterned metal/adhesive structure provides additional interstrata signal/power routing as well as alleviation of the wafer-alignment tolerance [20]. Planarity requirements are stringent, but appear to be reasonable. However, if the top wafers are bulk silicon wafers, the density of TSVs, which interconnect the bonded wafers (i.e., the top strata) and bring the I/Os to and from the stack, is nearly as limited as with the via-last 3D platform. As in the case of the via-last 3D platform, an SOI wafer is a better choice for the top wafer in the 3D stack. Silicon-on-insulator wafers can be thinned to less than 1 μm since the backside etch can stop at the BOX layer; therefore, short TSVs can be formed using BEOL damascene-patterning process as for bond vias.

Though the use of SOI wafers obviously makes the wafer thinning and TSV formation much easier (and hence low-cost), the choice to use SOI or bulk silicon wafer depends on starting wafer cost, overall processing protocols, components in the system (Si CMOS logic, SRAM, DRAM, NAND, SiGe RF, etc.), and IC manufacturer preferences. Further research and development is necessary to fully evaluate these promising platforms.

These 3D integration technology platforms open a door for a variety of applications [46], such as for optical interconnects [8, 9], memory stacks [47], memory-intensive microprocessors [48], 3D power delivery [6, 7], and software radio implementation using SiGe BiCMOS pipelined A/D converters [4, 5]. For example, in a 3D optical interconnect implementation an optical-only stratum can be used to distribute an optical clock across the die area, with 45° mirrors to direct the optical beam through the interlevel dielectrics in the CMOS strata [8, 9]. As a result, only photodetectors need to be added in the CMOS strata. In a 3D power delivery implementation, a DC–DC converter stratum (possibly with a separate stratum of high-quality passives) can be used to distribute prime power on a cellular basis directly to the IC chip [6, 7]. Therefore, the parasitics associated with long interconnects between the power conversion system and the IC chip in conventional power delivery system can be minimized; wide control bandwidth of high-switching frequency is possible, enabling dynamically scalable voltages; and power/ground package pin counts (I/Os) can be significantly reduced. We believe that power delivery with such wide control bandwidth point-of-load (PoL) converters is very promising for future 3D systems on a chip (SoC) implementation, including multicore and multifunction applications.

Looking forward, the first "3D killer applications" will be extremely high-density heterogeneous memory stacks (e.g., NAND, SRAM, DRAM, FRAM, and phase-change nonvolatile memory) and extremely high-resolution/low-cost imagers, followed perhaps by 3D integration of mixed signal/RF circuits with digital CMOS. Advanced handheld devices will also continue to drive the mix and match of various 3D approaches [49].

A variety of 3D approaches may be combined to offer more flexible integration with even higher functionality, signal integrity, or added value. For example, stacking discrete die (or devices, such as solid-state lasers for photonics or GaN transistors for power switch or microwave applications) onto a wafer stack provides a pathway for heterogeneous integration of compound semiconductor devices or analog circuits with digital circuits.

Once 3D integration technology is mature and the design and manufacturing infrastructure (such as ECAD tools, fabrication equipment, and standards) is in place, it is likely that more ICs will be designed for general purposes, because massive production of 3D ICs would lower the manufacturing cost for complex ICs such as microprocessors and SoCs. For instance, a general-purpose 3D multicore processor with extremely high-density memory, FPGA, and software in the nonvolatile memory in the same stack could be reprogrammed by FPGA or software for multiple purposes, or varying purposes, over a processor's lifetime.

10.10 Conclusions

Various evaluations indicate that adhesive wafer bonding is a mechanically and electrically robust option for 3D integration. Two viable 3D platforms, which are denoted as the "via-last" and "via-first" 3D platforms, are demonstrated. Various unit processes and characterization methods are developed for these two 3D platforms, and can also be applied to other 3D platforms with little modifications if needed. The via-last 3D process flow established BCB as a viable bonding adhesive and simplifies the bonding process; the via-first 3D process flow incorporates an interstrata redistribution layer, which introduces considerable flexibility. These 3D technology platforms are promising for heterogeneous integrations of future systems.

Acknowledgments Rensselaer 3D integration research programs were supported by DARPA, MARCO, and NYSTAR through the Interconnect Focus Center, as well as by SRC, EVGroup, IBM, Freescale, and SEMATECH. We gratefully acknowledge the contributions from many colleagues associated with the 3D group at Rensselaer and from our collaborators at the University at Albany, EVGroup, IBM, Freescale Semiconductor, and SEMATECH.

References

1. Niklaus F, Stemme G, Lu J-Q, Gutmann R (2006) Adhesive wafer bonding. Journal of Applied Physics (Applied Physics Review – Focused Review), 99, 031101(3): 031101-1–28
2. Lu J-Q, Cale TS, Gutmann RJ (2005) Wafer-level three-dimensional hyper-integration technology using dielectric adhesive wafer bonding. Materials for information technology: In: Zschech E, et al (eds) Devices, Interconnects and Packaging, Springer-Verlag (London) Ltd., London, pp 386–397
3. Burns J, McIlrath L, Keast C, Loomis A, Warner K, Wyatt P (2001) Three-dimensional integrated circuits for low-power, high-bandwidth systems on a chip. In: IEEE Intl. Solid-State Circuit Conf. (ISSCC 2001), pp 268–269
4. Gutmann RJ, Zeng AY, Devarajan S, Lu J-Q, Rose K (2004) Wafer-level three-dimensional monolithic integration for intelligent wireless terminals. J Semicond Technol Sci, 4(3): 196–203
5. Lu J-Q, Devarajan S, Zeng AY, Rose K, Gutmann RJ (2004) Die-on-wafer and wafer-level three-dimensional (3D) integration of heterogeneous IC technologies for RF-microwave-millimeter applications. In: Cho YS, et al (eds) Materials, Integration and Packaging Issues for High-Frequency Devices II, MRS Proc. vol 833, pp G6.8.1–G6.8.6
6. Lu J-Q, Sun J, Giuliano D, Gutmann RJ (2006) 3D Architecture for power delivery to microprocessors and ASICs. In: The 3rd International Conference on 3D Architectures for Semiconductor Integration and Packaging, Burlingame, CA
7. Sun J, Lu J-Q, Giuliano D, Chow P, Gutmann RJ (2007) 3D Power delivery for microprocessors and high-performance ASICs. In: 22nd Annual IEEE applied power electronics conference and exposition (APEC 2007), pp 127–133
8. Lu J-Q, Jindal A, Persans PD, Cale TS, Gutmann RJ (2003) Wafer-level assembly of heterogeneous technologies. In: Digest of International Conference on Compound Semiconductor Manufacturing Technology (GaAs MANTECH 2003 Scottsdale), pp 91–94
9. Persans PD, Ojha M, Gutmann RJ, Lu J-Q, Filin A, Plawsky J (2004) Optical interconnect components for wafer level heterogeneous hyper-integration. In: Carter RJ, Hau-Riege CS, Kloster GM, Lu TM, and Schulz SE (eds) Materials, Technology, and Reliability for Advanced Interconnects and Low-k Dielectrics, MRS Proc. vol 812, pp F6.11.1–F6.11.5
10. McMahon JJ, Lu J-Q, Gutmann RJ (2005) Wafer bonding of damascene-patterned metal/adhesive redistribution layers for via-first 3D interconnect. In: 55th IEEE Electronic Components and Technology Conference, pp 331–336
11. Lu J-Q, Kwon Y, Kraft RP, Gutmann RJ, McDonald JF, Cale TS (2001) Stacked chip-to-chip interconnections using wafer bonding technology with dielectric bonding glues. In: IEEE International Interconnect Technology Conference (IITC 2001), pp 219–221
12. Lu J-Q, Lee KW, Kwon Y, Rajagopalan G, McMahon J, Altemus B, Gupta M, Eisenbraun E, Xu B, Jindal A, Kraft RP, McDonald JF, Castracane J, Cale TS, Kaloyeros A, Gutmann RJ (2003) Processing of inter-wafer vertical interconnects in 3D ICs. In: Melnick BM, Cale TS, Zaima S, T. Ohta T (eds) Advanced Metallization Conference 2002 (AMC 2002), MRS Proc vol V18, pp 45–51
13. Ramm P, Bonfert D, Ecke R, Iberl F, Klumpp A, Riedel S, Schulz SE, Wieland R, Zacher M, Gessner T (2002) Interchip via technology by using copper for vertical system integration. In: McKerrow AJ, Shacham-Diamand Y, Zaima S, Ohba T (eds) Advanced Metallization Conference, pp 159–165

14. Lu J-Q, Kumar A, Kwon Y, Eisenbraun ET, Kraft RP, McDonald JF, Gutmann RJ, Cale TS, Belemjain P, Erdogan O, Castracane J, Kaloyeros AE (2000) 3D Integration using wafer bonding", Advanced Metallization Conference, eds., Edelstein D, Dixit G, Yasuda Y, T. Ohba T, MRS Proc. vol V16, pp 515-521
15. Lu J-Q, Jindal A, Kwon Y, McMahon JJ, Rasco M, Augur R, Cale TS, Gutmann RJ (2003) Evaluation procedures for eafer bonding and thinning of interconnect test structures for 3D ICs. In: IEEE International Interconnect Technology Conference, pp 74–76
16. Guarini KW, Topol AW, Ieong M, Yu R, Shi L, Newport MR, Frank DJ, Singh DV, Cohen GM, Nitta SV, Boyd DC, O'Neil PA, Tempest SL, Pogge HB, Purushothaman S, Haensch WE (2002) Electrical integrity of state-of-the-Art 0.13 mm SOI CMOS devices and circuits transferred for three-dimensional (3D) integrated circuit (IC) fabrication. In: Technical Digest of IEEE International Electron Devices Meeting, pp 943–945
17. Burns JA, Aull BF, Chen CK, Chen CL, Keast CL, Knecht JM, Suntharalingam V, Warner K, Wyatt PW, Yost DRW (2006) A wafer-scale 3-D circuit integration technology. IEEE Trans. Electron Devices, 53(10): 2507–2516
18. Chen KN, Lee SH, Andry PS, Tsang CK, Topol AW, Lin YM, Lu J-Q, Young AM, Ieong M, Haensch W (2006) Structure design and process control for Cu bonded interconnects in 3D integrated circuits. In: Technical Digest of IEEE International Electron Devices Meeting, pp 367–370
19. Patti R (2006) Three-dimensional integrated circuits and the future of system-on-chip designs. Proceedings of the IEEE, 94(6): 1214–1222
20. Lu J-Q, McMahon JJ, Gutmann RJ (2006) Via-first inter-wafer vertical interconnects utilizing wafer-bonding of damascene-patterned metal/adhesive redistribution layers. In: 3D Packaging Workshop at IMAPS Device Packaging Conference, Scottsdale, AZ
21. Gutmann RJ, McMahon JJ, Lu J-Q (2006) Damascene patterned metal/adhesive redistribution layers. In: Bower CA, Garrou PE, Ramm P, Takahashi K (eds) Enabling Technologies for 3-D Integration, , Mater Res Soc Symp Proc, 970, Warrendale, PA, paper # 0970-Y04-02
22. Enquist P (2006) Bond Interconnect (DBI) – Technology for scalable 3D SoCs. In: The 3rd International Conference on 3D Architectures for Semiconductor Integration and Packaging, Burlingame, CA
23. Lu J-Q, Kwon Y, Rajagopalan G, Gupta M, Bae DL, McMahon J, Hong CK, Kraft RP, Erdogan O, Belemjian PM, McDonald JF, Cale TS, Gutmann RJ, Eisenbraun E, Xu B, Castracane J, Kaloyeros A (2002) Fabrication of via-chain test structures for 3D IC technology using dielectric bonding on 200 mm wafers. In: McKerrow AJ, Shacham-Diamand Y, Zaima S, Ohba T (eds) Advanced Metallization Conference, pp 151–157
24. Lu J-Q, Jindal A, Kwon Y, McMahon JJ, Lee KW, Kraft RP, Altemus B, Cheng D, Eisenbraun E, Cale TS, Gutmann RJ (2003) 3D System-on-a-chip using dielectric glue bonding and Cu damascene inter-wafer interconnects. In: Mathad GS, Cale TS, Collins D, Engelhardt M, Leverd F, Rathore HS (eds) International Symposium on Thin Film Materials, Processes, and Reliability, ECS Proceedings PV2003-13, pp 381–389
25. Garrou P, Scheck D, Im JH, Hetzner J, Meyers G, Hawn D, Wu J, Vincent MB, Wong CP (2000) Underfill adhesion to BCB (Cyclotene^{TM}) bumping and redistribution dielectrics. IEEE Transactions on Advanced Packaging, 23(3): 568–573
26. Yang DQ, Martinu L, Sacher E, Sadough-Vanini A (2001) Nitrogen plasma treatment of the dow cyclotene 3022 surface and its reaction with evaporated copper. Appl Surf Sci, 177(1–2): 85–95
27. Price DT, Gutmann RJ, Murarka SP (1997) Damascene copper interconnects with polymer ILDs. Thin Solid Films, 308–309: 523–528
28. Adema GM, BerryMJ, Dibbs MG, Garrou P, Heistand RH, Huber BS, Mohler CE, Stokich TM, Townsend PH, Turlik I (1992) Cure management of benzocyclobutene dielectric for electronic applications. In: Proceedings of Society of Advanced Material Process Engineering (SAMPE) Electronics Conference 1992, pp 1–10, Baltimore

29. Mills ME, Townsend P, Castillo D, Martin S, Achen A (1997) Benzocyclobutene (DVS-BCB) polymer as an interlayer dielectric (ILD) material. Microelectron Eng, 33(1): 327–334
30. Niklaus F, Kumar RJ, McMahon JJ, Yu J, Lu J-Q, Cale TS, Gutmann RJ (2006) Adhesive wafer bonding using partially cured benzocyclobutene (BCB) for three-dimensional integration. J Electrochem. Soc, 153(4): G291–G295
31. Kwon Y, Seok J, Lu J-Q, Cale TS, Gutmann RJ (2006) Critical adhesion energy of benzocyclobutene (BCB)-bonded wafers. J Electrochem Soc, 153(4): G347–G352
32. Wimplinger M, Lu J-Q, Yu J, Kwon Y, Matthias T, Cale TS, Gutmann RJ (2004) Fundamental limits for 3D wafer-to-wafer alignment accuracy. In: Carter RJ, Hau-Riege CS, Kloster GM, Lu TM, Schulz SE, (eds) Materials, Technology, and Reliability for Advanced Interconnects and Low-k Dielectrics, MRS Proc. vol 812: pp F6.10.1–F6.10.6
33. Kwon Y, Jindal A, McMahon JJ, Lu J-Q, Gutmann RJ, Cale TS (2003) Dielectric glue wafer bonding for 3D ICs. In: McKerrow AJ, Leu J, Kraft O, Kikkawa T (eds) Materials, Technology, and Reliability for Advanced Interconnects and Low-k Dielectrics, MRS Proc. vol 766: pp 27–32
34. Kwon Y, Seok J, Lu J-Q, Cale TS, Gutmann RJ (2005) Thermal cycling effects on critical adhesion energy and residual stress in benzocyclobutene (BCB)-bonded wafers. J Electrochem Soc, 152(4): G286–G294
35. Kwon Y, Seok J, Lu J-Q, Cale TS, Gutmann RJ (2007) Critical adhesion energy at the interface between benzocyclobutene and silicon nitride layers. J Electrochem Soc, 154(6): H460–H465
36. Pozder S, Lu J-Q, Kwon Y, Zollne S, Yu J, McMahon JJ, Cale TS, Yu K, Gutmann RJ (2004) Back-end compatibility of bonding and thinning processes for a wafer-level 3D interconnect technology platform. In: IEEE International Interconnect Technology Conference (IITC 2004), pp 102–104
37. Gutmann RJ, Lu J-Q, Pozder S, Kwon Y, Jindal A, Celik M, McMahon JJ, Yu K, Cale TS (2003) A wafer-level 3D IC technology platform. In: Ray GW, Smy T, Ohta T, Tsujimura M (eds) Advanced Metallization Conference, pp 19–26
38. Lu J-Q, Kwon Y, Jindal A, McMahon JJ, T.S. Cale TS, Gutmann RJ (2003) Dielectric glue wafer bonding and bonded wafer thinning for wafer-level 3D integration. In: Bengtsson S, Hunt CE, Baumgart H, Suga T, Hesketh P (eds) Semiconductor Wafer Bonding VII: Science, Technology, and Applications, ECS Proc. V. 2003–19, pp 76–86
39. McMahon JJ, Niklaus F, Kumar RJ, Yu J, Lu J-Q, Gutmann RJ (2005) CMP Compatibility of partially cured benzocyclobutene (BCB) for a via-first 3D IC process. In: Kumar A, Lee JA, Obeng YS, Vos I, Johns EC (eds) Chemical-Mechanical Planarization—Integration, Technology and Reliability, MRS Proc vol 867: pp W4.4.1-6
40. Gutmann RJ, McMahon JJ, Rao S, Niklaus F, Lu J-Q (2005) Wafer-level via-first 3D integration with hybrid-bonding of Cu/BCB redistribution layers. In: Proceedings of SMTA International Wafer Level Packaging Congress (IWLPC), pp 122–127
41. Lu J-Q, Rajagopalan G, Gupta M, Cale TS, Gutmann RJ (2004) Planarization issues in wafer-level 3D integration. In: Boning DS, Bartha JW, Philipossian A, Shinn G, Vos I (eds) Advances in Chemical-Mechanical Polishing, Mater Res Soc Symp Proc, 816, Warrendale, PA, pp 217–28
42. Lu J-Q, Kwon Y, Rajagopalan G, Gupta M, McMahon J, Lee KW, Kraft RP, McDonald JF, Cale TS, Gutmann RJ, Xu B, Eisenbraun E, Castracane J, Kaloyeros A (2002) A Wafer-scale 3D IC technology platform using dielectric bonding glues and copper damascene patterned inter-wafer interconnects. In: IEEE International Interconnect Technology Conference, pp 78–80
43. Zhang J. Bloomfield MO, Lu J-Q, Gutmann RJ, Cale TS (2006) Modeling thermal stresses in 3-D IC interwafer interconnects. IEEE Trans. Semicond. Manuf., 19(4):437–448
44. Filippi RG, McGrath JF, Shaw TM, Murray CE, Rathore HS, McLaughlin PS, McGahay V, Nicholson L, Wang PC, Lloyd JR, Lane M, Rosenberg R, Liu X, Wang YY, Landers W,

Spooner T, Demarest JJ, Engel BH, Gill J, Goth G, Barth E, Biery G, Davis CR, Wachnik RA, Goldblatt R, Ivers T, Swinton A, Barile C, Aitken J (2004) Thermal cycle reliability of stacked via structures with copper metallization and an organic low-k dielectric. In: Proceedings of IEEE International Reliability Physics Symposium, pp 61–67

45. Pozder S, Jones R, Adams V, Li HF, Canonico M, Zollner S, Lee SH, Gutmann RJ, Lu J-Q (2007) Exploration of the scaling limits of 3D integration. In: Bower CA, Garrou PE, Ramm P, Takahashi K (eds) Enabling Technologies for 3-D Integration, Mater Res Soc Symp Proc, 970, Warrendale, PA, paper # 0970-Y02-01
46. Gutmann RJ, Lu J-Q (2008) Wafer-level three-dimensional (3D) integration for advanced CMOS systems. In: Iniewski K, (ed) Circuits for the Nanoscale: Communications, Imaging and Sensing, CRC Press pp 39–52
47. Zeng AY, Lu J-Q, Rose K, Gutmann RJ (2005) First-order performance prediction of cache memory with wafer-level 3D integration. IEEE Design Test Comput., 22(6):548–555
48. Zeng AY, Lu J-Q, Rose K, Gutmann RJ (2004) Wafer-level 3D manufacturing issues for streaming video processors. In: Proc. 15th Annual IEEE/SEMI Advanced Semiconductor Manufacturing Conference (ASMC 2004), pp 247–251, Boston, MA
49. Lu J-Q, Rose K, Vitkavage S (2007) Performance study and applications of 3D integration. In: Proceedings of the International Conference on 3D Architectures for Semiconductor Integration and Packaging, Burlingame, CA

Chapter 11
Direct Hybrid Bonding

Bart Swinnen, Anne Jourdain, Piet De Moor, and Eric Beyne

11.1 Introduction

The production of nonmonolithic three-dimensional (3D) systems by stacking components and interconnecting them by through-substrate vias (TSVs) is intrinsically limited to the stacking of "thin" dies. The restriction not only stems from the desire to shrink system height, it also is imposed by the limited capabilities of TSV processing. Most often the metallization process is limiting the aspect ratio (i.e., height/diameter) of the 3D via thus restricting the thickness of the TSV dies. Depending on the type of TSV technology employed, the final wafer or die thickness typically ranges from 100 μm down to 15 μm. Since dies or wafers of such thickness no longer are rigid, it is a requirement that the bond guarantees mechanical stability to the thin stacked die or wafer.

Pure dielectric bonding guarantees such stability simply by the fact that the thin die or wafer is bonded over its entire surface. This method, however, limits the options for electrical interconnection to a via-last approach enhancing the risk of system yield loss during via processing. This method, however is a key technology for chip embedding as described in [1–2].

In the case of metallic bonding, electrical interconnection is the prime function of the bond. Mechanical stabilization of the thin die requires introduction of dummy metallization, also in areas where no electrical interconnects are required. For TSV-based 3D systems this means that a patterned metal layer needs to be applied on the backside of a thinned wafer which may impose on the cost of the system or may limit the achievable interconnect density.

Compared to the above-discussed techniques, hybrid bonding is an outsider: it is a very recent approach and combines fixation of a thin wafer or die by means of a dielectric adhesive with the formation of a metallic interconnect. Figure 11.1 shows typical approaches to hybrid bonding as described in literature. The top figure is an

B. Swinnen
IMEC, Kapeldreef 75, B-3001 Leuven, Belgium
e-mail: swinnenb@imec.be

C.S. Tan et al. (eds.), *Wafer-Level 3D ICs Process Technology*,
DOI: 10.1007/978-0-387-76534-1_11, © Springer Science+Business Media, LLC 2008

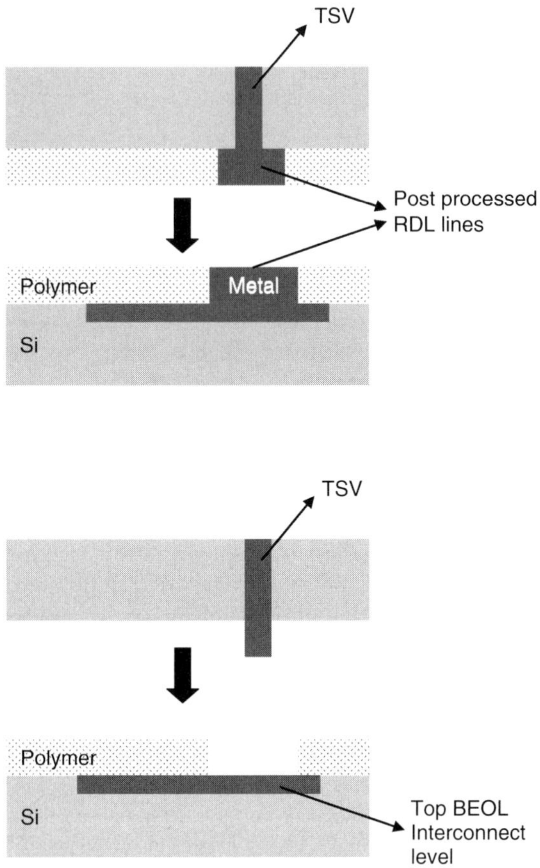

Fig. 11.1 Typical embodiments of the technique of hybrid bonding. (*Top*) Redistribution bonding: the bonded surfaces are patterned and planarized redistribution layers. (*Bottom*) Direct hybrid bonding: the TSV directly lands on the top metal layer of the back-end-of-line interconnect layers

example of redistribution layer bonding (RDL), first reported in 2005 by McMahon et al. [3]. Here a metal/dielectric layer, similar to a flip-chip redistribution layer is applied both on top of the bottom or landing wafer as well as on the backside of the thinned TSV wafer. Next, both redistribution layers are planarized by chemical–mechanical polishing (CMP) before the surfaces are joint and bonded.

As RDL is described in the previous chapter of this book, the current chapter focuses on the direct hybrid bonding approach. The bottom image in Fig. 11.1 shows a direct hybrid bonding approach as reported in 2007 by Jourdain et al. [4]. Here a patterned polymer dielectric layer is applied to the landing wafer only. The thinning process for the TSV die allows the vias to protrude from the wafer backside. During direct hybrid bonding, the TSVs fit into the openings in the dielectric on the bottom wafer and land on the top metal level of the landing wafer. The presence of topography during the direct bonding process does not occur in other approaches;

therefore, in this chapter we will also briefly describe the main features of the thinning process that precedes the direct hybrid bonding.

Both hybrid bonding methods are compatible with a via-first process flow, which limits the amount of processing that is required after the wafers or dies are stacked and reduces the risk of compound yield loss. The absence of litho on the back-surface of the thinned TSV wafers and the use of a compliant dielectric layer, eliminating the need for planarization processes, are the major features that differentiate direct hybrid bonding from RDL.

11.2 The Direct Hybrid Bonding Process

11.2.1 TSV Die Preparation

Three-dimensional integration using direct hybrid bonding starts with the preparation of the TSV wafers. In the approach pursued by IMEC, the process for via production is a single damascene process [5–6]. Three-dimensional vias are imaged and etched using stepper lithography and BOSCH etch processes. This is done immediately after processing of the contact layer and prior to processing the BEOL layers. In this way via diameters of 5 μm or less are easily achieved. The capabilities of barrier/seed processes today limit the aspect ratio of the vias to less than 5:1. Therefore a typical layout after complete BEOL processing of the TSV wafer is similar to the one shown in Fig. 11.2.

The bottom of the via is opened during the thinning process. First, most of the Si is removed from the wafer backside in a backgrinding process. The amount of Si

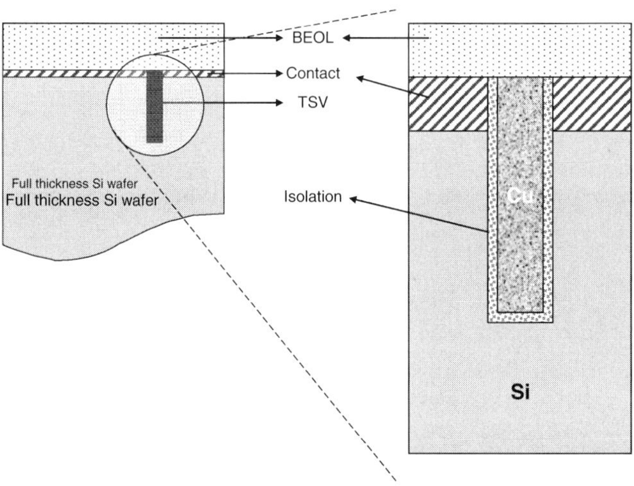

Fig. 11.2 (*Left*) Cross-sectional layout of a TSV wafer after full TSV and BEOL processing. (*Right*) Detailed cross-sectional layout of a blind TSV prior to wafer thinning

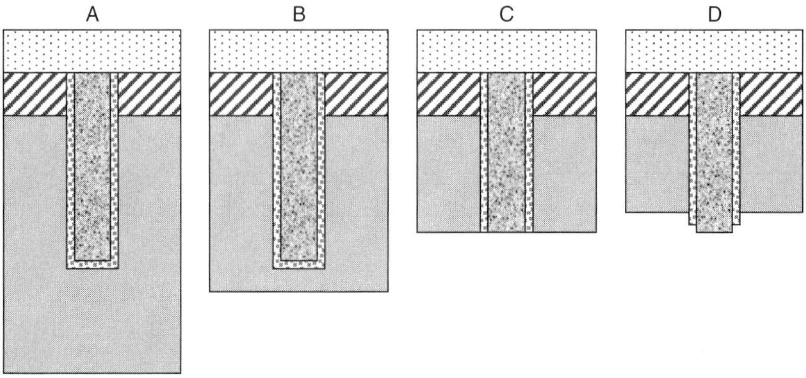

Fig. 11.3 Cross-sectional layout of a TSV (**a**) prior to wafer thinning, (**b**) after thinning by grinding, (**c**) after subsequent CMP, (**d**) after recess etch

removed is such that the via metallization is not exposed during this process. Typically the wafer thickness after this step exceeds the via depth by 3–5 μm: the wafer now looks as in Fig. 11.3b. The actual via opening is done in a subsequent CMP step that removes another 5–7 μm from the wafer backside: as shown in Fig. 11.3c the copper in the via is now exposed at the bottom of the wafer. In a last step, about 600–1000 nm of Si is removed in a dry-etch process. The etch serves as a stress relief process, and recesses the Si relative to the Cu-via surface so that the via protrudes above the Si-wafer surface. Due to the limited selectivity of the Si-etch process to via isolation material, the isolation layer is also slightly recessed. The via at this stage is depicted in Fig. 11.3d, while scanning electron microscopy (SEM) images are shown in the left part of Fig. 11.4. The right part of the same figure shows a detail of the TSV: one clearly distinguishes the central via metallization (Cu), the Cu-diffusion barrier (Ta), and the via isolation layer (SiO_2).

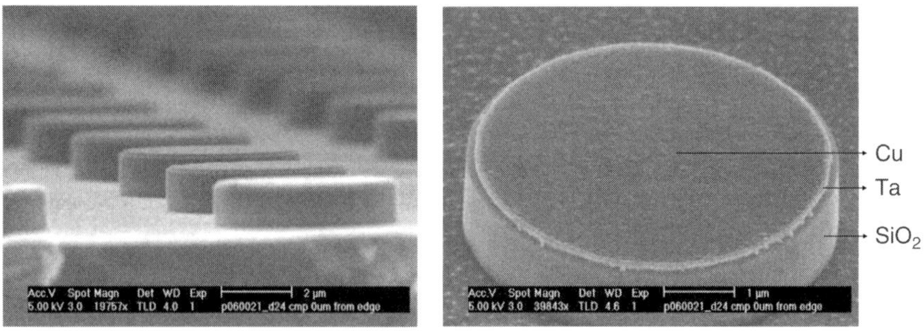

Fig. 11.4 (*Left*) Top-view SEM image of an array of 3D-vias after recess etching, (*right*) SEM of a single TSV after recess etch

11.2.2 Preparation of the Landing Surface: Dielectric Application and Patterning

In parallel to the fabrication of the TSV-wafer, the landing wafer is prepared. The top BEOL metal layer contains the Cu-landing pads to which the TSVs will interconnect. The next step in the preparation of the landing wafer is the application of a patternable dielectric glue layer. The dielectric glue layer used here is photosensitive benzocyclobutene (BCB), type XU-35133, with a target thickness after deposition of 1 μm or less. BCB Material properties can be found in [7]. The choice of BCB as the glue material is based on the well-known material properties of the currently available versions (series 4000) of the polymer. Indeed, its excellent mechanical resistance with shear strengths exceeding 20 MPa and its excellent bonding properties make BCB a first-rate candidate for hybrid bonding applications [8]. These BCB-variants furthermore are known to reflow at temperatures around 170°C.

The BCB is applied by spinning. After a semicure step, the photosensitive BCB layer is patterned using standard broadband 1X-photolithography. After developing the BCB layer, a short plasma descum removes incidental residues from the metal pads. The spinning and patterning processes are tuned to yield a BCB thickness that slightly exceeds the protrusion height of the TSV on the thinned wafer. At this stage of the process, the BCB is not yet fully cured and behaves as a tacky glue layer.

The patterning of the BCB layer frees up area on the landing wafer into which the BCB will reflow during the bonding process. As the adhesion of BCB to Cu in known to be poor, the BCB patterning mask is designed so that patterning results in BCB features occupying area only in between the Cu-landing pads. The top of Fig. 11.5 provides an optical top-view image of a selected area on the landing wafer: in the case shown here, the BCB patters are crosses of approximately $10 \times 10\ \mu m^2$. After the patterning process, the BCB can be quite rough as evidenced by the height profile map in the bottom of Fig. 11.5. This, however, does not pose any issues during further processing.

Next the TSV and landing wafers are brought in contact while keeping the system at a low temperature. At this stage of the process, the BCB features on the landing substrate touch the Si-surface of the TSV substrate. As the BCB thickness exceeds the Cu-via protrusion height, the metal features of top and bottom devices are not yet in contact (Fig. 11.6b). Next the temperature is ramped to the polymer reflow temperatures while the pressure on the top device is maintained. At temperatures from approximately 100°C–150°C, the BCB behaves as a nonconfined viscous liquid that is deformed under the applied pressure.

Figure 11.7 evidences the BCB reflow behavior. Figure 11.7a shows an optical image of a bare BCB surface immediately after patterning: the BCB forms a grid of rectangles around the Cu-landing pads on the wafer. Figure 11.7b shows a similar die after reflow at 100°C without application of force to the BCB: there is no evidence of BCB reflow under these conditions. The pattern shown in Fig. 11.7c is obtained by exposing the BCB to the same temperature treatment, but while simultaneously applying a force to the BCB. The force is applied by stacking a glass die on top of the BCB pattern and the picture is taken without removal of the

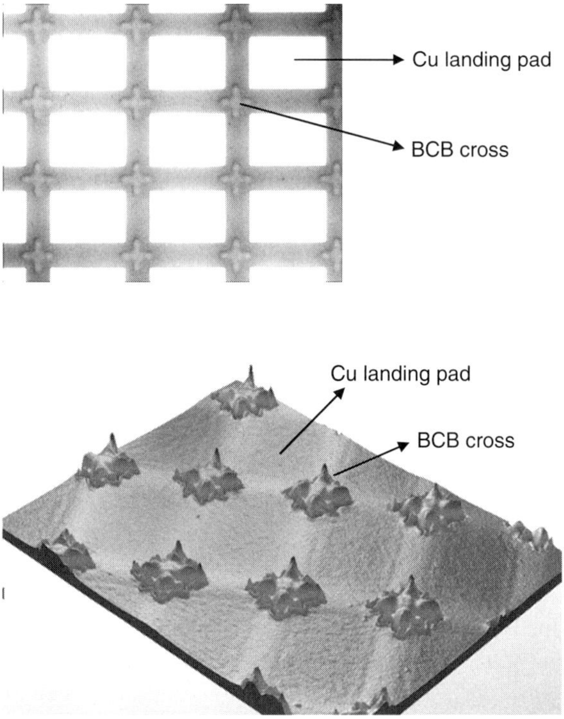

Fig. 11.5 (*Top*) Optical top-view image of a landing wafer after BCB application and patterning. (*Bottom*) VEECO height map after BCB patterning

glass cover. The texture of the BCB lines is no longer visible. A similar treatment at higher temperature results in the picture in Fig. 11.7d: the BCB lines are clearly reflown resulting in a BCB dot at the crossing of the BCB lines.

BCB deformation stops as the Cu-nails touch down on the Cu-landing pads (Fig. 11.6C). Whereas the temperature applied here is sufficient to reflow the

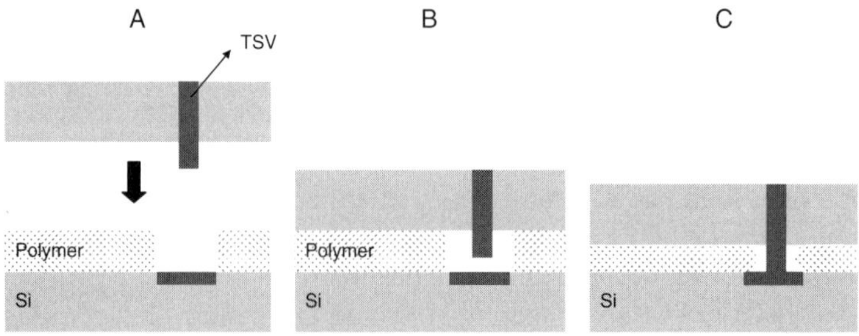

Fig. 11.6 Different stages of the hybrid bonding process. (**a**) TSV and landing dies or wafers are aligned. (**b**) Wafers or dies are brought in contact. (**c**) The polymer layer is compressed and reflows. The metallic interconnect is formed

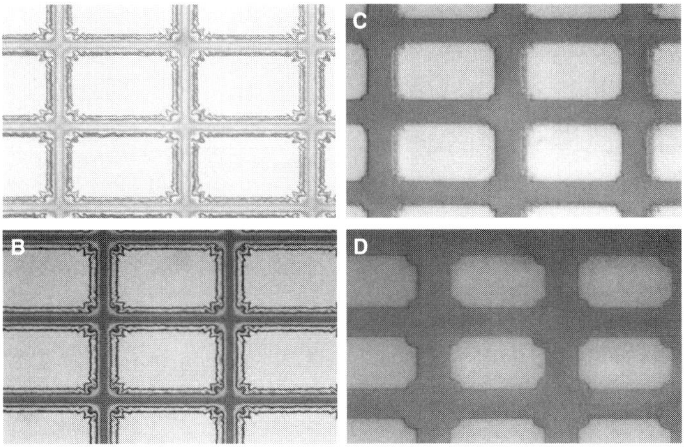

Fig. 11.7 Demonstration of BCB reflow. Optical imaes of BCB patterns: (**a**) immediately after patterning, (**b**) after reflow at 100°C without application of force, (**c**) after reflow at 100°C with simultaneous application of force, (**d**) after reflow above 100°C with simultaneous application of force

polymer, it is insufficient to form a reliable Cu–Cu interconnect. Therefore, the temperature is further increased to at least 300°C in order to realize the Cu–Cu interconnection between the TSV and the landing pad. At the same time the polymer glue layer is cured. Typical durations for the reflow and bonding processes are 15 min each. The force required during reflow and bonding steps are determined by the BCB and TSV densities, respectively.

Figure 11.8 displays an optical picture of a directly hybrid-bonded stack obtained in a die-to-die bonding mode. Cross-sectional views of the TSV areas in this stack are shown in Fig. 11.9. The structures shown in these pictures are elements of an electrically yielding via chain with 10 000 through-Si vias. The electrical characteristics of this via chain are identical to the ones reported by Swinnen et al. [2] for

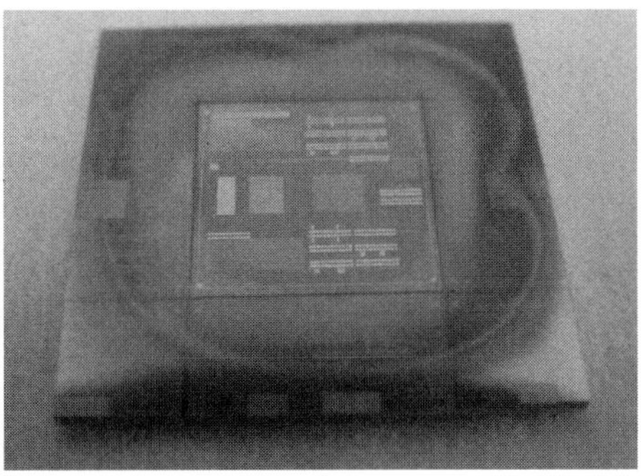

Fig. 11.8 Optical image of a directly hybrid-bonded die stack

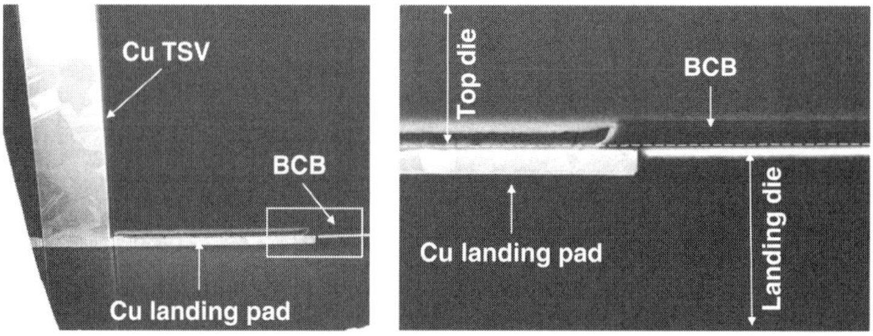

Fig. 11.9 (*Left*) Cross-sectional view of a TSV area in a directly hybrid-bonded die stack. (*Right*) Zoom in on the compressed BCB pattern

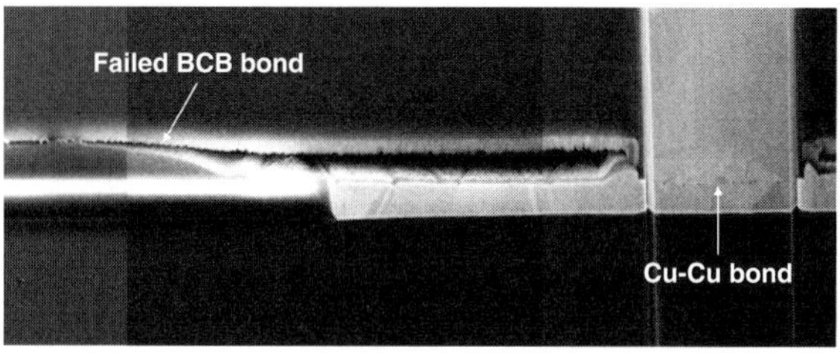

Fig. 11.10 Cross-section SEM image of a failed direct hybrid bond. The BCB layer is too thin to contact to the top die

a purely Cu–Cu bonded structure: this evidences that the BCB processing does not impact the electrical performance of the interconnects.

Figure 11.10 illustrates the importance of matching the thickness of the patterned BCB layer to the protrusion height of the Cu vias from the TSV die: it shows a cross-section SEM picture of a failed BCB bond. The Cu vias touched down on the landing pads before the BCB touched the TSV die: there is little or no contact between the dielectric glue and the top die.

11.3 Choice of Metal Bond Type and Dielectric

Literature descriptions of hybrid bonding integration schemes to date have all used different formulations of BCB as the dielectric bonding material and direct Cu-to-Cu bonding for the metallic interconnects. Whereas the integration schemes are not inherently limited to these materials, in practice, the choice of the dielectrics are mainly limited by the requirement that it should be compatible with the temperature budget and surface treatments required to realize the electrical interconnects.

For direct hybrid bonding approach, silicones are a so far unexplored class of dielectric materials that hold promise to counter BCB. The Cu–Cu interconnects can be replaced by intermetallic interconnections based on microbumps opening opportunities to lower the temperature budget required for realizing the electrical interconnect. Contrary to the case of RDL, the choice of both dielectric and metal are not restricted by the need for the planarization prior to bonding.

11.4 Collective Hybrid Bonding

As most other bonding methods, the direct hybrid bonding process in principle is compatible both with die-to-die and wafer-to-wafer bonding. Many organizations so far have preferred wafer-to-wafer bonding as it strongly outperforms die-bonding in terms of alignment accuracy capabilities and better alignment accuracy enables higher 3D interconnect densities. However, wafer-to-wafer 3D integration schemes inherently are limited to stacking dies of equal size. Moreover, when 2D device yield is limited, it leads to an important system yield loss. With these arguments, wafer-to-wafer stacking will probably be cost effective only in specific markets.

Die-to-wafer stacking on the other hand may be of more interest for the fabrication of heterogeneously integrated systems as it does not impose the requirement of equal die size. The method also is compatible with the selection of known good die (KGD) prior to stacking and therefore is of interest to cases where one of the components in the stacked system has limited yield. The cost of die-to-die or die-to-wafer stacking, however, for most bonding methods is limited by the throughput of the process. Especially when heat needs to be applied to achieve the bond, die-bonding is a lengthy process.

The introduction of the tacky polymer in the direct hybrid bonding scheme now offers a new possibility for die-to-wafer throughput optimization. The opportunity lies in the separation of pick-and-place and bonding operations. Figure 11.11 illustrates the concept: first TSV dies are aligned and placed onto the landing wafer (Fig. 11.11 (*left*)). The patterned and tacky dielectric that covers the landing wafer weakly bonds the stacked dies and fixes them during further handling. This operation can be performed at low temperatures and therefore its duration is determined mainly by the alignment process. Pick-and-place is repeated until the full wafer is populated. In a second stage, the fully populated wafer is moved to a wafer-level bonding toolwhere pressure and heat are applied to all stacked dies at once

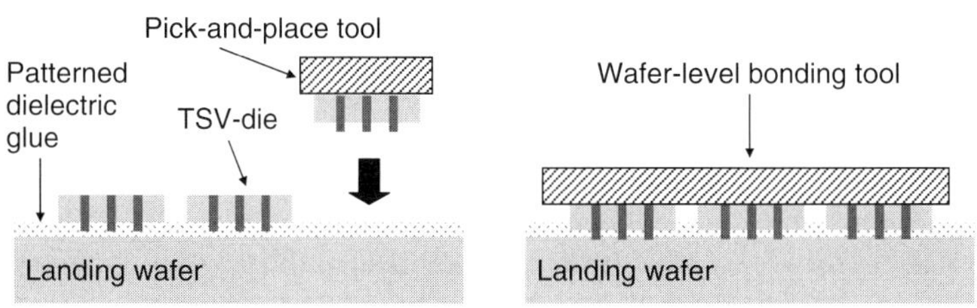

Fig. 11.11 Illustration of the collective hybrid bonding process. (*Left*) Die-to-wafer alignment and placement. (*Right*) Collective bonding of all dies placed on the landing wafer

(Fig. 11.11 (*right*)). Thus the dielectric reflow and the metallic interconnect bonding is performed for all stacked dies simultaneously. This process is called collective hybrid bonding.

11.4.1 Requirements for Pick-and-Place

The throughput enhancement achieved by segmentation of the hybrid bonding process into wafer population by pick-and-place followed by a collective bonding step implicitly assumes that pick-and-place can be done at temperatures low enough not to cure the polymer layer. Therefore, a sufficiently fast alignment method is needed. Whereas low-temperature pick-and-place on unpatterned BCB was implicitly demonstrated in the compliant bonding process demonstration, the temperature and time process windows for patterned BCB layers are still under investigation.

The tradeoff between throughput and alignment accuracy, however, remains an important concern for the pick-and-place process: high-throughput pick-and-place tools typically achieve alignment accuracies of about 10 μm only, which imposes a serious limitation on the achievable 3D interconnect density and rules out the method for logic stacking or for stacking of high-bandwidth memory on top of logic. Better alignment accuracies, down to 1 μm, are achieved by state-of-the-art flip-chip bonders. However, here the alignment accuracy comes at the expense of throughput due the lengthy alignment procedure. Therefore, the development of fast and highly accurate pick-and-place tools is an opportunity to the industry today.

11.4.2 Compliant Bonding

In the above collective hybrid bonding integration scheme, the stacked dies may eventually originate from one or more thinned source wafers. The method may even be used to assemble a system by stacking dies of a different type onto a single landing wafer. State-of-the-art thinning process can achieve target thickness accuracies of 1–2 μm. Therefore, it is possible that the individual stacked dies, differ in height by the same amount which is large as compared to the less than 1 μm thickness of the dielectric glue that has been used so far in this type of processes. These height differences can compromise the metallic bonding process if thick dies prevent the wafer-level bonding tool from applying sufficient force to the thinner dies in the assembly. Therefore, the bonding needs to be tolerant to small height differences across the stacked dies. In experiments, this has recently been achieved by including a high-temperature compatible polymer layer in the wafer-level bonder. Though these tool solutions are still in a developmental phase, a first process demonstration evidenced successful collective adhesive bonding of Si dies with thicknesses in the range of 670–700 μm to a Si wafer covered with an unpatterned 1-μm thick BCB layer [4].

11.5 Summary

The method of direct hybrid bonding enables simultaneous electrical interconnection and mechanical stabilization of stacked dies or wafer. The method has been demonstrated in a die-to-wafer mode using patterned dielectric polymer adhesive of less than 1-μm thickness in combination with a direct Cu-to-Cu interconnect resulting in electrically yielding via chains containing 10 000 through-Si vias. The approach opens the route to a collective hybrid bonding integration scheme in which dies are first aligned and placed on a landing wafer in a fast process, and next, collectively bonded at wafer level.

Whereas collective hybrid bonding is still in its infancy today, the method could be a first enabler for a cost-effective die-to-wafer integration process. Stacking of memory on top of logic or stacking of logic dies seem obvious market entry points for the collective hybrid bonding method.

References

1. Beyne E (2001) Technologies for very high bandwidth electrical interconnects between next generation VLSI circuits. In: International Electron Devices Meeting Technical Digest, IEEE, pp 371–374
2. Beyne E (2004) 3D Interconnection and packaging: impending reality or still a dream. In: Proceedings of the 2004 IEEE International Solid State Circuit Conference, IEEE, pp 138–145
3. McMahon JJ, Lu JQ, Gutmann RJ (2005) Wafer bonding of Damascene-Patterned metal/adhesive redistribution layers for via-first three-dimensional (3D) interconnect. In: Proceeding of the 55th Electronic Components and Technology conference, IEEE, pp 331–336
4. Jourdain A, Stoukatch S, De Moor P, Ruythooren W, Pargfrieder S, Swinnen B, Beyne E (2007) Simultaneous Cu–Cu and compliant dielectric bonding for 3D stacking of ICs. In: Proceedings of the International Interconnect Technology Conference, IEEE, pp 207–209
5. Beyne E (2006) The rise of the 3rd dimension for system integration. In: Proceedings of the International Interconnect Technology Conference, IEEE, pp 1–5
6. Swinnen B, Ruythooren W, De Moor P, Bogaerts L, Carbonell L, De Munck K, Eyckens B, Stoukatch S, Sabuncuoglu Tezcan D, Tőkei Z, Vaes J, Van Aelst J, Beyne E (2006) 3D integration by Cu–Cu thermo-compression bonding of extremely thinned bulk-Si die containing 10 μm pitch through-Si vias. In: International Electron Devices Meeting Technical Digest, IEEE, pp 371–374
7. http://www.dow.com/cyclotene/solution/index.htm
8. Jourdain A, Rottenberg X, Carchon G, Tilmans HAC, (2003) Optimization of 0-level packaging for RF-MEMS devices. In: Proceedings Transducers, 2003, Boston, pp 1915–1918

Chapter 12
3D Memory

Robert S. Patti

12.1 A Brief History of Memory

Electronic memory has been essential since the beginning of electronics. Virtually no logical or mathematical function can be implemented without some storage elements.

We will omit the history of electronic memory prior to integrated circuits (ICs), as this itself could fill a book. The first integrated memories, other than perhaps registers, were programmable read-only memory (PROM). Static random access memory (SRAM) devices soon followed PROMs. Dynamic random access memory (DRAMs) proved to be tricky to implement. Several companies tried 1T DRAM, but the first successful DRAM was the Intel 1103 3T DRAM. Within a few years, 1T DRAM appeared. This is the DRAM with which we are now very familiar, which dominates the entire volatile memory market. Figure 12.1 shows the evolution of memory. The evolution of DRAM is shown in Fig. 12.2.

Today's systems contain many more memory chips than processor chips. Even the processor chips themselves contain large numbers of memory registers and sizeable arrays of on-die memory. A state-of-the-art processor could easily be described as a memory device with some logic on it!

12.2 3D Memory By Many Means

12.2.1 3D Packaging

Three-dimensional ("3D") can mean many things. The first phase of 3D technology adoption is best described as 3D packaging. While some would point to single inline memory modules (SIMM) as 3D devices, a more reasonable first case would be the stacking of two dies in a single package. These early 3D devices were created

R.S. Patti
CTO, Tezzaron Semiconductor, Naperville, IL, USA
e-mail: rpatti@tezzaron.com

C.S. Tan et al. (eds.), *Wafer-Level 3D ICs Process Technology*,
DOI: 10.1007/978-0-387-76534-1_12, © Springer Science+Business Media, LLC 2008

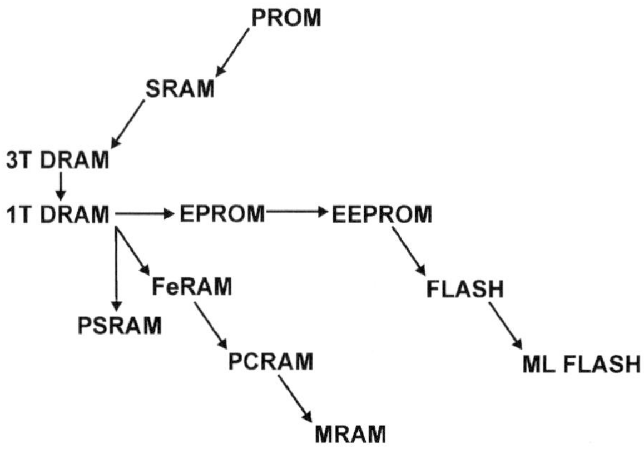

Fig. 12.1 The evolution of memory

by simply wire-bonding both dies to a common lead frame, doubling the memory capacity in the same circuit board space.

The next step in the evolution of 3D, and what the author would point to as the first real 3D packaging technology, would be the Irvine Sensors [1] memory stacks. These devices employed vertical metallization along the edges of a stack of memory dies. While these components had limited commercial success, they introduced

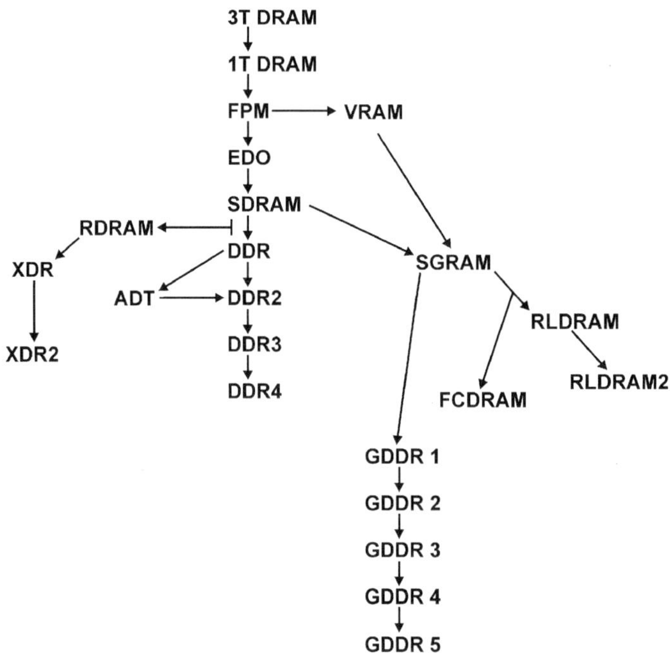

Fig. 12.2 The evolution of DRAM

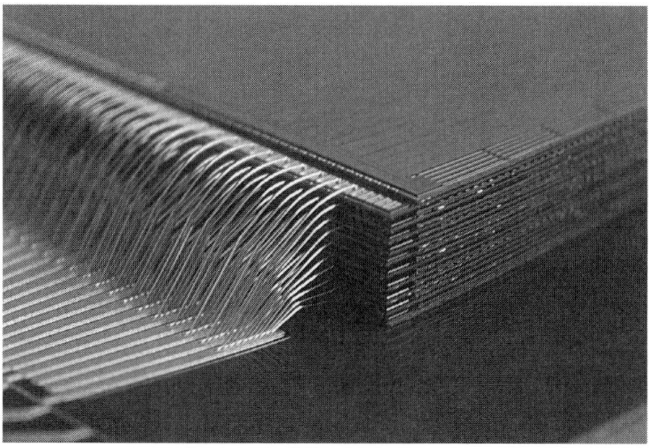

Fig. 12.3 Elpida's wire-bonded chip stack. Photo credit: Elpida Memory, Inc.

the world to 3D devices. From here, many variations of 3D packaging have been created. Today's system-in-package (SIP) technology is a direct outgrowth of Irvine Technology's initial stacking of memory devices.

System-in-package devices are used in high-volume manufacture and currently employed in most cell phones. Issues related to yield and known good die (KGD) are not fully solved, but they are understood and being addressed today. System-in-package enhancements such as package-on-package (POP) are nearing a long-sought goal: the ability to assemble SIP devices from multisourced off-the-shelf components.

Figures 12.3–12.5 are examples of 3D memory packages from leading companies like Elpida and Samsung [2, 3].

12.2.2 3D ICs

A subject of much discussion today is the definition of a 3D package versus a 3D IC. One's opinion often depends on one's capability. We will use this definition:

> A 3D IC contains at least 2 semiconductor layers and at least one layer of horizontal interconnect wiring between the semiconductor layers. Each semiconductor layer contains its own functional elements (transistors, diodes, etc.) with electrical junctions. The semiconductor layers may be built sequentially or separately. Elements on the semiconductor layers are connected vertically with TSVs or other vias, without employing the I/O drivers and associated electrostatic discharge (ESD) structures that are normally used to connect off-chip devices.

This definition should serve to identify most 3D ICs, although there will certainly be exceptions.

Some of the earliest 3D IC work incorporated amorphous silicon load transistors in low-power SRAMs. However, the first commercial 3D IC is generally recognized to be the PROM device created by Matrix Semiconductor (now part of SanDisk).

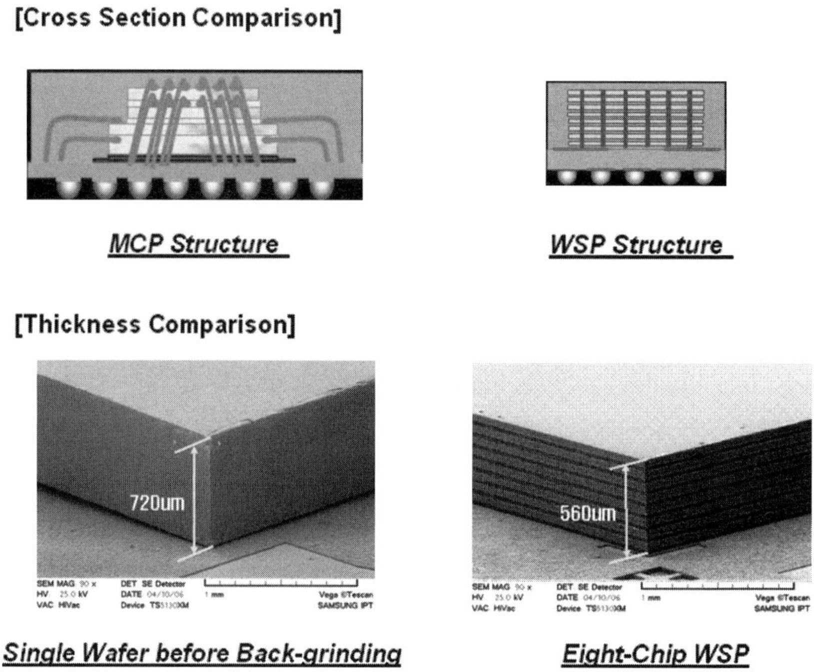

Fig. 12.4 Samsung's wafer-stacked package. Reproduced with permission from Kriss Lee, Samsung

This device was built by sequentially manufacturing vertical diode structures with embedded antifuses [4].

Another example would be the Infineon SOLID tin/copper process that was slated to premier as a 3D memory-to-controller device for use in smart cards. Unfortunately, the SOLID smart card project appears to have been cancelled. SOLID was the latest incarnation of Infineon's 3D IC technology. Their previous versions, using organic glues, were also cancelled before they reached commercial development.

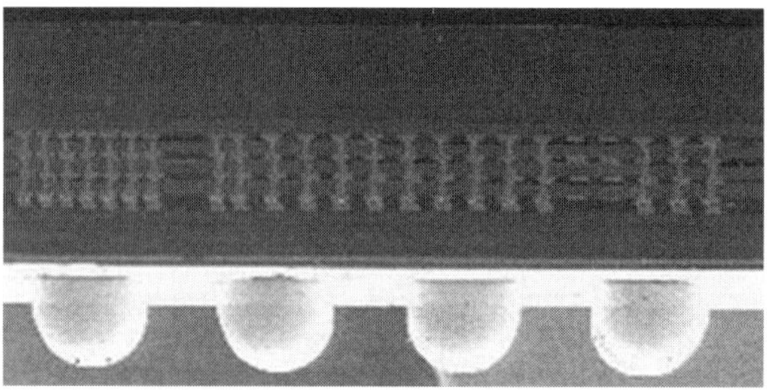

Fig. 12.5 Samsung's 3D DRAM Package with through-silicon vias (TSVs). Reproduced with permission from Kriss Lee, Samsung

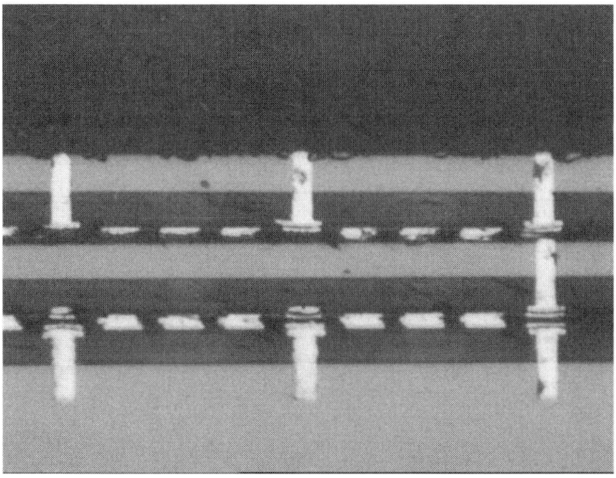

Fig. 12.6 A three-wafer stack showing seven Super-Via^{TM} interconnects

Many other promising 3D projects have also been abandoned. The fundamental reason is often a lack of joint buy-in and development between process engineering and design engineering. To create a successful 3D IC, process development and design development must proceed in tandem.

Tezzaron [5] built six different 3D IC designs as demonstration devices, including a memory-on-processor device and a two-layer SRAM. During the design of these devices, two different styles of vertical interconnect were developed for 3D ICs. The micrographs in Figs. 12.6 and 12.7 show both types. The second "Super-Contact^{TM}" type was ultimately used in the demo devices.

12.2.2.1 Separate Fabrication or Sequential Fabrication?

It is worth noting the difference between sequentially fabricated 3D IC layers and those that are created from separately fabricated layers that are stacked into a vertical assembly. The latter are often referred to as 2.5D ICs as opposed to the sequentially fabricated "true" 3D ICs. Sequentially fabricated devices never deal with more than

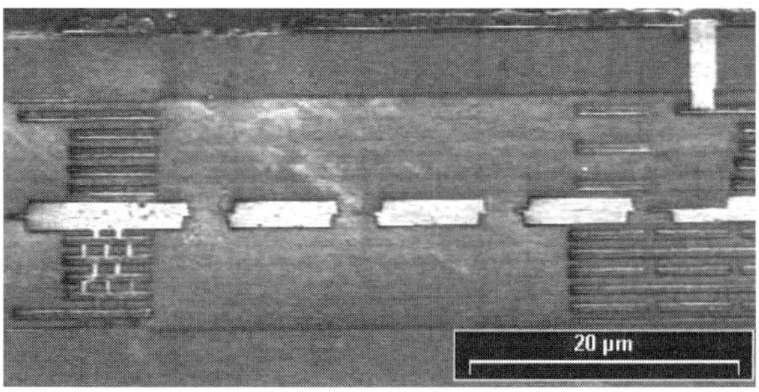

Fig. 12.7 A two-wafer stack showing one Super-Contact^{TM} at upper right

a single substrate during manufacture. These "true" 3D devices should have a significant advantage in the size and space requirements of the vertical interconnect.

12.2.2.2 Wafer Scale or Device Scale?

Another important distinction arises between 3D ICs manufactured at wafer scale and those built at device scale. Typically, 3D ICs are done at wafer scale (stacking whole wafers) and 3D packaging is done at device scale (stacking singulated dies), but the lines are blurred. Existing methods can produce both types of 3D devices at either scale.

12.3 DRAM

Three-dimensional integration can be applied to any type of memory. For the purposes of simplicity, this chapter addresses in depth only DRAM. Flash memory could easily be built the same way and have similar benefits. Indeed, hybrid devices of DRAM and Flash could also be made in a single stack. Three-dimensional SRAM could be built using similar methods, and would certainly have benefits, but the processes currently used for SRAM and the very high-speed nature of the existing devices mean that 3D SRAM offers fewer advantages than 3D Flash and 3D DRAM.

12.3.1 DRAM Manufacturing

Dynamic random access memory contains two very different types of electrical elements. Ideally, each type would be built in a different process. The core memory transistors must have very low leakage. On the other hand, the transistors that handle device I/O would perform best if they were made in a high-performance logic process like that used to build microprocessors; however, this cannot be done without compromising the core memory transistors. Table 12.1 summarizes the DRAM manufacturing issues.

Table 12.1 DRAM manufacturing issues

Memory elements	Desirable characteristics	Process requirements
Bit cells	Low leakage: – slow refresh – low power – low gate induced drain leakage	High-voltage Devices Vneg Well Thick Oxide
Sense amplifiers (sense-amps),Wordline drivers,Device I/O circuitry	High speed: – better sensitivity – better bandwidth – lower voltage	Low-voltage devices Copper interconnect Thin oxides

12.3.2 Simplified DRAM Overview

Dynamic random access memory is basically composed of arrays of memory cells surrounded by circuitry that writes data into the cells and reads (or senses) what is stored in the cells. Each memory cell consists of an access transistor and a capacitor to hold the stored voltage as long as possible.

The primary benefit of this very minimal design is its extremely tiny size. The standard for memory bit cell size is generally $8f^2$, where "f" is the minimum feature size of the process node. For example, at a 100-nm process node, the size of a DRAM memory cell should be $8 \times (100\,\text{nm})^2$ or $0.08\ \mu\text{m}^2$. Micron [6] produces a $6f^2$ memory cell, giving it a size advantage at the expense of more sophisticated processing technology.

The memory cell's access transistor couples the capacitor to the bitline. When the memory is written, the capacitor in the memory cell is either charged or discharged. During this operation the access transistor gate is driven to a super voltage, today about 1 V above V_{DD}. This voltage is internally generated in most DRAMs and allows the maximum amount of voltage to be stored on the capacitor. After the memory cell capacitor is charged (or discharged), the access transistor is turned off by driving the gate to ground, or even below ground. This effectively isolates and stores the bit for later retrieval. Figure 12.8 shows a simplified DRAM circuit.

12.3.3 DRAM Construction Issues

Because no capacitor can hold charge indefinitely, DRAM memory cells must be periodically "refreshed" (sensed and rewritten) to preserve the memory contents. The primary objective in a DRAM process technology is to minimize leakage and prolong the time between refresh cycles. Typically, the storage capacitor in a DRAM memory cell has a capacitance of 20–25 fF and a refresh period of 64 ms.

There are two basic capacitor types: trenched capacitors and stacked capacitors. The trenched capacitor is made by digging a very narrow hole into the silicon substrate, about 7–10-μm deep. The hole is coated with a dielectric and then filled with heavily doped polysilicon. The difficulty with this approach is the very high aspect ratio of the hole; common values range from 30-1 to 50-1. The stacked capacitor structure typically forms a cup-shaped device above the access transistor. The 3D nature of this type of capacitor also has its challenges. The cup may be as much as 3-μm high; this leaves a very uneven surface that must be filled and smoothed so that upper metal layers can be put in place.

The charge stored on the capacitor must be retained well enough to be correctly sensed 64 ms later. The stored charge is nominally close to either V_{DD} or ground. When sensed, the value is referenced to $V_{DD}/2$. Any charge greater than $V_{DD}/2$ is "1", less than $V_{DD}/2$ is "0". This is simple so far, but now for some ugly reality: the bitline connected to the memory cell (and its capacitor) has its own capacitance. The wire capacitance, plus capacitance from the other 128 to perhaps 1024 access transistors, plus the input capacitance of the sense amplifier itself – all this parasitic

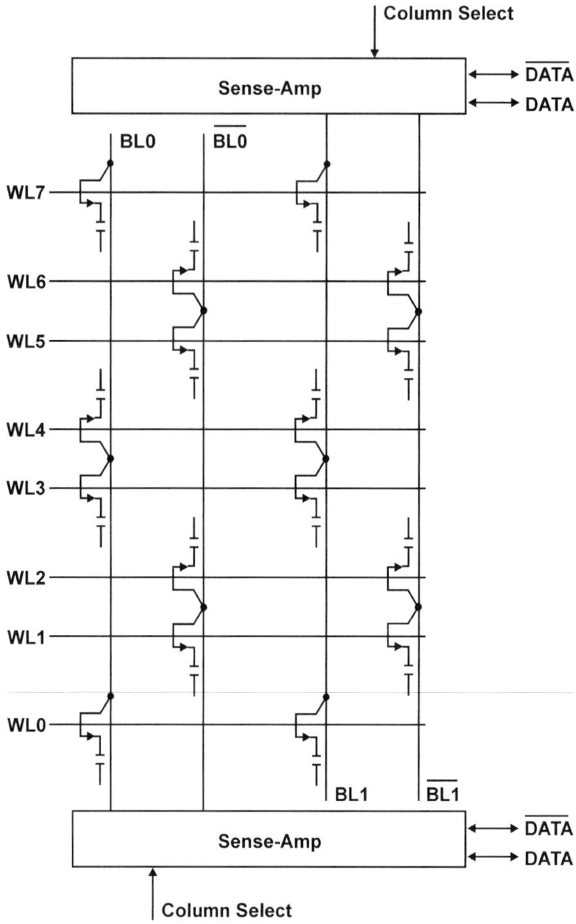

Fig. 12.8 A simplified DRAM circuit

capacitance totals ~200 fF. The result is that during an access to the memory cell its charge is distributed and balanced with the bitline. Hence, only about 10% of the memory cell capacitor's original voltage is present on the bitline. For a double-data-rate three (DDR3) memory this means a best-case bitline deflection of 75 mV, assuming the memory cell capacitor was fully charged to V_{DD} (1.5 V) and the bitline was perfectly precharged to $V_{DD}/2$. The 75-mV signal is further reduced by leakage from the original voltage, by low power supply voltage, by internal noise, and by offset in the sensing circuit.

To help deal with internal noise, bitlines are used in pairs. The sense-amp compare is done between an active bitline and an inactive bitline immediately adjacent to the one being sensed. These are typically annotated as bit and bitbar. (An interesting side effect of this architecture is that every other column in a memory array stores inverted data. If a DRAM contains all 1 s or all 0 s, half the capacitors in the memory cells actually hold V_{DD} while the others hold 0 V.) Bitline pairing is a significant help, but noise can be quite large due to the thousands of simultaneously firing sense-amps. Instantaneous currents can be measured in amperes! A fair estimate of noncancelling internal noise effects is 25 mV.

Offset in the sense-amp is caused primarily by transistor mismatch that comes from random manufacturing variations. As devices shrink, transistor mismatch increases. Historically, the mismatch was strongly related to transistor gate area variation, but a new effect can be seen as devices get smaller than 100 nm. This is the clustering of implant material, which can increase the offsets of adjacent transistors by an additional 50%. With this effect, a sense-amp can have an offset as high as 25 mV.

If we assume a worst-case power supply voltage of 1.4 V, then our stored charge and deflection are scaled to about 70 mV. Subtracting our noise and offset allowances leaves us with a 20-mV deflection or a 200-mV stored charge margin. This 200-mV margin is vulnerable to the deleterious effects of leakage. If we assume leakage to be a linear current sink, 20 fF leaking 20 fA would lose 1 V in 1 s. Thus, with a 200-mV margin, we can afford to leak about 200/64 $\times$ 20 fA or 63 fA. Of course, this ignores overdrive at the sense-amp (the overdrive requirement should reduce the available leakage margin by 25–50%), but it at least sets an outside limit.

There are four important leakage sources. The primary source is junction leakage. The drain of the access transistor where it connects to the storage capacitor has a reverse-biased pn junction. This will typically account for 80–90% of the leakage at an operating temperature of 85°C or higher. The next largest source is the subthreshold leakage across the access transistor. The threshold of the access transistor is kept high, typically in the 600–700-mV range. In addition, many DRAM designs actually drive the wordline, and thus the gate of the access transistor, below ground. Pumping the substrate or bulk negative with respect to ground lowers the subthreshold leakage by raising the effective Vt of the transistor, but this can increase the junction leakage at the drain and may also aggravate gate-induced drain leakage (GIDL). Normally, bitlines are kept at $V_{DD}/2$, which helps to limit this leakage, but if the same row is continuously being accessed, such as happens when a page is open, the leakage can increase.

Junction leakage and subthreshold leakage account for virtually all of the leakage that a modern DRAM sees today. The other two sources are just starting to become issues. One well-known issue is oxide leakage. The access transistor can leak into or out of the storage capacitor if the oxide is thin on that transistor. Thus DRAM access transistors have always had very thick gate oxides and they continue to do so. (This has an undesirable side effect of high transistor on resistance.) It is also important to note that the capacitor also has leakage across the dielectric. This dielectric is typically oxide, although high-κ dielectrics will likely be used in storage capacitors starting at ~65 nm. An associated effect of the thin oxide is GIDL which can be as large as the oxide leakage. Various drain implant techniques are being developed and employed to minimize this, but it can be seen in 65-nm logic processes and below. In logic processes, where leakages are typically reported in nanoamperes, femtoampere leakage is generally ignored.

From the discussion above, it becomes quickly apparent that everything that is done to make stored data more persistent comes at the expense of speed or size.

12.3.4 Array Efficiency

Array efficiency is a measure of the die area used by memory arrays versus the area used by wordline drivers, sense-amps, input/output (I/O), and other overhead circuitry. Over the years, constant improvements in DRAM design have increased the array efficiency. Some designs approached 80% efficiency at the time that synchronous DRAM (SDRAM) was introduced. Recently, that trend has suffered a significant reversal. On-die termination and lower driver impedances have allowed significant increases in bandwidth, but they consume much more die area. The addition of DLLs, burst hardware, and programmable pipelines for performance tuning has also added area. The number of bits on a bitline, which had been increasing, has reversed in trend; as voltage is decreasing, so are the bits per bitline. This means that more sense-amps are required for the same number of memory cells. Also, as technologies shrink, some electrical elements cannot keep pace, so they consume a larger proportion of the silicon area. Sense-amps, for example, must remain somewhat larger in order to control input offset error. Similarly, items like charge pumps, internal regulators, bandgaps, and other analog elements do not scale as well as memory cells. The net effect is that array efficiency is rapidly worsening. DDR4 is likely to continue the negative trend, as its data link cyclical redundancy check (CRC), further voltage reduction, and other planned features will take an even larger proportion of real estate on the die. Table 12.2 is a summary of typical array utilization.

12.3.5 The Memory Wall

Prior to the 1980s, memory improved in lock step with processors; whenever the speed of a processor doubled, the speed of the memory also doubled (according to Bob Johnson, former VP of Engineering at Burroughs and longtime computer architect, after whom the Johnson Counter was named). Since that time, DRAM has seen significant increases in the number of bits on a given device, but its performance improvement has been disappointing in terms of latency. Latency is defined

Table 12.2 Typical array utilization

DRAM type	Array utilization
SDRAM	78%
DDR1	70%
DDR2	47%
DDR3	38%
DDR4	Perhaps less than 30%

as the time that elapses between the presentation of a new address to the memory and the return of data from that address. During the last 25 years, latency for commodity DRAM improved from 200 ns to 45 ns – a factor of less than 5 – while processor performance increased by a factor of 100 000.

As this huge performance gap developed, monumental design and architecture refinements were invented to mask the problem. Multilevel cache memories, multi-threading central processing units (CPUs), out-of-order execution, and speculative execution are examples of the ingenious efforts to bridge the chasm between CPU and memory speeds. In spite of the speed gap, the availability of huge DRAM memories enabled the development of very large applications and the incredible user interfaces that we all depend on today. Small memories, no matter how fast, would not allow the existence of Windows (or Mac OS, if you prefer).

However, that was then – this is now. We have nearly exhausted our bag of design tricks to hide the poor latency of memory. Turning the bandwidth knob is essentially the last enhancement available, and even that is approaching its limits. Bandwidth has doubled with each of the last few memory generations, but each generation is taking longer to release. DDR2 managed to achieve 1066 Mega Transfers per second (MT s^{-1}); DDR3 is planned to reach 1600 MT s^{-1} and may extend to 2133 MT s^{-1}; but the next doubling (from 1066 to 2133) will likely take 3 or 4 years. DDR4 is planned to obtain a 3.2 GT s^{-1} rate, but that is perhaps 7 years away.

A paper early in 2007 suggested that, at minimum, multicore processors in 2010 will need several data channels 128-bits wide, running at 1600 MT s^{-1} (mega-transfers per second) – one such data channel for each CPU core. It follows that an eight-core microprocessor would require 1024 bits of data bus providing 1.6 GT s^{-1} for reasonable memory subsystem performance. Other than the large number of I/O pins on the CPU and the nearly impossible routing on the motherboard, there is the small matter of the 40 W+ of power just to support the data bus. The whole memory interface could use 55 W or more. While this might be doable, certainly the next generation of 16-core processors will not be able to scale. The inevitable conclusion is that change is required. Memory must address latency and find a new method to increase bandwidth.

12.3.6 Repair and Redundancy, Test, and Reliability

Virtually all DRAM made today has zero yield when initially tested. Repairs are made by activating redundancies built into the arrays. At an early production node, the yields even after repair might be as little as 20% or 30%. These numbers rise significantly as the node matures. Historically, the final tested yield of DRAM devices has been greater than 90%. But as the geometries shrink, the sheer number of transistors and capacitors increases, and each successive generation of memory devices has more and more complicated peripheral circuitry; as a result, ultimate yields have shrunk. A DDR3 memory at 1 Gb density may reach only a 70% or 75% yield.

There are two basic types of defects in deep submicron devices. The first class of defects, usually caused by particle contamination, damages several memory cells at a time. Because today's dense memory arrays have rows and columns in very close proximity, the defect caused by a single 1-μm particle can span several rows and/or columns – a close cluster of flaws affecting many bits in a small area. Built-in redundancies are designed to compensate for this, within the limitations presented by available wiring density and simple area costs. Typically, substitutions are made in groups of four or eight rows of bits and two or four columns.

The second class of defects damages a single transistor or capacitor. A single-bit defect may originate as a particle, a material contaminant, or a simple random variation. Keep in mind that there are billions of devices on a memory die. The fact that we can build devices with parts per billion of failure is amazing by itself. The granularity of repair, however, means that a single defective bit may force the substitution of as many as 8 K bits. Improving the granularity would require more control and complexity and more interconnect requirements, but the memory array size is already limited by the pitch of metals. Just as DRAM logic speed suffers greatly from the process compromises that improve data retention, more complicated redundancy creates additional latency issues for which there is no budget. Simply adding more repair bits is not an answer, either. First, it affects cost by adding to the die area; second, DRAM (like all logic today) has issues with wire delay. Wire resistance is increasing as well as capacitance. Time-of-flight requirements mean that repair elements must have close proximity to the originals.

Testing itself is expensive. Leading-edge memory testers that process 256 or even 512 memories in parallel can easily cost millions of dollars. Each memory cell must be tested to ensure it can be written and read back as both "1" and "0". Additionally, leakage between adjacent memory cells can cause neighborhood pattern-sensitive faults (NPSF) that give rise to peculiar problems. For example, one memory cell might fail only when storing "1" while both adjacent memory cells are storing "0". There are 192 different patterns related to each cell in an array. If memories were exhaustively tested to check all possible patterns under worst-case conditions, it would take hours to test each part. Even with highly parallel testing, this is cost prohibitive. Manufacturers spend a great deal of time and effort understanding how their devices tend to fail, so they can significantly reduce the number of required tests. Even so, testing is time-consuming and costly; a 1-Gb DRAM can spend several minutes on a tester. Indeed, for today's mature high-yielding memories (e.g. 64-Mb SDRAM), test is the single largest cost.

Proper process screening and test are generally assumed to ensure reliability; we expect very few ICs to fail in the lifetime of a product. Unfortunately, the large number of transistors and their near-atomic scale are accelerating device wear-out. Soon enough, we may come to expect device wear-out within the usable lifetime of any IC. This is especially true for memories, as they typically contain more active elements (transistors and capacitors) than even the most sophisticated processors – and a typical system usually contains dozens more memories than processor chips.

Table 12.3 Inter-related effects of 3D IC DRAM

Characteristic	3D benefit
• Power	10%–80% lower
• Clock speed	10%–50% faster
• Latency	0%–87% lower
• Cost	–50% to +15%

Table 12.4 Independent effects of 3D IC DRAM

Characteristic	3D benefit
• Density	200%–500% higher
• Weight	20%–400% lower
• Soft error rates	Generally better
• New functionalities	Significant
• Specifications, e.g., operating temperature	Much better

Given this, a compelling need is arising to have devices "repair" themselves in the field. Today's memories are designed for factory repair; the decisions of how and what to repair are made by large pieces of test equipment and their associated computers. To provide field reparability, all these capabilities would need to be added to the humble memory device. The DRAM process, because of its low transistor performance, is hardly suitable for manufacturing built-in test hardware. Instead, error-correction code (ECC) circuitry built into CPUs and chipsets may be needed in order to mask the memory errors.

12.3.7 3D Benefits in DRAM

The preceding text provides at least a basic understanding of DRAM and the issues that currently limit its manufacturing and performance. What can 3D do to improve DRAM and meet future memory needs?

Generally, 3D packaging provides an increase only in circuit density. This benefit normally comes at an increase in cost and a decrease in yield. Three-dimensional ICs, however, can provide a wide array of benefits. The anticipated impact of 3D IC DRAM is summarized in Tables 12.3 and 12.4 below. Note that the values in the first table are interrelated and will cause design tradeoffs. For example, the very best cost and the very lowest latency cannot be achieved simultaneously, although both cost and latency can improve compared to an equivalent 2D component.

12.4 Building 3D DRAM

This section outlines one approach to building 3D IC DRAMs, based on the capabilities of the Tezzaron wafer-scale 3D process.

First, the DRAM circuitry is separated into layers, dividing the elements by their manufacturing considerations. Memory cells will be fabricated on dedicated wafers in a low-leakage DRAM process. The rest of the DRAM elements (sense-amps, wordline drivers, I/O, etc.) will be fabricated on different wafers in a high-performance logic process. This process separation takes advantage of the best of both logic and DRAM processes.

As shown in Fig. 12.9 below, several layers of memory cells can be stacked on a single logic layer. An interesting empirical result is that the best fit for matching the logic wafer to the memory wafer, in terms of matching the area requirements of the layer constituents, is to build the logic wafer in a process that is a half node larger. For example, 110-nm DRAM wafers would best match 130-nm logic wafers.

Separating the elements to this degree requires a great deal of vertical integration, so the vertical interconnect (TSV) must be capable of achieving a tight pitch. In Tezzaron memories, bitline multiplexing and local subwordline decoding relax the pitch somewhat. The resulting pitch requirement is eight times the bitline pitch in current 110-nm memories. Tezzaron uses an $8\,f^2$ memory cell DRAM process today for its memory wafers, with a bitline pitch twice the node (0.22 μm) and wordline pitch four times the node (0.44 μm). This translates into a required TSV pitch of $8 \times 220\,\text{nm}$ or 1.76 μm. In future memories, a factor of $16x$ will likely be necessary. While TSVs used in the memory wafers must be at this tight pitch, the actual bonding pitch can be made larger by fanning out the signals prior to the wafer-to-wafer interconnect. The fanout can be done either before the bond layer or on the bond interface layer itself.

Bitline multiplexing creates one design side effect: the need to sequentially access four separate bitlines in the memory array. This adds some complexity to the sense-amp but, for the other improvements, this is a small cost.

12.4.1 Latency

Bitlines and wordlines are typically fabricated in polysilicon. Because of the high resistance of this material, the RC propagation delay can be several nanoseconds. To

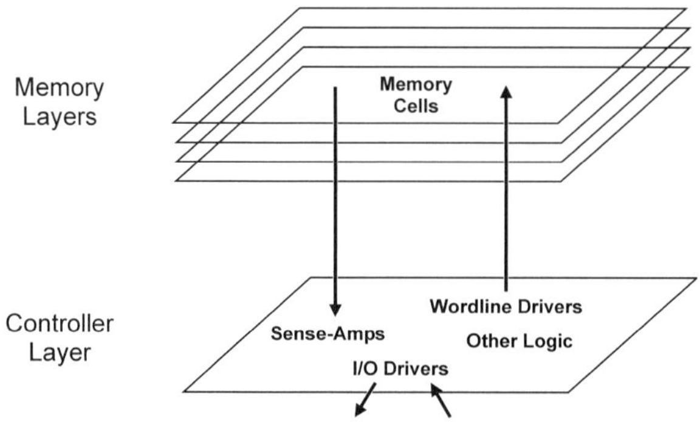

Fig. 12.9 Process-separated DRAM layers

improve this, Tezzaron cuts the length of these wires in half, proportionately reducing both the number of bits on each wire and also the total capacitance on each wire. Propagation delay is proportionate to the square of the wire length. Theoretically, halving the bitline length should return a $4x$ improvement in speed; in practice, the result is a $3x$ improvement. This is a great start.

Further direct speed gains are realized by implementing the sense-amp and wordline drivers in a high-performance logic process. These elements achieve a $3x$–$5x$ improvement over their DRAM process counterparts. The secondary memory data buses, the I/O pipeline, and I/O itself tend to have additional gains.

The bottom line is a latency reduction of 66%–87% over a typical 2D DRAM. Note that if the memory is stacked four layers high, as in the diagram, there is exactly the same number of sense-amps and wordline drivers, proportional to the number of memory bits, as in the original 2D device – but with vastly improved performance.

12.4.2 Power

Shorter bitlines and wordlines reduce the necessary access power by as much as 40%. In addition, logic functions that are implemented in a logic process rather than the original DRAM process are 20%–50% more power efficient. At the I/O level, the additional performance available in the p-channel devices allows a 20% reduction in the I/O capacitance. This directly lowers signaling power while reducing the size of the pad predrivers and the power they consume.

If desired, the power savings could be "spent" on faster operation. Lower latency allows faster cycle times, so the 3D DRAM could be run at about a 35% faster clock rate and consume the same power as an equivalent 2D DRAM.

12.4.3 Reliability

Half-length bitlines improve the signal level by more than 75%. This gives better data integrity – which, if desired, can be "spent" to extend the refresh period or to operate at a higher temperature.

Tezzaronfurther enhances its 3D DRAM reliability by building its Bi-STAR engine into each chip as shown in Table 12.5. Bi-STAR^{TM} performs ECC, dynamic memory management, and on-chip test and repair. These functions consume additional silicon area, but they provide large positive tradeoffs (as discussed in the next section on cost). ECC obviously adds the benefit of soft error masking, but it also works to locate and diagnose hard failures. Bi-STAR circuitry audits the entire memory at every power-up, performing hundreds of memory tests and using built-in redundancy to replace any of a variety of possible hard failures. Beyond power-up, Bi-STAR can replace bad bits or rows "on the fly" during normal operation. This feature can extend a DRAM's mean time between failures (MTBF) by an order of magnitude.

Table 12.5 Bi-STAR^{TM} reparability (Data from Tezzaron)

Reparable hard failures
• Bad memory cells • Leaky bits • Shorted bitlines • Shorted wordlines • Bad sense-amps • Bad CAMS • Bad line drivers • Bad secondary bus drivers

Although Bi-STAR could be built into a normal 2D DRAM device, it would not be practical. The normal DRAM process would retard much of the critical timing path, raising the latency and lowering the bandwidth. Further, Bi-STAR performs its power-on memory tests in ~64 ms on a 3D device. To accomplish this feat, it uses comparator-per-bitline testing, a technique that tests hundreds of thousands of memory cells in parallel. To enable comparator-per-bitline testing in a 2D DRAM, the memory array size would increase by as much as 100%.

12.4.4 Cost

Overall, the combined cost savings derived from 3D can be as high as 50%. Certainly, some cost is added as each layer is bonded and thinned, and as TSVs must be added to all but one of the layers in the stack. This said, the cost reductions far outweigh the added stacking costs. The savings come from four major sources: memory wafer processing, array efficiency, testing, and yield.

12.4.4.1 Memory Wafer Processing

The processing of memory-only wafers for a 3D stack can cost 30%–40% less than normal 2D DRAM wafers as shown in Table 12.6. Because there are no logic elements, memory-only wafers can be produced in as few as 15 mask layers – half the number of steps used by leading-edge DRAM processes.

12.4.4.2 Array Efficiency

Memory wafers contain only memory cells, TSVs, and a small amount of decoding. The array efficiency is close to 90%. This compares to 38% for a 2D DDR3 memory. Of course, 3D carries the overhead of the logic wafer in the stack, but even taking this into account, the array efficiency of a two-layer stack composed of one memory wafer and one logic wafer is ~45%. A stack with four memory layers and a single logic layer can reach 72%, and a stack with eight memory layers can reach 80%.

Table 12.6 DRAM manufacturing steps (Data from Tezzaron)

2D DRAM wafer: cells and supporting circuitry	3D DRAM wafer: cells	3D DRAM wafer: supporting circuitry
3 Well masks	1 Well mask	3 Well masks
3 Oxide masks	1 Oxide mask	2 Oxide masks
5 Implant masks	2 Implant masks	4 Implant masks
2 Poly masks	2 Poly masks	1 Poly mask
1 Contact mask	1 Contact mask	1 Contact mask
5 Metal masks	3 Metal masks	7 Metal masks
1 Passivation mask	0 Passivation mask	0 Passivation mask
3 Capacitor masks	3 Capacitor masks	0 Capacitor masks
0 Stacking masks	2 Stacking masks	2 Stacking masks
23 Masks	15 Masks	20 Masks

The process outlined here is one of many currently in use. "Typical" DRAM processes can use from 15 to 35 mask steps.
The number of stacking masks will vary by process; Tezzaron uses two per wafer.

12.4.4.3 Testing

Testing can make up 25% of a DRAM's final device cost. Tezzaron's Bi-STAR can reduce test cost by 80% while providing a far more exhaustive battery of testing than is reasonable at the factory today.

12.4.4.4 Improved Yield

Three-dimensional memory provides enhanced reparability. First, the shorter rows and columns enable a finer-grained repair so that each repair consumes fewer of the available redundant memory cells. Also, redundant elements can be shared from layer to layer within a 3D stack. As an example, if a repair requires more replacement rows than are available in the local layer's array, they can be borrowed from any other layer. For clustered defects caused by particles, the ability to use spares on other layers significantly improves reparability. Other abilities, such as utilizing a bit cache on the logic layer, can further enhance the recoverability of highly defective memory wafers. The dominant failure mechanism in a 3D DRAM becomes the nonredundant areas of the logic layer. This can be as little as 3% of the entire silicon area of the device. In all, yields in excess of 95% are possible in 3D devices that use leading-edge technology.

Bi-STAR can also increase the yield of an SIP. If a memory device has Bi-STAR, it can perform a complete at-speed test and repair even after assembly into an SIP device. Additional defects that might develop from the assembly process, especially those thinning-related defects widely reported in the industry, are easily addressed with Bi-STAR. Known good die does not directly allow a method for full at-speed test, nor does it provide a post-assembly repair methodology. Three-dimensionality enables Bi-STAR, and this in turn enables much higher memory content in SIPs without the yield issues currently associated with multiple memory dies. Table 12.7 is a summary of Bi-STAR technology.

Table 12.7 Summary of Bi-STAR™ technology (Data from Tezzaron)

Bi-STAR™ summary
• Innovative way to improve the yield of highly parallel structures such as memory • Basis: integration of intelligent self-test, self-repair • Performs greater level of testing than normally available during normal chip-level or wafer-level testing • Tests and compares >300 000 nodes or bits/clock cycle; more than 1000 times faster than any external memory tester

12.4.5 Other Benefits of 3D Memory

A number of other benefits can be realized with 3D DRAMs. Lower soft error rates result from built-in ECC as well as from thinner substrates, which provide less interaction volume for incident particles or energies. Density improvements are obvious, as each additional layer of memory in the 3D stack adds a mere 12 μm in height. Also, the lower capacitance at the I/O pins can be directly translated into additional bandwidth by improving signal integrity and allowing higher speeds.

12.5 Tezzaron 3D Process Flow

Described here is the Tezzaron 3D process used to create 3D DRAM chips. This is a wafer-to-wafer process using a type of TSV called Super-Contact™. Tezzaron also performs chip-to-wafer bonding and can use a second type of TSV called Super-Via™, but the process outlined here is best for DRAM. The sequence shown builds two levels of memory cells onto a control wafer.

The process begins on each wafer after the transistors have been created but before the contact metal. On each wafer, Super-Contacts are etched through the oxide and into the silicon substrate, approximately 6-μm deep. The walls are lined with SiO_2/SiN, then each Super-Contact is filled with tungsten and finished with chemical–mechanical polishing (CMP) as shown in Fig. 12.10. This completes the unique processing requirements at the wafer level.

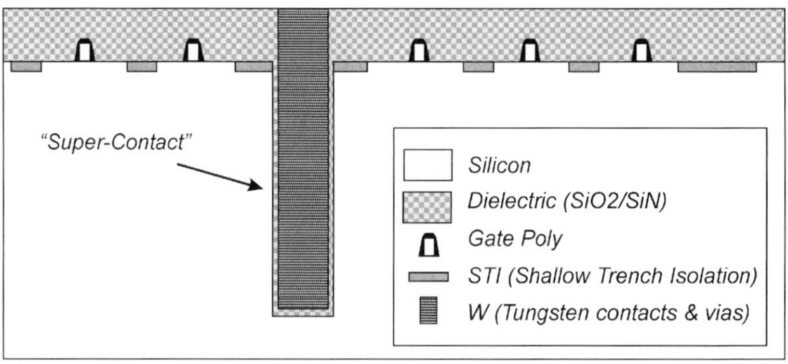

Fig. 12.10 Cross-section of wafer showing one Super-Contact

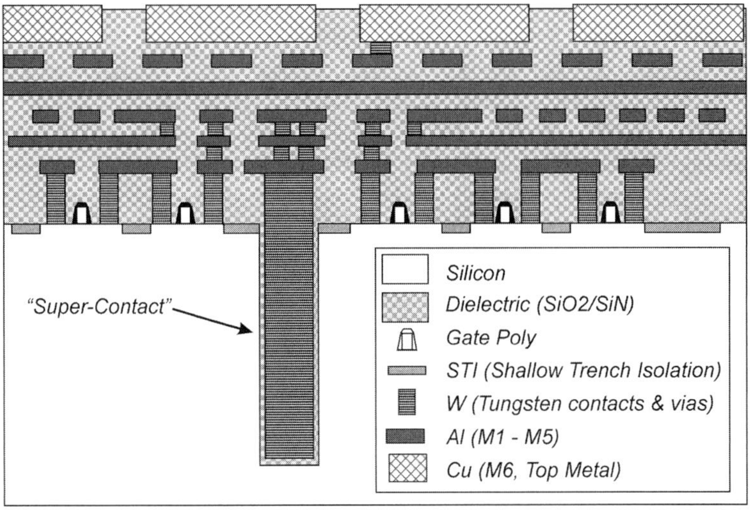

Fig. 12.11 Cross-section of a wafer ready for bonding

After the Super-Contacts are in place the wafers are finished with their normal processing, which can include a combination of aluminum and copper wiring layers. The last layer on each wafer must be copper in order to provide bondpoints for stacking. The oxide surfaces are then slightly recessed between the copper bondpoints. See Fig. 12.11.

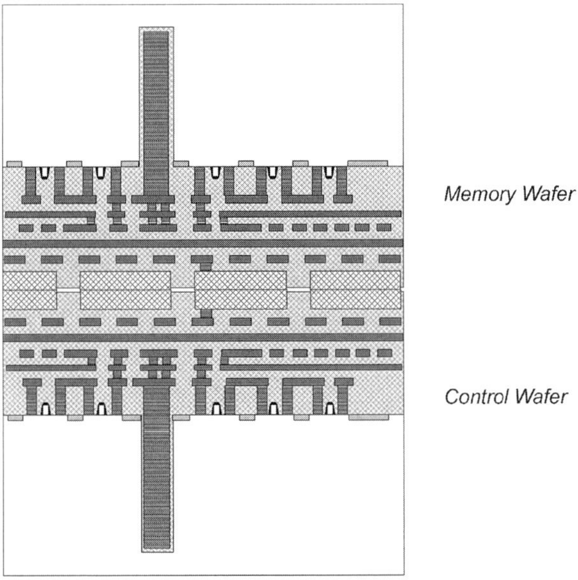

Fig. 12.12 Two-wafer stack, memory cell wafer on top of control wafer

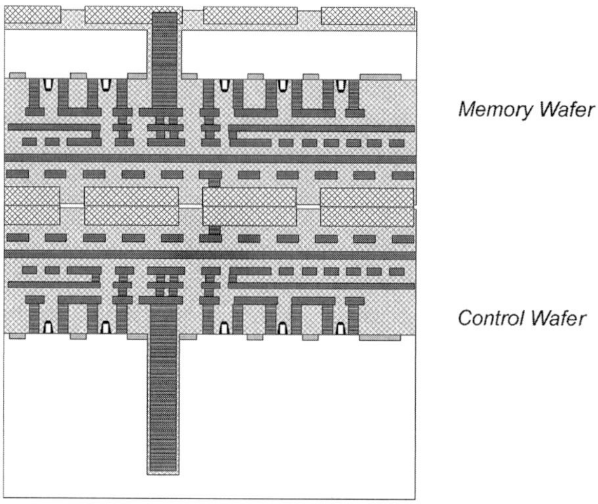

Fig. 12.13 Two-wafer stack ready for addition of third wafer

In Fig. 12.12, the first memory wafer is aligned face-to-face with the control wafer and bonded in a copper thermal diffusion process that takes place at approximately 375°C.

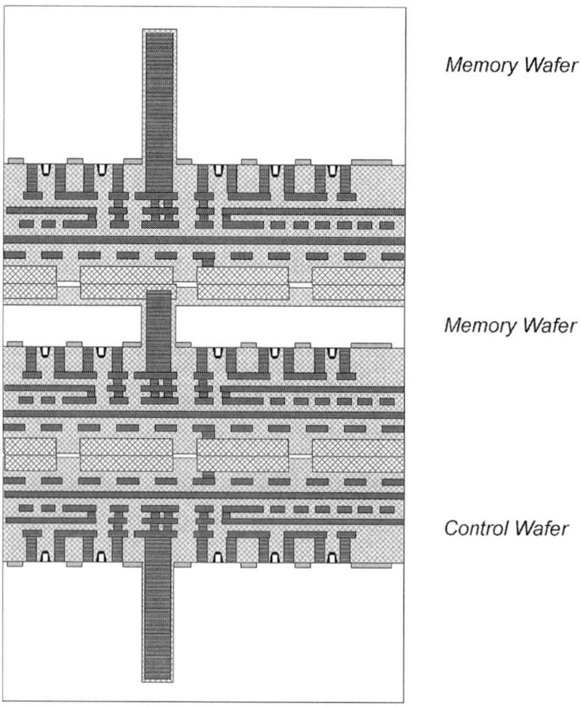

Fig. 12.14 Three-wafer stack, two memory cell wafers on a control wafer

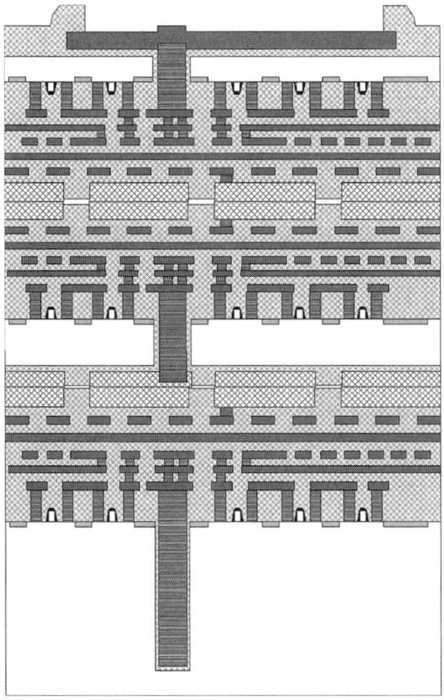

Fig. 12.15 Finished three-wafer stack

After bonding, the back of the memory cell wafer is thinned to the bottom of its Super-Contacts. This leaves a substrate thickness of about 5 μm. Thinning is done with a combination of wafer grinding, CMP, and etching. The backside of the thinned wafer is covered by an oxide, then a single damascene copper process creates bondpoints for subsequent stacking. The oxide surface is slightly recessed between the copper bondpoints as before. This is depicted in Fig. 12.13.

The second memory cell wafer is added to the stack with its face bonded to the back of the first memory cell wafer, using the same techniques as the first bond. See Fig. 12.14.

The stack is then inverted so that the control wafer (the first wafer) is on top for final processing. The control wafer undergoes the same thinning process used before, stopping on its tungsten Super-Contacts. Instead of a copper damascene process for bondpoints, an aluminum layer is deposited for normal wire bonding. This is shown in Fig. 12.15

12.6 Embedded Memory versus Stacked Memory

The ever-increasing abundance of transistors in electrical products has given rise to ever-increasing integration. In many cases, integration requires the joining of devices that have dissimilar processing requirements. Most modern silicon-on-chips

(SOCs) would clearly benefit by being built in multiple different semiconductor processes.

A good example is embedded memory. On the plus side, embedded memory can implement a very wide bus for increased bandwidth. Also, as discussed earlier, building DRAM in a logic process can improve memory latency. On the negative side, adding DRAM or Flash memory to a logic die adds extra masking steps to the process and produces memory with significantly less density than a dedicated memory process. In the specific case of DRAM, embedded arrays have only 1/5 or 1/6 as many bits per square millimeter.

One alternative to embedded memory is 3D packaging. In many cell phones today, DRAM is chip-stacked onto a logic device. This solution avoids the extra processing steps and keeps the memory arrays at high density, but it does not provide the extra bandwidth or reduced latency of embedded DRAM. In addition, the use of separate chips requires complete I/O drivers with ESD structures, just as if the chips were mounted side-by-side.

A better solution is to create a memory device that is designed specifically for intimate 3D assembly. There are a variety of techniques that perform die-to-wafer assembly with very fine pitch and thousands of interconnects between a die and a host wafer. Die attach processes available from Tezzaron, Ziptronix, Cubic Wafer, and a number of other sources can achieve vertical interconnect pitch of 25 μm or less. If both a memory device and a host device were designed to take advantage of this, they could be intimately connected as a single IC (no I/O drivers or ESD) with a very wide bus for maximum bandwidth. At least one specification already exists for this type of attachment: the Intimate Memory Interconnect Specification (IMIS), available at 3D IC Alliance website (http://www.3D-IC.org). This represents a first attempt at providing off-the-shelf solutions for SOC developers to implement 3D solutions. The ultimate goal is to build 3D systems with the cost and simplicity of separate IC components while retaining SOC-like density and performance.

Tezzaron already offers a device called Octopus™ that conforms to the IMIS specification. This may be the first foray into this new field of intimate memory, but other well-known semiconductor companies are looking at similar solutions. Intel recently described a hypothetical CPU with 1 Gbit of DRAM stacked between the CPU die and the package substrate. Given the issues with memory latency and bandwidth, it is obvious that intimate processor/memory solutions are bound to happen. It is not a question of "if", but of "when".

12.7 3D Futures

Three-dimensionality has a great deal of potential to change the way designers approach new chips and even entire new systems. Various studies show that 3D ICs can provide anywhere from 10% to 1000% improvement in performance. The key difference in these studies seems to be the level of fundamental redesign that is envisioned. To realize the most significant gains, 3D will require a clean new look at each design. This is perhaps the largest reason that 3D adoption has taken so long. Three-dimensionality does not work well as an evolutionary progression; process changes and design changes of relatively radical scale must be undertaken simultaneously. As described in this chapter's section on DRAM, 3D enables many improvements, but success requires design changes that are possible only in 3D. An incremental adoption of either process-only or design-only would not yield a viable commercial device.

Another fascinating potential in 3D is the mixing of dissimilar materials, also known as heterogeneous integration. This opens new memory possibilities that cannot be enabled by any other method. Because 3D can stack materials that do not commonly coexist with CMOS processing, there are opportunities for hybrid construction of magnetoresistive RAM (MRAM), ferroelectric RAM (FeRAM), phase-change memory, carbon nanotubes, and any number of other exotic technologies. Separating the memory cell technology from the sense, drive, and I/O circuitry may be fundamentally enabling to many of these new memory materials.

Memory must move closer both physically and architecturally to the computing engines it serves. Just as important, memory, which has had virtually no performance gain compared to logic, must be accelerated to feed those computing engines at appropriate speeds. Because 3D integration is the only way to adequately address both of these issues, it is likely to become the norm for memory as well as for logic within a very few years.

References

1. http://www.irvine-sensors.com
2. http://www.elpida.com
3. http://www.samsung.com
4. Herner SB et al (2004) Vertical p-i-n polysilicon diode with antifuse for stackable field-programmable ROM. IEEE Elec Dev Lett 25:271–273
5. http://www.tezzaron.com
6. http://www.micron.com

Chapter 13
Circuit Architectures for 3D Integration

Nisha Checka

13.1 Introduction

The recent electronics revolution has been fueled by the decades-long trend of exponential growth in circuit performance through device scaling. In current and future technologies, simple device scaling does not result in the same performance improvements. For deeply scaled technologies, optimizing interconnect performance is of equal importance as device performance. Following the International Technology Roadmap for Semiconductors (ITRS) specifications for scaled interconnect, worst-case and even average-case interconnect performance decreases with each technology generation [11]. Because the performance improvement achievable through Moore's law scaling is bottoming out, new technologies to achieve performance enhancement are being explored. Three-dimensional integration is one such technology.

Three-dimensional integration offers several performance improvements for electronic systems. First, 3D integration offers a greater device density for a given footprint area. Essentially, more functionality can be packed into a given area. Second, long, global wires can be replaced with shorter, local wires by exploiting vertical interconnect used to connect device layers. This results in lower power dissipation and shorter timing delays. Finally, a 3D technology permits the integration of heterogeneous technologies. Using a wafer-bonding approach to 3D integration, each device layer is fabricated independently of every other device layer allowing each system to be fabricated in a different technology. Circuits that are part of system-on-a-chip (SOC) applications can be fabricated in the device technology that yields optimal performance and then subsequently integrated together using a 3D integrated technology. Because each subsystem is fabricated in the optimal technology, optimal overall system performance can be achieved.

The implication of each benefit of 3D integration on circuit architectures is the subject of this chapter. Section 13.2 describes SOC applications. Section 13.3

N. Checka
Massachusetts Institute of Technology, Cambridge, MA, USA
e-mail: nchecka@ll.mit.edu

C.S. Tan et al. (eds.), *Wafer-Level 3D ICs Process Technology*,
DOI: 10.1007/978-0-387-76534-1_13, © Springer Science+Business Media, LLC 2008

describes the impact of 3D integration on two key types of digital systems. Section 13.4 describes how 3D integration can be used with nanoscale technologies. Finally, section 13.5 describes how 3D integration compares to planar packaging technologies.

13.2 Three-Dimensional SOC

Three-dimensional integration can be used to integrate disparate technologies in a 3D stack. As a result, it exhibits much potential for SOC applications. To implement an SOC in a standard planar technology, a single-wafer fabrication technology is chosen for all subsystems. Typically, the optimal technology is different for each subsystem.

For example, consider the SOC depicted in Fig. 13.1a. This SOC contains the full gamut of technologies: a display, DC–DC converters for power management, a micro battery, microelectromechanical system (MEMS) sensors, and data converters to convert the sensor data for further processing by the microprocessor which requires memory for data storage. The SOC also contains a radio frequency (RF) link to send and receive data and a digital signal processor (DSP) to process the received data. A high-speed technology such as GaAs would yield better RF performance than the low-power digital complementary metal-oxide

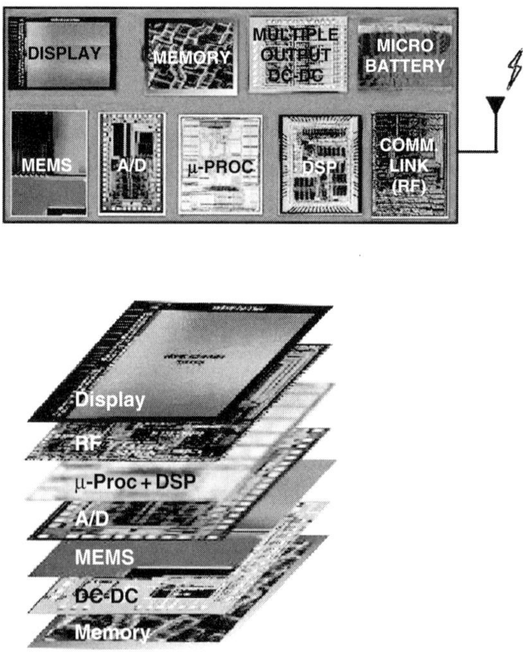

(a)

(b)

Fig. 13.1 Silicon-on-a chip implementations: (**a**) 2D SOC; (**b**) 3D SOC

semiconductor (CMOS) process that is the preferred fabrication technology for the digital subsystems. A standard planar technology would require a single fabrication technology; thus, the individual subsystems operate suboptimally compromising overall system performance.

Three-dimensional integration allows for the integration of heterogeneous technologies. Each subsystem could be fabricated in its optimal technology and subsequently bonded together to create the 3D integrated stack shown in Fig. 13.1b. This results in simultaneously achieving optimal subsystem performance as well as optimal overall system performance. Parasitic coupling issues as a result of vertically integrating the subsystems in close promixity can arise. This is the subject of the next section. Packaging approaches can also be used to implement heterogeneous technology SOCs. A comparison between packaging approaches and monolithic 3D integration approaches is discussed in Section 13.5.

13.2.1 Parasitic Coupling

Cost, performance, and portability drive the push for increased system integration on a single chip. Such integration necessitates the coexistence of analog and digital systems on the same chip. The integration of analog and digital systems on a single chip not only results in a smaller form factor but also large performance improvements. However, this integration is not without challenges [5]. Parasitic interactions induced by switching digital circuits can adversely affect the performance of the analog subsystem through propagation via the shared substrate. Three-dimensional integration can mitigate this parasitic coupling. The analog and digital subsystems could be fabricated on separate substrates breaking the resistive connection through the substrate. However, coupling between the subsystems can still occur vertically through the bonding interface.

Because the subsystems are vertically stacked, several coupling paths arise. First, consider face-to-face stacking where the two device layers are integrated such that the interconnect of each device layer faces each other. This is depicted in Fig. 13.2a. In this case, the interconnect of one subsystem can couple noise to the interconnect of the other subsystem. Furthermore, the interconnect of device layer 1 can induce noise into the substrate of device layer 2; however, this effect is mitigated by interconnect shielding and the large separation distance to substrate 2. In face-back stacking, the two device layers are integrated such that the interconnect of the bottom device layer faces the substrate of device layer 2 as shown in Fig. 13.2b. In this case, the interconnect of device layer 1 induces noise in the substrate of device layer 2. Because the interconnect of device layer 1 is closely spaced to substrate 2, the coupling can be quite significant.

To determine the amount of vertically coupled noise, simulations using HFSS [1], a 3D field solver, were performed [4]. Because the coupling in a face-to-face configuration is largely dependent on the interconnect pattern and the amount of shielding from surrounding interconnect, this structure was not simulated.

Two sets of simulations were performed on the face-back structure to determine the effect of the bonding material on the coupling mechanism. MIT's 3D integration

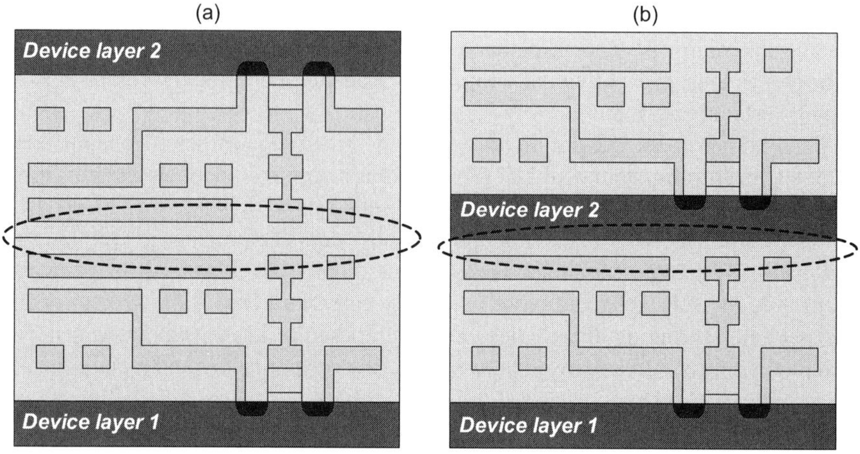

Fig. 13.2 Device layer configurations. (**a**) Face-face; (**b**) Face-back

technology uses copper as the bonding interface [23], whereas the MIT Lincoln Laboratory technology uses oxide as the bonding material [3]. One set of simulations assumed a floating copper bonding interface while the other assumed oxide as the bonding material. The family of curves generated for each set of simulations correspond to a change in the geometry of the test structure as shown in Fig. 13.3a and b.

Figure 13.4 shows the results of the simulation. The use of a copper bond provides an additional 10–20 dB of isolation (depending on the geometry) over an oxide bond. The improved isolation from the copper bond is due to the role of fringing fields. Field lines originating on the metal line of the bottom device layer terminate on the copper bond rather than on the substrate. This reduces the coupling between the metal line and the top substrate. In the structure with the oxide bond, the field lines terminate on the top substrate. More fringing fields terminate on the bond as the bond area increases resulting in improved isolation. In the extreme, the bond can have an area equal to the entire die area. If copper is used as the bonding material, the bond can be grounded and thus acts as a ground shield between the two substrates [18]. The curve labeled "ground" in Fig. 13.4 shows the ground plane

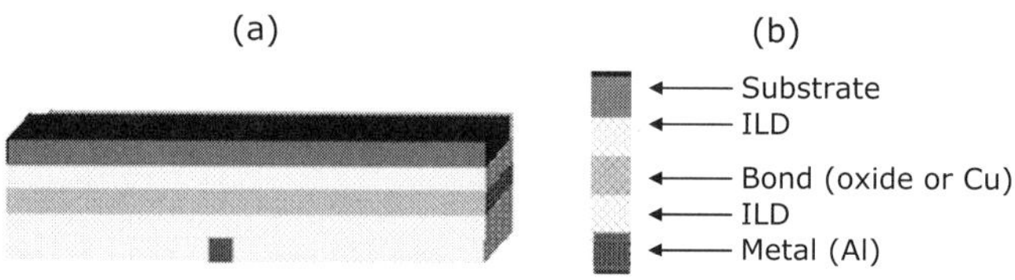

Fig. 13.3 Test structures used in HFSS simulations. (**a**) Three-dimensional test structure where all geometries are larger than the interconnect; (**b**) Three-dimensional test structure where all geometries are of the same size

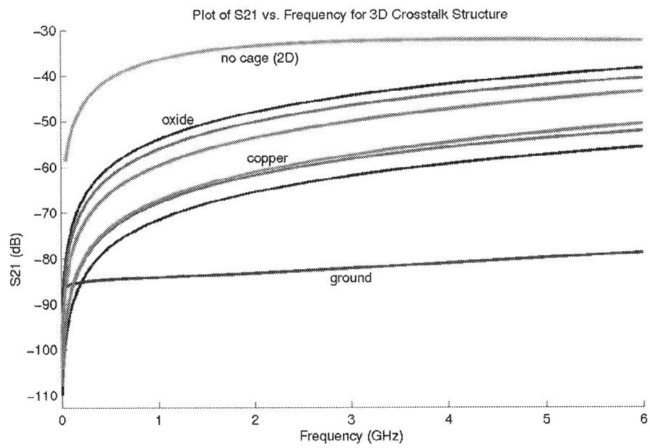

Fig. 13.4 Isolation of the various test structures

isolation. In this case, as much as 85 dB of isolation can be achieved resulting in near-perfect isolation between the two layers. This isolation can also be achieved in a 3D integrated technology using oxide bonding by using one of the device metal layers as the ground plane.

13.3 Digital Systems

By using a 3D integrated technology, some digital architectures achieve greater performance enhancement than others. These architectures best leverage the improvements that 3D integration provides. Such applications fall into one of two categories. The first class of applications exploits increased device density to achieve enhanced performance. The second class of applications exploits the shift in interconnect distribution through the use of vertical interconnect. Interconnect typically dominates the performance of this class of circuits. Each application is discussed further in this section.

13.3.1 Imager Systems

Three-dimensional integration offers significant improvements in circuit wirelength metrics. The additional device layers increase the number of nearest neighbors for a given transistor. Thus, every transistor, gate, and module has increased wiring bandwidth. Classes of circuits that heavily rely on device density to achieve enhanced performance exploit the device density advantage offered by 3D integration. Active focal plane architectures are one such class of circuits. Imagers consist of photodetection elements and readout circuits. The readout circuits are typically implemented in a charge-coupled device (CCD) or CMOS technology and usually consist of data converters to interface to the photodetection elements, some signal amplification circuits, as well as some circuits to manage noise levels. Imagers implemented in a

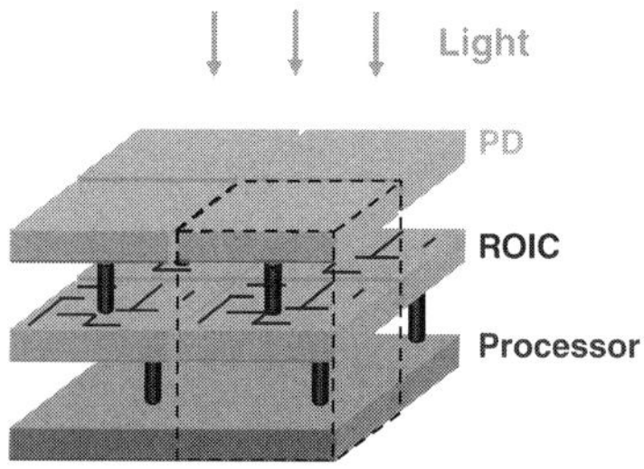

Fig. 13.5 Cross-section of a 3D imager. Courtesy of V. Suntharalingam

3D integrated technology allow the readout circuitry to be placed in close proximity to the photodetection elements [21]. Placing the signal processing behind every pixel reduces the total power and bandwidth requirements of the imager. Figure 13.5 is a schematic diagram of a 3D imager.

Three-dimensional packaging approaches such as flip chip are typically used for commercial imagers and involve bump-bonding independently optimized photodetectors and readout circuits into a stack [2]. The two main limitations of packaging approaches are that the stack is limited to two layers and because of the large size of the bump bond, the pixel size is quite large. Because the 3D vias are smaller in size than 3D packaging connections and can be packed more densely, smaller pixel sizes are achievable with 3D integration. In addition, the number of device layers is theoretically unlimited using 3D integration.

13.3.2 Interconnect-Dominated Systems

In current and future high-performance electronic systems, interconnect has a significant impact on overall system performance. In fact, for current technology nodes, the delay of a medium-length wire exceeds the delay of a typical gate. For example, consider a fan-out-of-four (FO4) inverter (i.e., an inverter driving four identical inverters). The delay of the FO4 inverter reduces with technology scaling. However, the delay through an intermediate-level 1-mm wire increases exponentially with technology scaling because of the increase in interconnect resistance. Because of scaling, the cross-sectional area of the wire is reduced increasing interconnect resistance. Figure 13.6 compares FO4 gate delay with interconnect delay. This comparison indicates that current architectures designed with regard to the logic gates and their devices will not scale as required for future generations.

Three-dimensional integration can alleviate this interconnect delay problem. A 1-mm-long wire typically represents global interconnect that spans much of the die width. By partitioning the circuit and stacking the two partitions on top of each other, the long wire that formerly spanned the die width is replaced with a short, vertical interconnect, typically a via. Rahman et al. found that for a wide class of 3D integration technologies the wirelength distribution shifts in response to an increase in the number of device layers as shown in Fig. 13.7 [17].

The leftward shift in the distribution indicates that longer, global wires are placed with shorter, local wires. This is the mechanism by which 3D integration can

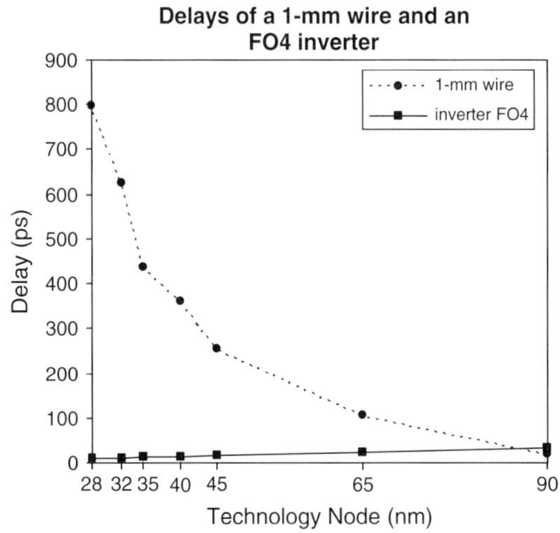

Fig. 13.6 Comparison of the delays associated with 1-mm wire and that of a FO4 inverter. Data projections from the 2006 ITRS Roadmap [11]

improve the performance of interconnect-dominated systems. The level of performance improvement achievable is highly dependent upon the vertical interconnect used. Most 3D integration technologies use through-wafer vias on the order of 1 μm in diameter [3, 10, 23].

The parasitics of the 3D via greatly affect the performance improvement achievable. If the resistance and capacitance of the via are large, then the performance improvement diminishes [7]. Figure 13.8 shows the total wirelength versus the number of device layers as a function of via cost for a benchmark circuit. Via cost is defined as the via capacitance relative to the capacitance of 1μm of metal wire. Thus, a via cost of 1 represents a via that has equivalent parasitics as a 1-μm metal line. As via cost increases, the improvement achievable with 3D integration decreases. Comparing a circuit using either one or two device layers, the wirelength

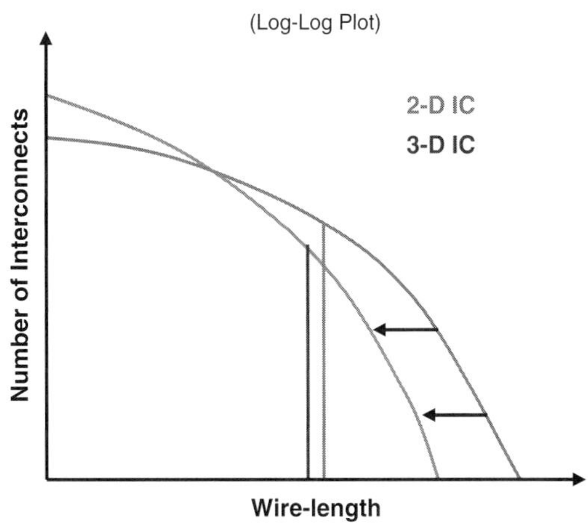

Fig. 13.7 Wirelength distribution of a typical circuit as a function of the number of device layers

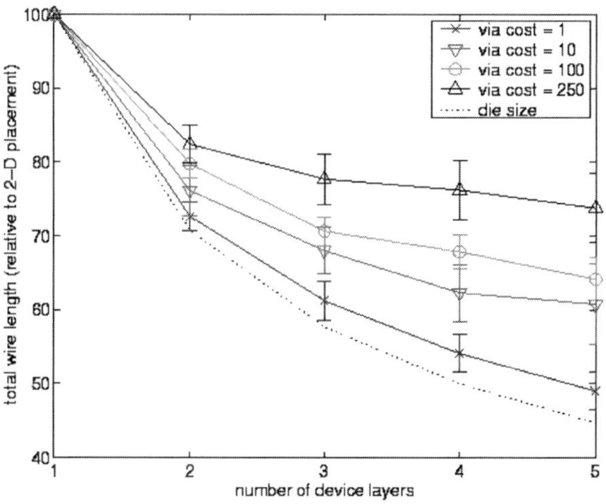

Fig. 13.8 Total wirelength vs. the number of device layers as a function of via cost [6]. Courtesy of S. Das

reduction can be as much as 28% for a via cost of 1 to as little as 18% for a via cost of 250. This is because it becomes more "expensive" to use a 3D via as its parasitics increase.

The shift in the wirelength distribution as a result of additional device layers most significantly affects interconnect power dissipation. For this reason, interconnect-dominated circuits benefit from 3D integration.To quantify the energy savings obtained by using additional device layers, Das et al. examined two test circuits, a 64-bit multiplier-accumulator circuit (MAC) [15] and a fast Fourier transform (FFT) datapath circuit [24], using their 3D place and route tool, PR3D [8], and subsequently simulated the circuit at the transistor level using Synopsis Nanosim. For the MAC, the interconnect is responsible for 67% of the total energy consumed [6]. Figure 13.9 shows how the interconnect energy consumption scales with the number of wafers for the MAC. For the FFT datapath circuit, the interconnect consumes roughly 44% of the total energy.

For the MAC, the addition of one device layer reduces interconnect energy consumption by 30%. The additional device layers reduce energy consumption by 24% in the FFT datapath circuit. Because the interconnect consumes a greater fraction of the total energy for the MAC circuit, the additional device layer results in greater improvements in energy consumption than the FFT circuit. This analysis shows that circuits that are more dominated by interconnect have more potential to benefit from 3D integration.

The field-programmable gate array (FPGA) is another circuit that can benefit from 3D integration. An FPGA consists of an array of configurable logic blocks (CLBs) connected together using complex routing. Kusse showed that interconnect overwhelmingly dominates FPGA power consumption consuming at least 65% of the total power [13]. A substantial amount of work on low-energy FPGAs focuses on

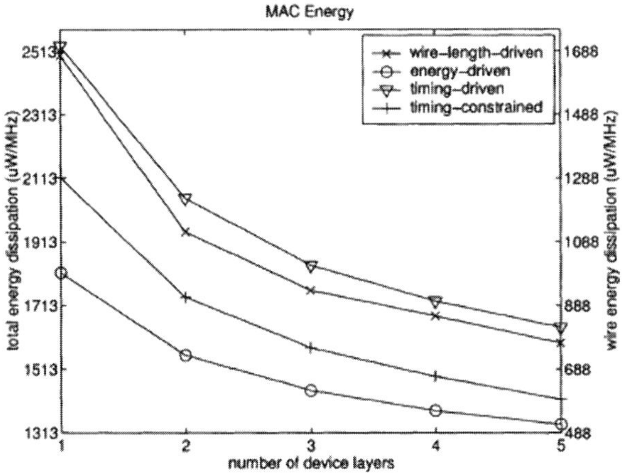

Fig. 13.9 Energy consumption for MAC as a function of the number of device layers [6]. Courtesy of S. Das

reducing the impact of interconnect. Since 3D integration reduces the need for long wires and can reduce the total wirelength, a 3D implementation of an FPGA can offer substantial improvements in speed, power, and functionality. Rahman found that FPGA interconnect delay could be reduced by as much as 45% for short wires and as much as 60% for long wires [16].For a 3D integrated FPGA using two device layers, power dissipation can be reduced by almost 35% [16].

Field-programmable gate arrays (FPGA) can not only benefit from the shift in interconnect distribution provided by 3D integration but can also benefit from the increased device density. For an FPGA implemented using a single device layer, each CLB has eight nearest neighbors. The addition of a single device layer can increase the number of nearest neighbors for a given CLB to 17. A 3D-integrated FPGA has more CLBs within a given footprint area and could potentially implement functions that cannot be realized in a standard 2D FPGA.

13.4 Nanoscale/Microscale Integration

Many believe that a radical paradigm shift from lithography-based fabrication is required to overcome the challenges associated with CMOS scaling. Such a shift entails assembling devices from fundamental units (i.e., nanometer-scale molecules) rather than defining devices through lithographic patterning. This procedure is most likely only feasible to build two terminal devices. Because of their limited functionality, much work has focused on building hybrid CMOS/nanodevice circuits [14, 19, 20]. In such hybrid systems, metal-oxide semiconductor field-effect transistors (MOSFETs) are used for signal restoration, long-range communications, and input/output (I/O) while nanodevices are used for information storage, interconnect, and some signal processing [20].

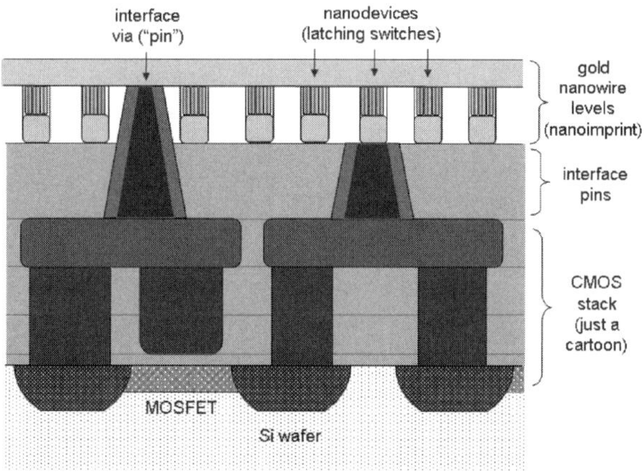

Fig. 13.10 Cross-section of the CMOL structure. From [25] © 2007 The Institution of Engineering and Technology

Several CMOS/nanodevice hybrids exist such as CMOS/molecular hybrid (CMOL) and field-programmable nanowires interconnect (FPNI). CMOL devices are chosen as the subject of discussion in this work. In the CMOL technique, CMOS devices are used to implement the more difficult logic functions while nanowires and junctions are used for signal routing and wired-OR logic "CMOL" [14, 19]. Figure 13.10 shows the cross-section of the CMOL structure.

In CMOL circuits, crossing nanowires separated by a molecular layer form junctions, which can be configured as electronic devices [19]. The nanowires connect to the CMOS circuits via metallic pins on the CMOS surface.

The most prevalent application of CMOL technology is implementing reconfigurable logic circuits such as FPGAs. As discussed in section 13.3.2, FPGAs consist of arrays of configurable logic cells that are interconnected using programmable interconnect. In a CMOL FPGA cell, the basic cell consists of two pass transistors and an inverter connected to the nanodevice using two pins [25]. Figure 13.11 shows the CMOL FPGA cell. The CMOL FPGA cell can be configured to implement NOR and NAND functions. Thus, any logic function can be implemented by connecting several cells together.

By uniformly distributing pins across the CMOS chip and having one nanowire connection per pin, CMOL obtains maximal signal bandwidth between the CMOS chip and the nanolayers resulting in higher-density circuits than using pure CMOS.

Despite the virtues of CMOL, fabrication of CMOL circuits is challenging. Because CMOL pins are only a few nanometers in diameter, they are particularly difficult to fabricate. Height control of CMOL pins presents another challenge. Pins of different heights are used to connect to either the top or bottom nanowires. Because it is difficult to accurately control the height of the pin, the wrong nanowire is often connected.

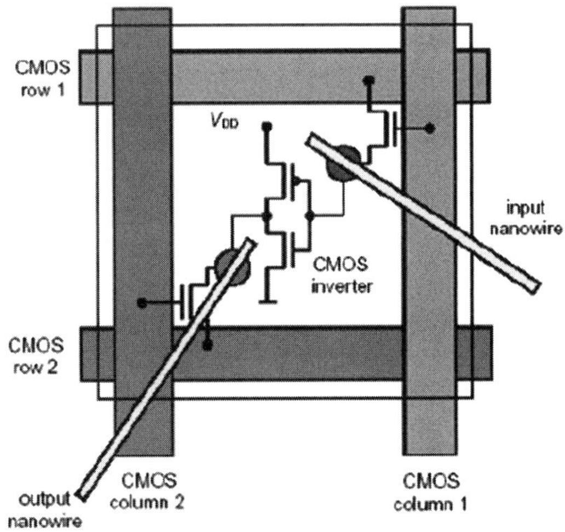

Fig. 13.11 CMOL FPGA cell. From [25] © 2007 The Institution of Engineering and Technology

If the CMOL pins could be eliminated, CMOL fabrication would become more feasible. Such a task is possible by leveraging 3D integration to build a new CMOL structure. In the 3D CMOL structure, two 2D CMOL structures are bonded in a face-to-face configuration as shown in Fig. 13.12.

In the 3D cell, the pins of the 2D CMOL stack are replaced with standard vias. The 3D CMOL cell is equivalent to two 2D CMOL cells stacked vertically. Thus, each 3D CMOL cell contains four pass transistors, two inverters, four vias, and four nanowires for the same footprint. The pass transistors, inverters, and vias are placed to maintain the same density as a 2D CMOL cell; however, nanowire density doubles allowing for more flexibility in connections [25].

Because each 3D CMOL cell is essentially two 2D CMOL cells, a logic design using 3D CMOL will occupy half of the footprint of a design using 2D CMOL circuits. A memory design using 3D CMOL will also occupy half of the footprint of

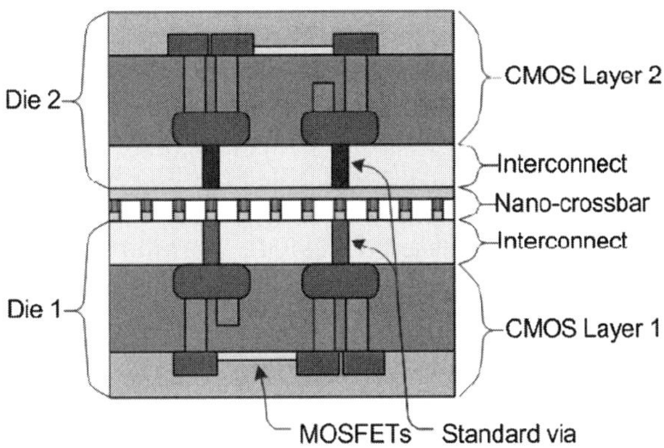

Fig. 13.12 3-D CMOL structure. From [25] © 2007 The Institution of Engineering and Technology

the 2D CMOL design because each 3D cell contains twice the number of nanowires that can be configured as memory elements.

In general, very little difference in delay and power is observed when comparing 3D CMOL designs to 2D CMOL. The delay path is not significantly altered when using the 3D cell; thus, delay is largely unaffected. Similarly, the number of devices and switching behavior is similar for a 3D design resulting in little change in power consumption [25].

13.5 Comparison Between 3D Integration and Planar Packaging

Three-dimensional packaging techniques such as multichip module (MCM-V) and flip chip offer many of the same benefits as a 3D technology. In a MCM-V package, individual die are fabricated and subsequently bonded to a printed circuit board (PCB). Input/output pads are wirebonded to the connections on the surface of the PCB. The separate PCBs are then interconnected via a high-bandwidth interconnect backplane that serves as the communication interface between dies [12]. This technology increases the effective density. Wirelengths can be reduced; however, delay and power dissipation typically do not improve since long wires are replaced with a wirebond connection and a line through the backplane. These interconnects are neither high performance nor low latency when compared with standard metal interconnect. Using ultrathin and multilayer packaging approaches, the interconnect backplane can be made higher performance. However, all chip-to-chip interconnect connections are located at the periphery of the die. Table 13.1 summarizes the parasitics associated with two of the main packaging technologies as well as that of a via used to connect device layers (i.e., 3D via) in a 3D integrated technology.

In a flip-chip technology, individual wafers are connected using solder bumps that are not limited to the periphery of the die and can be placed anywhere. However, solder bumps tend to be rather large and thus limit the density of chip-to-chip interconnections. The size of the 3D via is substantially smaller than both a wirebond and a solder bump; thus, the parasitics are significantly smaller. The small size of the 3D via allows for dense connections anywhere across the die area. Because the parasitics of the chip-to-chip interconnect are substantially smaller than for other packaging technologies, a system fabricated using 3D integration will exhibit lower delays and lower power consumption than its packaging counterpart.

Table 13.1 Parasitics comparison for a wirebond technology, flip chip, and 3D integration. Data for wirebond from [22]. Data for flip chip from [9]

	Wirebond	Flip chip	3D
Resistance	122 mΩ	1.2 mΩ	1.7 mΩ
Lead capacitance	25 fF	< 1 fF	–
Lead inductance	2.6 nH	0.2 nH	0.05 nH
Pitch	100 μm	200 μm	4 μm

References

1. Ansoft HFSS. http://www.ansoft.com/products/hf/hfss/
2. Bai Y, Bernd SG, Hosack JR, Farris MC, Montroy JT, Bajaj J (2004) Hybrid CMOS focal plane array with extended UV and NIR response for space applications. In: Proceedings of the SPIE, Bellingham, WA, pp 83–93
3. Burns JA, Aull BF, Chen CK, Chen C-L, Keast CL, Knecht JM, Suntharalingam V, Warner K, Wyatt PW, Yost D-RW (2006) A wafer-scale 3-D circuit integration technology. IEEE Trans. Electron Devices 53:2507–2516
4. Checka N (2005) Substrate noise analysis and techniques for mitigation in mixed-signal RF systems. Ph.D. Thesis, Massachusetts Institute of Technology
5. Checka N, Wentzloff DD, Chandrakasan AP, Reif R (2005) The effect of substrate noise on VCO performance. In: Proceedings of the IEEE RFIC, Long Beach, CA, pp 523–526
6. Das S (2004) Design automation and analysis of three-dimensional integrated circuits. Ph.D. Thesis, Massachusetts Institute of Technology
7. Das S, Chandrakasan AP, Reif R (2003) Three-dimensional integrated circuits: performance, design methodology, and CAD tools. In: Proceedings of the IEEE Computer Society Annual Symposium On VLSI, pp 13–18
8. Das S, Chandrakasan A, Reif R (2003) Design tools for 3-D integrated circuits. In: Proceedings of the ASP-DAC, pp 53–56
9. Flip chips reference. http://www.flipchips.com
10. Gutmann RJ, Lu J-Q, Devarajan S, Zeng AY, Rose K (2004) Wafer-level three-dimensional monolithic integration for heterogeneous silicon ICs. In: Dig. Papers Meeting on Silicon Monolitic Integrated Circuits in RF Systems, pp 45–48
11. ITRS roadmap. http:// public.itrs.net
12. Kelly G, Morrissey A, Alderman J, Camon H (2000) 3-D packaging methodologies for microsystems. IEEE Trans Adv Pack 23:623–630
13. Kusse E (1997) Analysis and circuit design for low power programmable logic module. M.S. Thesis, University of California at Berkeley
14. Likharev KK, Strukov, DB (2005) Introducing molecular electronics. Springer, Berlin
15. Open cores. http://www.opencores.org
16. Rahman A (2001) System-level performance evaluation of three-dimensional integrated circuits. Ph.D. Thesis, Massachusetts Institute of Technology
17. Rahman A, Reif R (2000) System-level performance evaluation of three-dimensional integrated circuits. IEEE Trans. VLSI Syst 8:671–678
18. Reif R, Checka N, Chandrakasan AP (2006) Multi-layer integrated semiconductor structure having an electrical shielding portion. U.S. Patent No. 10749096
19. Snider G, Williams RS (2007) Nano/CMOS architectures using a field-programmable nanowire interconnect. Nanotechnology 18:1–11
20. Strukov D, Likharev KK (2006), A reconfigurable architecture for hybrid CMOS/nanodevice circuits. In: Proceedings of FPGA, Monterey, California, pp 131–140
21. Suntharalingam V, Berger R, Burns JA, Chen C, Keast CL, Knecht JM, Lambert RD, Newcomb KL, O'Mara DM, Rathman DD, Shaver DC, Soares AM, Stevenson CN, Tyrrell BM, Warner K, Wheeler BD, Yost D-RW, Young DJ (2005) Megapixel CMOS image sensor fabricate in three-dimensional integrated circuit technology. In: ISSCC Dig. Tech. Papers, San Francisco, California, pp 356–357
22. Swaminathan M, Smith LD (1994) CBGA vs. CQFP: electrical performance comparison and tradeoff study. In: Proceedings of the IEEE EPEP, Monterey, California, pp 10–13
23. Tan CS, Reif R (2005) Silicon multilayer stacking based on copper wafer bonding. Electrochem Solid-State Lett 8:G147–G149
24. Wang A, Chandrakasan A (2005) A 180-mV subthreshold FFT processor using a minimum energy design methodology. IEEE J Solid-State Circuits 40:319–319
25. Wang W, Haruehanroengra S, Tu D, Liu M (2007) 3D CMOL: 3D integration of CMOS/nanomaterial hybrid digital circuits. IEE Micro Nano Lett 2:40–45

Chapter 14
Thermal Challenges of 3D ICs

Sheng-Chih Lin and Kaustav Banerjee

14.1 Introduction

During the past few decades, complementary metal-oxide semiconductor (CMOS) technology scaling along Moore's law has been the classical solution for the semiconductor industry to meet the ever-increasing demand for lower cost and higher performance [1–4]. However, in the nanometer regime, the pace of the transistor scaling has been slowing down due to the challenges and hindrances of severe short-channel effects, increasing variability, and power/thermal problems [5–7]. Also, due to the increase of functionality (transistor counts) of a planar (single active-layer) integrated circuit (IC), the complexity of interconnecting the devices increases dramatically and requires a large number of metal layers. Consequently, performance improvement from transistor scaling cannot be fully exploited and has gradually been constrained by interconnects [8].

Under this scenario, three-dimensional (3D) integration has been proposed as a promising technology to overcome the bottleneck of interconnects in future advanced nanoscale ICs [9, 10]. Three-dimensional integration scheme involves monolithic stacking of multiple active layers and leads to a considerable reduction in the number and average lengths of the longest global wires seen in traditional planar (2D) chips by providing shorter "vertical" paths for connection (Fig. 14.1). Besides the benefits of interconnect performance [9–11], it is conceivable that heterogeneous integration of designs in CMOS and various non-silicon technologies (SiGe, GaAs, InP, etc.) will be easier to realize in a single chip by the 3D architecture (different active layers) than by existing 2D (planar) chips [9].

Although 3D technology promises significant benefits, thermal issues are expected to offset the gains from this technology due to the degradation of performance and reliability [9, 13]. It is obvious that the heat problem will become worse due to the dramatic increase of power density when a 2D chip is redesigned into a 3D structure with the same functionality (identical power dissipation in a smaller

S.-C. Lin
University of California, Santa Barbara, CA, USA
e-mail: sclin@ece.ucsb.edu

C.S. Tan et al. (eds.), *Wafer-Level 3D ICs Process Technology*,
DOI: 10.1007/978-0-387-76534-1_14, © Springer Science+Business Media, LLC 2008

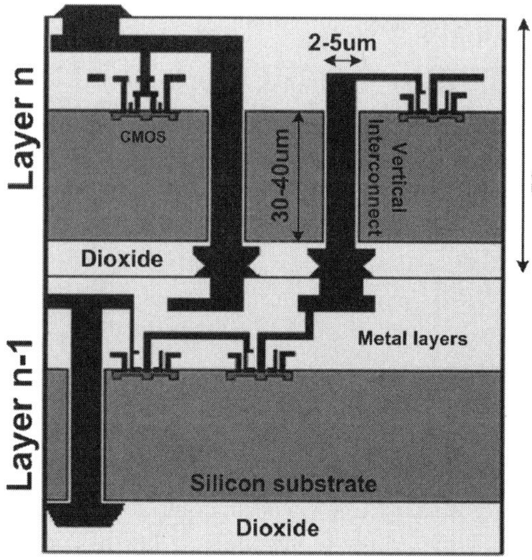

Fig. 14.1 Cross-sectional view of a vertically stacked *n* active-layer IC interconnected by through-wafer vias (TWV). From [12] © 2006 ACM, Inc. Included here by permission

size). Similarly, heat and thermal problems are expected to be exacerbated for 3D applications when many planar (2D) designs are integrated by stacking one layer on top of another [13–15]. Due to the low thermal conductivity of the dielectrics between active layers [16], heat generated by the stacked active layers located away from the heatsink are difficult to transfer, and hence, lead to temperature gradient in the vertical direction of a 3D chip. Therefore, heat and thermal considerations are imperative for determining the practical applicability of 3D technology and for evaluating various 3D design options.

Traditionally, thermal infrared (IR) imaging system has been used for acquiring thermal profiles of 2D ICs [17, 18]. This system offers limited resolution of substrate thermal profiles and is not suitable for 3D designs with multiple layers. Similarly, integrated thermal sensors are commonly employed to ensure that hot-spots do not exceed the specified maximum temperature criteria in high-performance ICs. However, only a limited number of sensors can be integrated into each active layer due to the routing and pin-out constraints. Most importantly, these techniques can only provide thermal profiles after fabrication that is not practical for early design optimization. Hence, an accurate chip-level thermal profile estimation methodology is necessary, especially for 3D applications.

In this chapter, we will focus on the thermal challenges of 3D (multilayers) ICs and discuss their implications for thermal management and their mitigation methods. First, impact of heat on device and interconnect reliability issues are briefly described. Then, an analytical die temperature model of a multilayer IC will be reviewed. In addition, the origin of various electrothermal couplings between chip power, substrate (die) temperature, operating frequency, and supply voltage will be discussed. Subsequently, an accurate chip-level leakage-aware methodology for 3D IC thermal profile estimation is illustrated that self-consistently takes various electrothermal couplings into consideration with a realistic package thermal model that comprehends different packaging layers and noncubic structure of the package. Finally, implications of temperature profiles generated by the proposed methodology for 3-D IC power estimation and thermal management are discussed.

14.2 Thermal Effects in 3D ICs in the Nanometer Regime

While continued scaling of CMOS technologies provides substantial benefits in the form of higher transistor packing density, higher circuit performance, and lower cost of ICs, power consumption and power densities (Watts per unit chip area) have been increasing steadily [5, 6]. Moreover, as CMOS has scaled from generation to generation, power dissipation has historically increased proportionately to increasing transistor density and switching speeds. However, with the minimum feature size of the transistor entering the nanometer regime (<100 nm), leakage power has become a significant fraction of the overall chip power [7]. Also, most leakage mechanisms are strongly temperature dependent. This strong coupling between temperature and leakage can cause further increase in total power dissipation. When several active layers are stacked (3D architecture), the heat and thermal problems are expected to be exacerbated [13, 15].

14.2.1 Impact of Heat on Device and Interconnect Reliability

Elevated substrate temperature is widely known to have a strong impact on the performance and lifetime of devices and interconnects under “field”, “accelerated testing”, and “burn-in” conditions. Higher temperature increases the risk of damaging the devices and interconnects (since major back-end and front-end reliability issues including electromigration (EM), time-dependent dielectric breakdown (TDDB), and negative-bias temperature instability (NBTI) have strong dependence on temperature), even with advanced thermal management technologies [19–21]. Moreover, due to the increase in the number of interconnect levels and introduction of low-κ dielectric materials with poor thermal conductivity, chip-level thermal problems have become even worse [13, 16, 22]. Hence, there is a critical need to accurately estimate the silicon substrate thermal gradients and temperature profile for the development and thermal management of future generations of all high-performance ICs, including 3D chips.

14.2.2 Analytical Average Die Temperature Model

A schematic diagram of a 3D IC with n active layers is illustrated in Fig. 14.2a. Each active layer (including one chip layer and one metallization layer) is separated by the glue layer. The 3D IC can be represented by a first-order equivalent thermal circuit as shown in Fig. 14.2b where P and T denote the power dissipation and the average die temperature of each active layer [13].

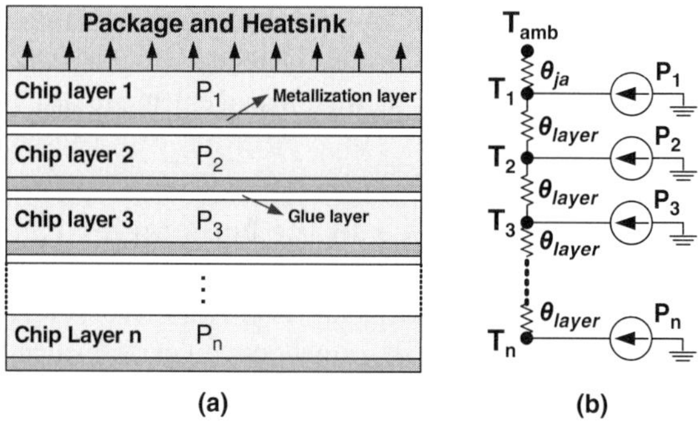

Fig. 14.2 (**a**) A 3D (n layers) IC. Thermal solutions (package and heatsink) are attached to the substrate of chip layer 1. (**b**) Equivalent thermal circuit used for evaluating the temperature of the 3D stack. P and T denote the power dissipation and the average die temperature of each active layer, respectively. T_{amb} is the ambient temperature. θ_{ja} and θ_{layer} represent the junction-to-ambient and layer-to-layer effective thermal resistance, respectively

Preliminary analysis for estimating the average die temperature of a 3D IC can be carried out by employing the equivalent thermal circuit. Assuming the direction of heat flow is from layer n to layer 1 and to the heatsink, the die temperature of the first layer can be estimated by the following:

$$T_1 = T_{\text{amb}} + \theta_{\text{ja}} \cdot \left(\sum_{k=1}^{n} P_k \right), \tag{14.1}$$

where T_{amb} represents the ambient temperature and θ_{ja} denotes the effective junction-to-ambient thermal resistance. Similarly, the temperature rise (above T_{amb}) of each active layer can be calculated by the following analytical expression, where θ_i is θ_{ja} (when $i = 1$) and θ_{layer} (when $i > 1$):

$$\Delta T_j = \sum_{i=1}^{j} \left[\theta_i \cdot \left(\sum_{k=i}^{n} P_k \right) \right]. \tag{14.2}$$

According to the first-order equivalent thermal circuit in Fig. 14.2b, heat can only flow toward T_{amb} (package and heat-sink). Thus, the highest temperature is determined by the temperature of the uppermost (nth) layer. From [13], the temperature rise of the uppermost layer is expected to increase with the square of the number of active layers ($\sim n^2$) under assumptions of identical power dissipation in each layer and identical thermal resistance between adjacent layers. Hence, it is clear that heat and thermal problems are expected to exacerbate (highest temperature increases quadratically) as the number of active layer in a 3D IC increases [13].

Although the first-order analytical model comprehends the thermal couplings between different active layers, the analysis simply employs a 1D equivalent thermal circuit with constant and average power dissipation at each layer that arbitrarily

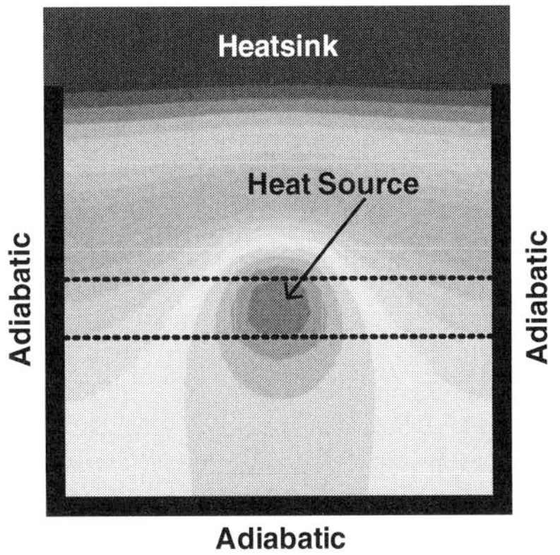

Fig. 14.3 Cross-sectional view of a steady-state temperature profile. The power dissipation of the active layer (defined by two dotted lines) is nonuniform (only one heat source). The boundaries are assumed to be adiabatic except the top (attached to heat-sink)

imposes the direction on heat transfer. This assumption (average power for each active layer) results in increasingly higher temperatures at the layers which are further away from the heat-sink and can mislead the estimation of temperature in a multiple-layer (3D) IC. As shown in Fig. 14.3, when the power dissipation is nonuniform, temperature could be lower at a point that is away from the heat-sink. Moreover, the first-order 1D analysis also ignores the electrothermal couplings within each active layer (e.g., correlations between temperature and power) which become critical in nanoscale designs [23]. Moreover, as the power dissipation of the entire 3D IC increases, a more detailed consideration of thermal solutions (e.g., package and heat-sink in Fig. 14.2a) must be taken into account (described in the following subsections).

14.2.3 Origin and Significance of Electrothermal Couplings

Typically, switching power and leakage power are the two major contributors to total chip power dissipation. The short-circuit component is relatively small and can be considered as a constant factor of total power [24, 25].

The switching power results from the charging and discharging of circuit capacitances between different voltage levels and increases with the chip frequency and supply voltage. The leakage power, especially subthreshold leakage, used to be negligible, but is rapidly becoming the dominant contributor to the total chip power because it is highly temperature sensitive (being thermionic emission based) (Fig. 14.4a) and exacerbates with technology scaling (Fig. 14.4b). Note that gate leakage (tunneling based) is temperature independent and can be mitigated by gate engineering [26]. Also, the junction (diode) leakage is relatively small as compared to subthreshold leakage [27].

The subthreshold leakage increases significantly due to the fact that supply voltage (V_{dd}) scaling necessitates threshold voltage (V_{th}) scaling to maintain a required performance according to International Technology Roadmap for Semiconductors (ITRS) prediction (Fig. 14.4c).

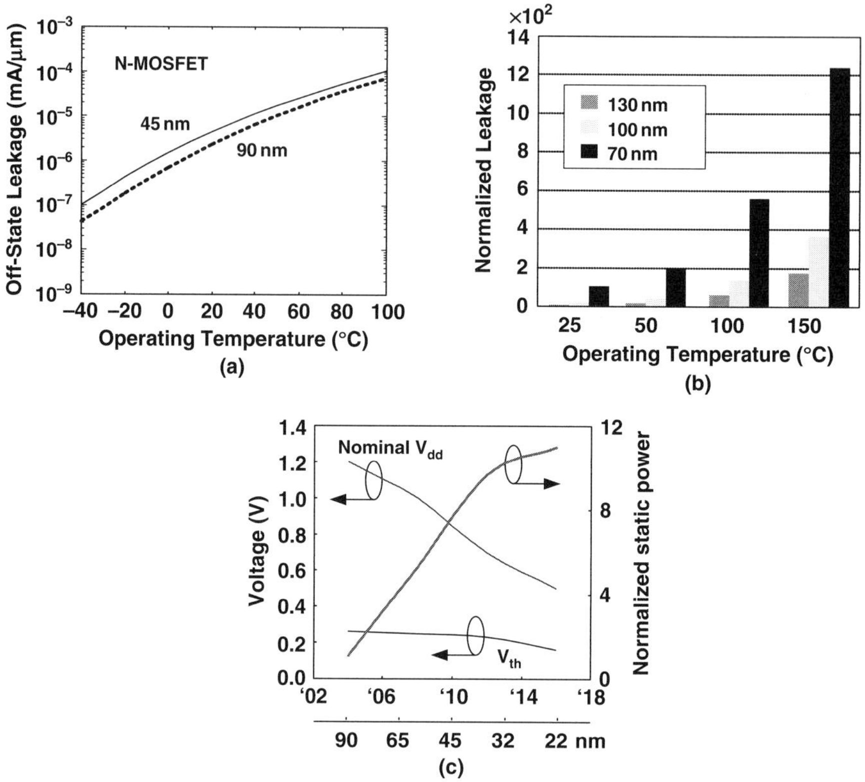

Fig. 14.4 (**a**) Transistor off-state leakage current for N-metal-oxide semiconductor field-effect transistor (N-MOSFET) (45- and 90-nm effective channel lengths) based on BSIM3 models as a function of operating temperature. (**b**) Leakage power dissipation of an NMOS device for different technology nodes based on BSIM3 models showing the impact of temperature. The leakage power dissipation is normalized with respect to (w.r.t) the value at 130-nm node at 25°C. (**c**) Nominal supply voltage, threshold voltage, and static power based on ITRS'04, as technology scales. From [28] © 2007 IEEE

In addition, elevated temperature lowers the threshold voltage of the transistor, and thus increases the leakage further [29]. Moreover, since the gap between the wavelength of light for optical lithography and the polysilicon gate length is increasing (Fig. 14.5a) [30], device channel length exhibits a significant amount of within-die variations [31], which in turn, leads to a significant impact on the distribution of leakage as shown in Fig. 14.5b.

The performance itself depends on temperature due to the dependence of the transistor on-current on operating temperature. Although the threshold voltage decreases at higher operating temperature and partially offsets the performance degradation resulting from the lower carrier mobility, the transistor on-current still decreases at higher operating temperatures (Fig. 14.5c).

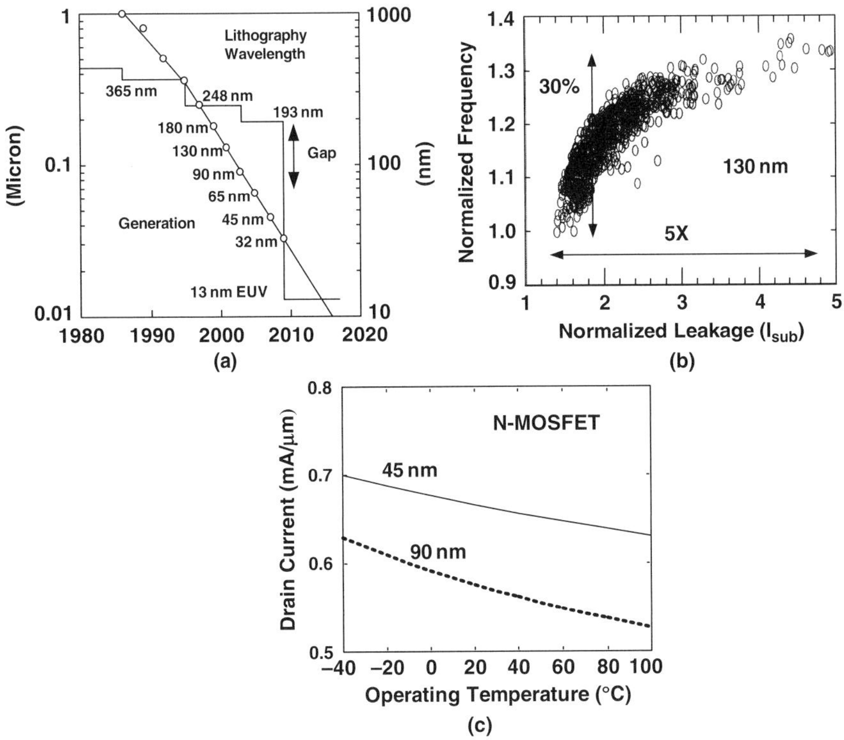

Fig. 14.5 (**a**) Increasing gap between polysilicon gate length and lithographic wavelength for different technology nodes [30]. (**b**) Distributions of frequency and standby leakage current for different microprocessors on a single wafer. (Courtesy: S. Borkar, Intel) (**c**) Transistor drive (drain) current for N-MOSFET (45- and 90-nm effective channel lengths) based on BSIM3 models as a function of operating temperature. From [28] © 2007 IEEE

The increase in total chip power consumption causes higher die temperature, which further increases subthreshold leakage. Therefore, a strong feedback loop builds up, leading to various electrothermal couplings [23], which had been inconspicuous in earlier generation of ICs. Fig. 14.6 illustrates such electrothermal couplings between performance, power dissipation, supply voltage, threshold voltage, and die temperature.

14.3 Self-Consistent Temperature Estimation for 3D ICs

As elevated and nonuniform temperature in a 3D IC extensively impacts the reliability, performance, and thermal management, acquiring accurate temperature profile of each active layer is necessary in the early design stage (before the 3D chip is fabricated). In this section, a self-consistent 3D temperature profile estimation methodology is presented. The method incorporates the electrothermal couplings,

$$performance\ (I_{on}) = f_1(V_{DD}, V_{th}, T_j, L_{nom})$$
$$P_{switching} = \alpha C V_{DD}^2 F$$
$$P_{subthreshold\_leakage} = f_2(V_{DD}, V_{th}, T_j, L_{nom})$$
$$P_{chip} \approx P_{switching} + P_{subthreshold\_leakage}$$
$$V_{th} = V_{th\_nom} + \Delta V_{th}$$
$$\Delta V_{th} = f_3(T_j, L_{nom}, t_{ox}, X_j, V_{DD}, \Delta_L)$$
$$T_j = f_4(P_{total}, T_{amb}, \theta_{ja})$$

(a)

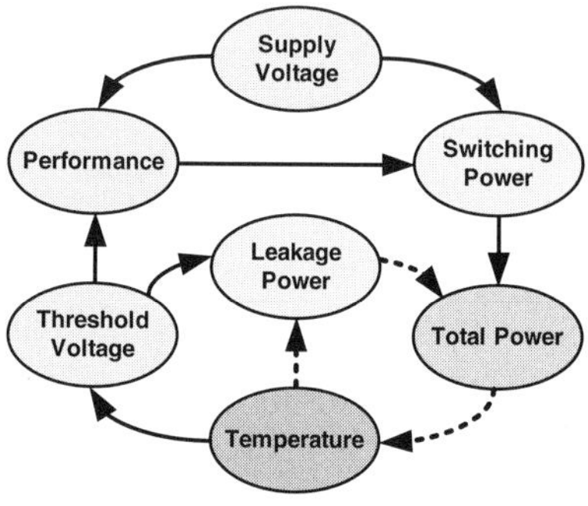

(b)

Fig. 14.6 (**a**) Models for various metrics are expressed in functional format. Couplings are indicated using broken lines. L_{nom} is the nominal gate length, α is the switching activity, C is the total load capacitance, F is the operating frequency, t_{ox} is the gate oxide thickness, and X_j is the junction depth. (**b**) Electrothermal couplings between different design metrics. As technology scales, the couplings between total power, leakage, and temperature (shown by dotted arrows) become increasingly prominent. From [28] © 2007 IEEE

as well as a realistic package thermal model to improve the accuracy of the thermal profile estimation and it is implemented via one of the widely used efficient algorithms for solving heat diffusion equations.

14.3.1 Typical Chip Package Structure and Heat Transfer Mechanisms

Due to the increase in silicon junction temperature for nanometer-scale technologies, packaging has been transformed from playing the traditional role of a protective mechanical enclosure to a sophisticated thermal management platform [32, 33]. Fig. 14.7 illustrates a cross-sectional view of a typical package structure of a planar high-performance IC including a "flip-chip land grid array" package and a socket that interfaces with the printed circuit board. The die is mounted on a package substrate (carrier).

Along the main heat transfer path as shown in Fig. 14.7, the die and the package substrate are attached to an integrated heat spreader (IHS). The IHS, with a relatively larger area than that of the die, spreads the nonuniform heat from the die region to the top of the IHS. This improves the heat flux from a smaller die area to a larger surface that serves as the mating surface for the heat-sink. Since the surface of these three major components (die, IHS, and heat-sink) are never smooth enough to have a perfect contact, they are bonded together with a thermal interface material (TIM) applied between them. The TIM improves the poor thermal conductivity caused by surface roughness (conductivity of TIM is much larger than that of air) and thus enhances the overall thermal performance of the packaging stack-up and cooling mechanisms.

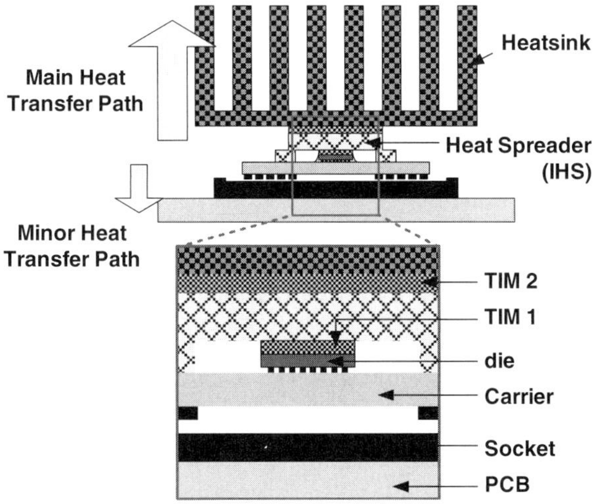

Fig. 14.7 A typical package assembly (drawing not to scale). Although the package structure is specific for a planar (2D) high-performance IC, the package can be applied to 3D ICs as well. From [28] © 2007 IEEE

There is a second heat transfer path from the die to the printed circuit board, through the interconnect and dielectric layers, input/output (I/O) pads, and carrier as shown in Fig. 14.7. The thermal resistance of this path (from junction to the printed circuit board) is normally several orders of magnitude higher than that of the major heat transfer path [34]. Therefore, this path can be neglected in the analysis because of the small fraction of heat it can transfer.

Heat is a form of energy that can be transferred as a result of temperature difference by three different modes: (1) conduction, in which heat passes through the matter itself, (2) convection, in which heat is transferred by relative motion of portions of the heated body, and (3) radiation, in which heat is directly transferred between distant portions of the body by electromagnetic radiation. The effect of radiative heat losses can be neglected (effects of heat conduction and convection are considered) since its influence is negligible when forced convection is employed in most high-performance ICs [35]. The silicon die is the main source of heat generation. Heat can be exchanged and transferred by conduction within the entire packaging stack-up and by convection at the surface of the heat-sink.

14.3.2 Full-Chip Package Thermal Model

Practical packaging structures typically employ the heat spreader and heat-sink with larger dimensions (compared to the die) to improve the thermal performance of the main heat transfer path (the realistic package thermal model is shown in Fig. 14.8a). In practice, the area of the heat spreader and heat-sink are at least $9x$ and $30x$ larger than the area of the die, respectively. Note that not only does the packaging structure

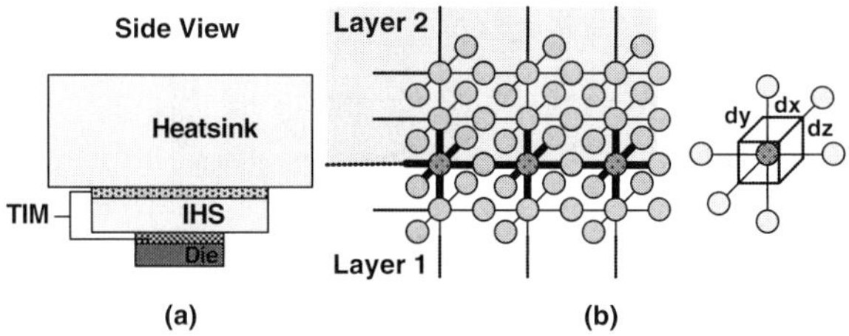

Fig. 14.8 (**a**) Side view of a realistic package thermal model indicating different dimensions for each layer. The thickness of different layers and the dimension of the layers are not drawn to scale. Note that the die could be planar or 3D IC. (**b**) Sketch of the discretization of the thermal packaging stack-up. Each node (*circle*) represents a discretized cell with a temperature value (T). Each discretized cell has six adjacent cells connected by edges (*lines*). Relationships between two adjacent cells are governed by (14.5) or (14.6) depending on heat transfer mechanisms. Effective thermal conductivity of cells between two adjacent layers (*darker nodes*) can be determined by (14.7) since the dimensions of a discretized cell are equal (i.e., $dx = dy = dz$). From [28] © 2007 IEEE

involve different materials with different thermal properties but also their dimensions with respect to the silicon die are different, which will significantly influence the heat transfer as well as the substrate thermal profile. The cubic package thermal mode, on the other hand, refers to a model in which all different package layers have identical areas and dimensions.

The temperature profile cannot be solved analytically due to the presence of complex geometry and complicated boundary conditions. Thus, numerical approaches will be employed for thermal profile estimation.

The fundamental physics of heat transfer in a chip is governed by the following 3D heat conduction equation and is subject to heat convection as the boundary condition [36]:

$$\rho C_p \frac{\partial}{\partial t} T(x, y, z, t) = \nabla \cdot [k(x, y, z, t) \nabla T(x, y, z, t)] + g(x, y, z, t) \tag{14.3}$$

$$k(x, y, z, t) \frac{\partial}{\partial n_i} T(x, y, z, t) = h\,[T(x, y, z, t) - T_{\text{amb}}] \tag{14.4}$$

where ρ is the density of the material (kg/m^3), C_p is the specific heat of material (J kg^{-1}°C), T is the temperature (°C), k is the thermal conductivity of the material (W m^{-1}°C), g is the internal heat generation (W m^{-3}), n_i is the outward direction normal to the boundary surface, h is the convective heat transfer coefficient (W m^{-2}°C), and T_{amb} is the temperature of the ambient air surrounding the package measured at a specified distance sufficiently far away from the surface of the entire package. Note that k is a measure of the ability of the material to conduct heat. Although it varies with temperature, the variance is relatively small within the range of operation [36]. Hence, a constant value of k is employed for each material in the packaging structure at the nominal temperature in the analysis. Also, for each layer, the thermal conductivity is identical in all directions (i.e., the material of each packaging layer is considered to be isotropic and homogeneous).

The aforementioned partial differential equations and boundary conditions can be rewritten as (14.5) and (14.6) where the temperature (T) is a function of the position (x,y,z) and time (t).

$$\frac{\partial T}{\partial t} = \left(\frac{k}{\rho C_p}\right)\left(\frac{\partial^2 T}{\partial x^2} + \frac{\partial^2 T}{\partial y^2} + \frac{\partial^2 T}{\partial z^2}\right) + \frac{p}{\rho C_p} \tag{14.5}$$

$$\frac{\partial T}{\partial n_i} = \frac{h}{k}\left[T - T_{\text{amb}}\right] \tag{14.6}$$

Electrothermal couplings are incorporated into the thermal model and the parameter p in (14.5) is a function of temperature, time, and the position within the die. Unlike the constant quantity g in (14.3), the parameter p represents the heat generation including electrothermal couplings and is recalculated at each evaluation step in a self-consistent manner.

The entire thermal packaging stack-up (packaging material layers) is discretized based on a typical high-performance package structure as Fig. 14.7. Relationships between discretized cells are governed by the heat partial differential equations and boundary conditions shown in (14.5) and (14.6). Physical thermal parameters, such as thermal conductivity, density, and specific heat of different layers, depend on material properties. Note that the dimensions of a discretized cell are chosen to be equal (i.e., dx = dy = dz). Thus effective thermal conductivity (k_{eff}) of cells between two adjacent layers, as represented by darker nodes in Fig. 14.8b between layer 1 and layer 2, can be simply determined by (14.7).

$$\frac{2}{k_{\text{eff}}} = \left(\frac{1}{k_1} + \frac{1}{k_2}\right), \tag{14.7}$$

where k_1 and k_2 represent the thermal conductivity of material in layer 1 and layer 2, respectively. A perfect thermal contact between the TIM layer and the adjacent materials is assumed since TIM is applied between two different layers to reduce the thermal contact resistance caused by surface roughness.

14.3.3 Numerical Approach and Methodology Overview

Partial differential equations (PDEs) of the general form shown in (14.8) are classified as parabolic PDEs (where ϕ is a function of x, y, z, and t) [36, 37] and can be solved using the finite difference approximation by two well-known approaches: explicit and implicit methods.

$$\frac{\partial \varphi}{\partial t} = \alpha \left(\frac{\partial^2 \varphi}{\partial x^2} + \frac{\partial^2 \varphi}{\partial y^2} + \frac{\partial^2 \varphi}{\partial z^2} \right) \tag{14.8}$$

The explicit method is simple and straightforward [36, 37]. The explicit method calculates the state of a system at the next time step from the state of the system at the current time. However, in many cases, time steps must be very small to maintain stability; this results in long computation time for a steady-state analysis. In order to overcome the aforementioned disadvantages of the explicit method, the implicit method considers both the current state and the state at the next time step [36, 37] and the stability can be maintained over much larger values of time step. However, this method is more complicated to set up and massive matrix manipulations require a considerable amount of computation memory and runtime for each time step.

The alternating direction implicit (ADI) method is a widely used algorithm for the numerical solution of parabolic PDEs involving multiple spatial variables [38, 39]. The advantage of applying this method arises from transferring a multiple dimensional parabolic PDE into a succession of 1D problems. Therefore, no large-scale matrix has to be computed, and it is easy to implement. Thus, the ADI method is employed as the core algorithm to solve the heat PDEs for achieving higher computation efficiency. It is important to note that although other computationally efficient methods exist, choosing any one of them over the others does not affect the accuracy of results.

In order to accurately estimate on-chip thermal gradients and the power dissipation profile, a self-consistent temperature profile estimation methodology is proposed with the capability of incorporating precise layout geometry and the power dissipation of individual circuit blocks in a chip [40].

Fig. 14.9 illustrates the overview of the methodology for substrate temperature profile estimation. The chip is partitioned into a mesh according to the information provided by the layout geometry and power distribution map. Nominal power dissipation (including switching and leakage power) for each functional block is used as initial value according to its activity, depending on specific circuit implementation and application. Note that for a 3D IC, each active layer will have dif-

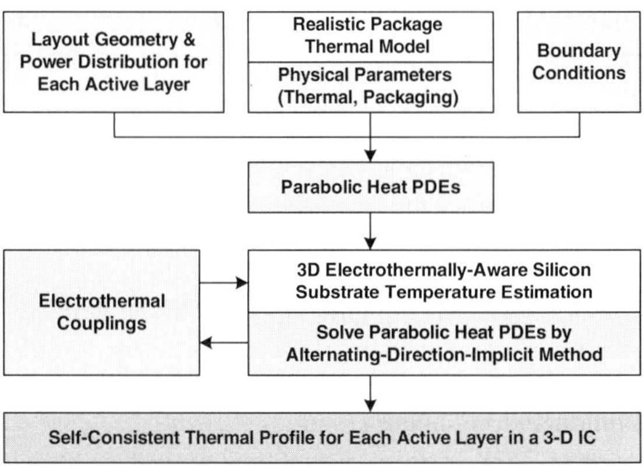

Fig. 14.9 The electrothermally aware methodology for silicon substrate temperature profile estimation. From [40] © 2007 IEEE

ferent layout geometry and power distribution. Physical parameters such as specific heat, thermal conductivity, and heat transfer coefficient depend on specific packaging material properties and applied cooling techniques. The full-chip realistic package thermal model is then incorporated, which comprehends both vertical and lateral heat transfer paths. Boundary conditions are determined by the operating environment. The simulator uses layout geometry, nominal power dissipation, boundary conditions, and physical thermal/packaging parameters as initial values to formulate PDEs and then solves these equations in a self-consistent manner using the ADI method for every mesh element. The algorithm converts a multiple-dimensional parabolic PDE into a succession of 1D linear equations. The electrothermal couplings are also embedded in the core of the simulator that simultaneously estimates temperature-dependent quantities for each simulation step. Once the difference of the temperature evaluation between two steps is within a certain range, the evaluation stops and the steady-state temperature profile is obtained. However, if the temperature exceeds the maximum criteria (defined by reliability constraints) for certain extreme cases due to poor packaging solutions or high power dissipation, the evaluation will terminate and thermal runaway will be reported.

The key aspect of the proposed approach as compared to traditional methods is illustrated in Fig. 14.10. Although the entire thermal profile can be obtained by the traditional evaluation, the traditional method is apparently misleading because it ignores the correlation between power and temperature. While one might think

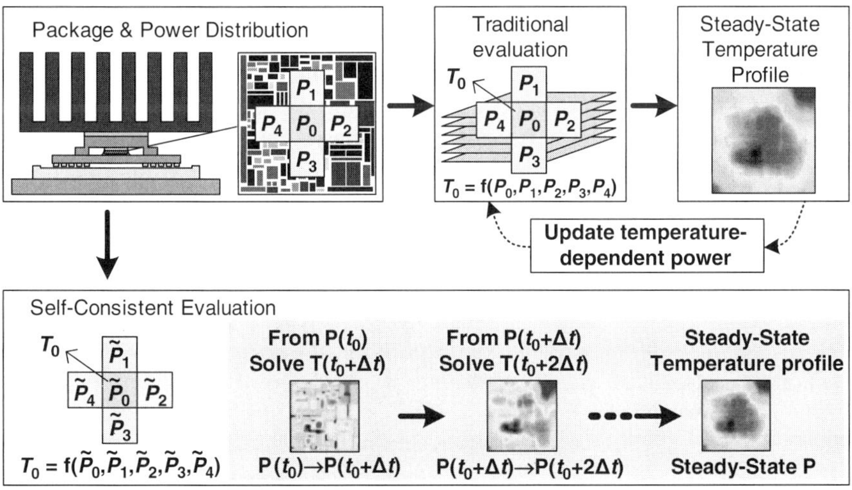

Fig. 14.10 Diagram showing the difference between traditional evaluation and the proposed self-consistent substrate thermal profile estimation methodology. Due to the strong interdependence of temperature and leakage power, temperature at the central block (T_0) is not simply a function of the nominal power dissipation within and adjacent to the center block (P_0, P_1, P_2, P_3, P_4) as per traditional analysis. Nominal power distribution should be updated self-consistently with the temperature evaluation (e.g., P_i is updated to $\tilde{P}_i$). The proposed method evaluates the temperature by incorporating the correlation between power and temperature at each time step (Δt). Note that Δt is chosen to be much larger than the electrical switching time of gates or logics in the area of concern. From [40] © 2007 IEEE

of applying the traditional evaluation iteratively by updating the temperature-dependent power (as shown by the dotted arrows), however, this dramatically increases the computation time. In addition, once the steady-state temperature is evaluated without considering the electrothermal couplings, the iterations (as shown by the dotted arrows) based on inaccurate information is meaningless. On the other hand, the proposed self-consistent approach evaluates the steady-state temperature profile by employing the ADI method such that the correlation between the power and the temperature can be incorporated at each time step. Hence, the self-consistent method inherently generates a more accurate power profile, which can then be used to generate an accurate temperature profile by efficient PDE solvers.

14.3.4 Setup and Implementation: An Example of a 2D IC Thermal Profile Estimation

A design with a die size of $10 \times 10\,\text{mm}^2$ (discretized into 100×100 grids) and with power densities per functional block is shown in Fig. 14.11. The power dissipation of the chip or each functional block depends on the application (workload, activity, etc.). However, in this analysis, the power distribution map is known. The nominal total power consumption of the chip at ambient temperature (45°C) is 96 W (nominal active power = 93.1 W, leakage power = 2.9 W). The short-circuit component is relatively small; therefore it is neglected for simplicity. The physical and thermal properties of all packaging layers are evaluated according to a practical packaged high-performance microprocessor [40].

In order to demonstrate the importance of incorporating electrothermal couplings and realistic package thermal model for estimating the substrate temperature profile, four different simulation scenarios are compared using the design shown in Fig. 14.11. Although the results of the proposed methodology have not been verified against direct measurements, the method simply ensures the self-consistency

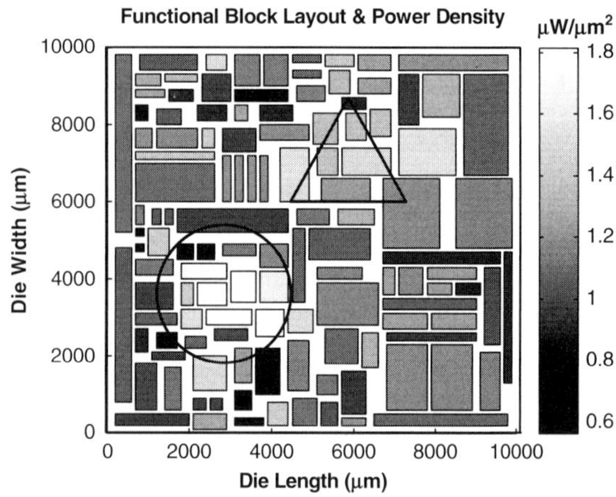

Fig. 14.11 Functional block layout of a chip. Power densities associated with functional blocks are also shown. The circle encloses a region where blocks have the highest power density. The triangle encloses the functional blocks that have higher leakage power dissipation than all other blocks. From [40] © 2007 IEEE

between power and temperature during each iteration of the PDE solver, which has been validated against an industrial-quality computational fluid dynamics (CFD) software [41]. The same heat equations are employed and the inclusion of the electrothermal couplings does not change the fundamental equations governing thermal transport via heat conduction and convection but provides an algorithm to self-consistently solve the temperature and leakage power. Hence, once the core of the solver has been validated against the CFD, the results of the methodology can be trusted even with the inclusion of the electrothermal couplings.

Although the results are specific to the aforementioned 2D IC, the conclusions are more generic. It can be observed that there is a region indicated by a circle in Fig. 14.11 where blocks have highest power density. In addition, there is a region indicated by a triangle where blocks have 10 times leakage power dissipation with respect to the values of other functional blocks. However, the average power density of the circuit blocks in the triangle is around 60% of the average power density value in the circle.

Fig. 14.12 and Fig. 14.13 represent the silicon substrate temperature profiles generated under four different scenarios, respectively:

1. Traditional method + cubic package thermal model
2. Traditional method + realistic package thermal model
3. Self-consistent method + cubic package thermal model
4. Self-consistent method + realistic package thermal model

Note that all temperature profiles are shown using a constant temperature range (56–66°C) for ease of comparison in Fig. 14.12 and Fig. 14.13.

The impact of electrothermal couplings on the substrate temperature evaluation can easily be observed by comparing Fig. 14.12b and Fig. 14.13b, which both employ the realistic package thermal model (Fig. 14.8a) and the same cooling conditions. The substrate thermal profile (Fig. 14.12b) is generated using a traditional thermal simulator without considering electrothermal couplings. The highest temperature (hot-spot) is approximately 64.23°C and is located in a region with the highest power density (indicated by a circle in Fig. 14.11). However, a different substrate temperature profile (Fig. 14.13b) is obtained by employing the proposed self-consistent methodology. From the temperature profile in Fig. 14.13b, two hot-spots can be observed: one in the region with the highest power density and the other in the region with a higher percentage of leakage power. Unlike the traditional evaluation, the highest temperature is around 63.81°C and is located in the region with a higher percentage of leakage power (indicated by the triangle in Fig. 14.11). Note that the self-consistent methodology comprehends the couplings between power (active and leakage) and temperature. The steady-state power dissipation (active and leakage) is self-consistent with the temperature and may not be equal to the nominal power dissipation.

As explained in [28], regions with higher switching power density do not necessarily yield a higher temperature due to the various electrothermal couplings. Although the highest temperature values are similar in Fig. 14.12b and Fig. 14.13b, the temperature profile obtained by the self-consistent evaluation shows an

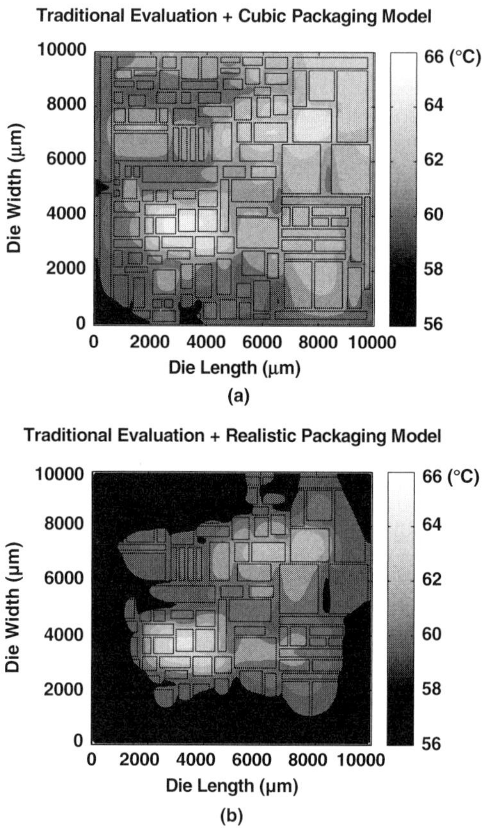

Fig. 14.12 Silicon substrate temperature profile generated by traditional evaluation without considering electrothermal couplings. (**a**) A cubic package thermal model is employed. Only one hot-spot can be observed. T_{max} is approximately 65.49°C and located in the region with higher power density. $T_{max} - T_{min}$ is approximately 8°C. (b) A realistic package thermal model is employed. Only one hot-spot can be observed. T_{max} is approximately 64.23°C and located in the region with higher power density. $T_{max} - T_{min}$ is approximately 11°C. From [40] © 2007 IEEE

additional hot-spot and thus a different temperature distribution. The traditional estimation is clearly misleading in terms of hot-spot count, location, and the overall spatial temperature profile as it neglects the electrothermal couplings between power dissipation and temperature.

The impact of employing two different package thermal models for the cooling path on the temperature profile estimation can be observed by comparing Fig. 14.13a and Fig. 14.13b. For fair comparison, the layout, power density distribution, and discretization of the die are kept identical. In addition, the physical and thermal properties of each packaging layer material are kept constant in both models. Fig. 14.13a shows the estimated substrate temperature profile by using a cubic (unrealistic) package thermal model. Although the electrothermal couplings are considered, unrealistic package thermal model underestimates the lateral heat spreading of packaging layers (particularly in IHS and heat-sink), and thus results in a higher maximum and average substrate temperature. However, it is also important to note that although the maximum temperature is lower, the temperature gradient from the hot-spot to the edges of the chip is higher while employing the realistic package thermal model (e.g., T_{max} is 65.69°C in Fig. 14.13a and 63.81°C in

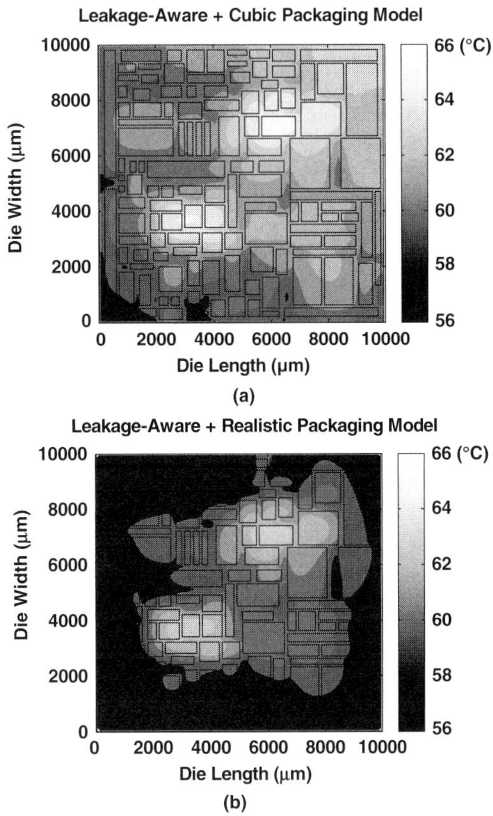

Fig. 14.13 Silicon substrate temperature profile generated by the proposed self-consistent method. (**a**) A cubic package thermal model is employed. Two hot-spots can be observed. The highest temperature (T_{max}) is approximately 65.69°C and located in the region with higher percentage of leakage power. $T_{max} - T_{min}$ is approximately 8°C. (**b**) A realistic package thermal model is employed. Two hot-spots can be observed. The highest temperature (T_{max}) is approximately 63.81°C. $T_{max} - T_{min}$ is approximately 11°C. From [40] © 2007 IEEE

Fig. 14.13b; $T_{max} - T_{min}$ in Fig. 14.13a and Fig. 14.13b are about 8°C and 11°C, respectively). Due to the use of larger heat spreader and heat-sink in the realistic package thermal model, better lateral heat spreading leads to lower maximum temperature but to even lower temperatures at the edges of the chip. This, in turn, is expected to impact the physical design issues such as partitioning and placement schemes for high-performance ICs including multicore designs.

14.3.5 3D IC Thermal Profile Estimation: Analysis and Implications

In [9, 42, 43, 44], several possible applications for this revolutionary 3D technology have been explored. One of the most promising applications is that of integrating a processor-and-memory system on a single 3D chip. Preliminary thermally aware performance analysis of the 3D processor-memory hierarchy (assuming an average temperature for each active layer) is performed with different benchmarks at different processor frequencies [12]. The impact of thermal constraint on performance of the processor-memory hierarchy is summarized in Fig. 14.14.

For the application, which is highly memory-intensive (e.g., benchmark: *mcf*), the execution time per instruction (t_{exe}) is lower for a 3D system (Fig. 14.14a) while for the application that is less memory-intensive (e.g., benchmark: *twolf*), the difference in execution time between 2D and 3D system is negligible (Fig. 14.14b). Moreover, when the system is constrained by a maximum allowable temperature (which arises from reliability concerns), the maximum allowable frequency (f_{max}) of a system is limited. For memory-intensive systems (Fig. 14.14a), even though thermal considerations place a lower limit on the f_{max} in 3D, better performance can still be achieved as compared to the 2D system running at a higher frequency.

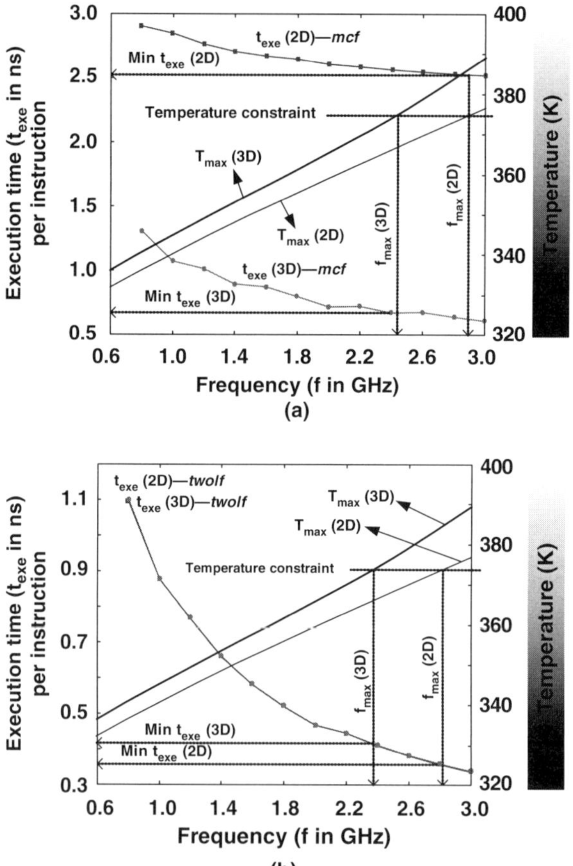

Fig. 14.14 Execution time per instruction (t_{exe}) and maximum average temperature as a function of operating frequency for 2D and 3D chip (processor-memory hierarchy). (**a**) Highly memory-intensive application (*mcf*). (**b**) Less memory-intensive application (*twolf*). From [12] © 2006 ACM, Inc. Included here by permission

This is because the 2D system cannot overcome the memory interface bottleneck. On the other hand, for applications that are not memory intensive (Fig. 14.14b), the system performance is not dominated by memory accesses. Hence, under this scenario, the 2D system, which has a higher limit of f_{max}, has a system performance better than the 3D system, which is constrained to operate at a lower frequency [12]. Note that this analysis employs an average temperature for active layers and this implies the temperature will be higher at the layer which is away from the heat-sink. However, when detailed temperature profile is taken into consideration, the average temperature model may mislead the temperature estimation and the performance analysis.

As the traditional planar (2D) technology has already been threatened by power and associated thermal problems, the success of 3D integration not only depends on the development of processing technologies but also requires thorough and accurate estimation of thermal profiles in a 3D IC.

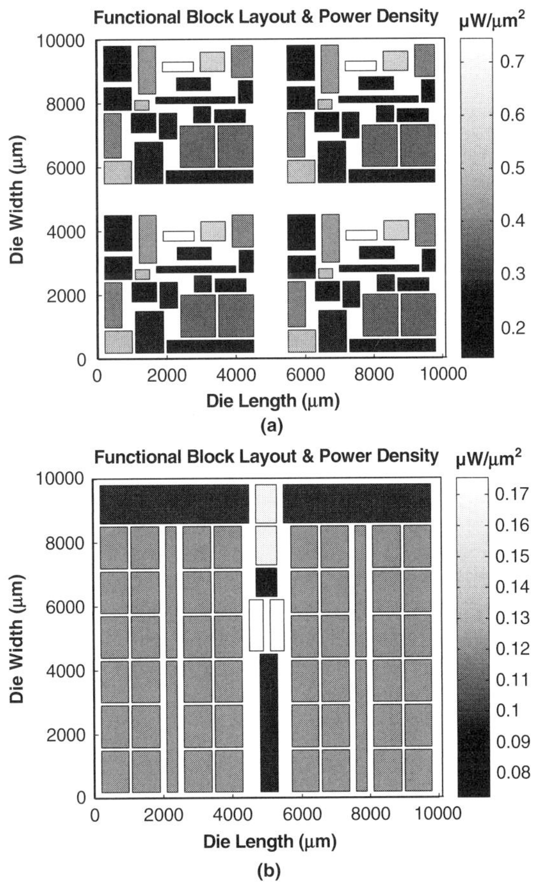

Fig. 14.15 Functional block layout of two additional active layers (first layer shown in Fig. 14.11) in a 3D IC. Power densities associated with functional blocks are also shown. (**a**) The nominal total power consumption of the layer at ambient temperature (45°C) is 20 W (nominal active power = 19.6 W, leakage power = 0.4 W). (**b**) The nominal total power consumption of the layer at ambient temperature (45°C) is 10 W (nominal active power = 9.8 W, leakage power = 0.2 W)

Chip-level thermal and reliability issues of planar (2D) IC designs can be comprehended by employing the aforementioned thermal profile estimation methodology while considering the packaging and electrothermal couplings. From Fig. 14.7, the major heat transfer path of a planar (2D) IC is clearly from the active layer to the thermal packaging and heat-sink. However, due to the presence of different power dissipation and distribution of different active layers in a 3D IC, the direction of the heat flow significantly depends on the arrangement and the placement of the active layers.

A generic 3D IC with three active layers is considered in this subsection (one active layer is shown in Fig. 14.11 and the power density maps of the other two active layers are shown in Fig. 14.15). Note that the thickness of each active layer in a 3D IC (around 50 μm, in Fig. 14.1) is much smaller than that in a planar (2D) IC (several hundred microns) for practical integration, assuming the power dissipation of active layers in Fig. 14.15 is 20 W and 10 W, respectively. Typically, in order to reduce the thermal resistance between active layers and heat-sink, layers with higher power dissipation will tend to be placed closer to the heat-sink than layers with less power dissipation. Thus, in this scenario the layer shown in Fig. 14.11 will be attached directly to the package structure and followed by Fig. 14.15a and subsequently by Fig. 14.15b. Note that besides the boundary that is attached to the heat-sink and exposed to the ambient, all other boundaries are considered to be adiabatic in the analysis.

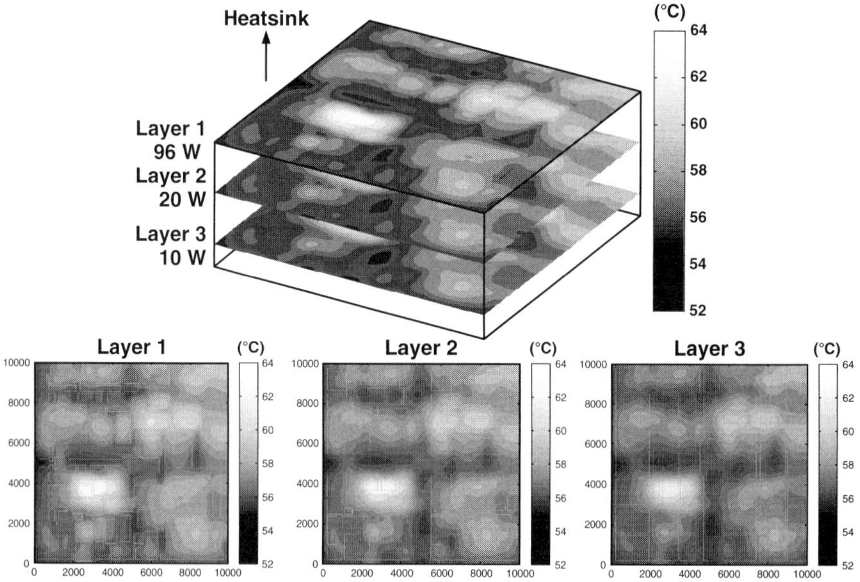

Fig. 14.16 Temperature profiles of the active layers in the 3D IC generated by the proposed self-consistent method. Similar to **Fig. 14.2a**, layer 1 is directly attached to the thermal solutions (e.g., heatsink)

With the same packaging and environment conditions as in the previous subsection, steady-state temperature profiles of the layers in the 3D IC can be estimated by using the self-consistent methodology (Fig. 14.16). The maximum temperature for each active layer is 64.19°C (layer 1), 63.89°C (layer 2), and 63.47°C (layer 3), respectively. Although power dissipation of layer 2 and layer 3 is much lower than layer 1, steady-state temperature profiles of layer 2 and layer 3 are influenced and raised by layer 1. As discussed in Section 14.2, employing first-order analytical thermal model for 3D IC analysis with average power and temperature of each active layer certainly misleads the temperature estimation and overestimates the maximum temperature of layers that are away from the heat-sink.

14.4 Implications and Opportunities for 3D IC Thermal Management

Unlike planar (2D) ICs, thermal management for 3D ICs requires thorough considerations not only for each active layer (2D level) but also the correlative impacts between active layers (3D level).

At the 2D level, power-reduction techniques and thermal management for conventional planar (2D) ICs, including device-, circuit-, and architecture-level techniques, can be directly employed to reduce power dissipation and thermal gradient. For CMOS technologies, device short-channel effects [29], which lead to higher subthreshold leakage, have been shown to be improved via substrate engineering. For instance, vertically nonuniform doping (retrograde channel profile) enhances inversion layer mobility because of the lower surface doping [45, 46], while laterally nonuniform channel implants (halo doping) reduce threshold voltage roll-off by compensating 2D charge-sharing effects in short-channel transistors [47–49]. Transistor gate-tunneling leakage, which increases with the ever-thinning silicon dioxide gate dielectric [50], can be alleviated by replacing the thin silicon dioxide by a thicker insulating material with higher dielectric constant (high-κ). In addition, a metal gate electrode was also used to replace the poly-silicon gate to have a better control of the threshold voltage [51, 52]. Similarly, at the circuit level, low-power design methodologies include dual- or multi-V_{dd} and V_{th} schemes as well as adaptive body-biasing techniques can be applied [53]. Transistor gating is also considered for low-power or power-constrained designs. For instance, the clock-gating technique is used to reduce clock tree power dissipation [54]. Power gating and sleep transistor insertion techniques reduce leakage by turning off idle circuitry [55]. Thermally aware placement schemes (within-layer) are also available in the literature to optimize performance and operating temperature [56, 57]. Moreover, in [58], a 3D IC thermal placement method using an iterative force-directed approach is presented. At the architecture level, pipelining and parallel (including multi-core) structures are often implemented in low-power designs. The throughput can be maintained at a lower V_{dd} by parallel implementation. Also, applying pipelining can reduce power consumption while the switching rate and V_{dd} are reduced [59]. Note that these methods reduce power consumption at the cost of area, performance, or noise margin penalty.

Besides the aforementioned techniques, chip cooling has always been considered as an effective knob for power and thermal management [60]. Conventional cooling techniques and thermal management for planar (2D) ICs cannot be directly applied for 3D IC thermal management but require a holistic consideration including all active layers and thermal solutions (packaging, etc.). For instance, the boundary conditions of one active layer are determined by its adjacent layers, and hence, the aforementioned 2D-level (within-layer) placement schemes are required to comprehend the layer-to-layer effects.

As shown in the aforementioned example (Fig. 14.15 and Fig. 14.16), layer 1 has the highest power dissipation and the highest temperature occurs in layer 1 even though this layer is closest to the heat-sink. However, temperature profiles and the

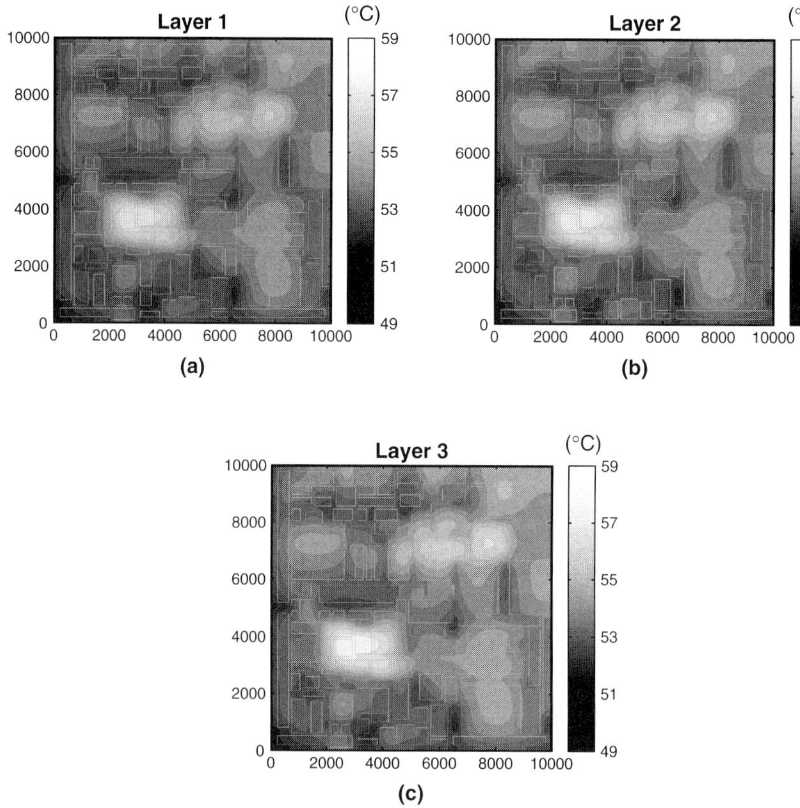

Fig. 14.17 Temperature profiles of the active layers in the 3D IC generated by the proposed self-consistent method. All layers are identical with the same power distribution and dissipation (one-third as compared to that of **Fig. 14.11**). Layer 1 is closest to the heat-sink. In this analysis, it is assumed that all boundaries are adiabatic besides the one attached to the heat-sink and exposed to the ambient. The maximum temperature is (**a**) 58.02°C, (**b**) 58.51°C, and (**c**) 58.77°C

maximum temperature of the active layers in a 3D IC will change when employing different arrangement schemes of layers. Here we discuss an example with three identical active layers stacked in a 3D IC. The power distribution of the active layers is similar to Fig. 14.11 but the power dissipation of each functional block in the active layer is one-third as compared to that of Fig. 14.11 (note that stacking three high-power dissipation layers with the same thermal solutions leads to thermal runaway). Fig. 14.17 shows the steady-state temperature profile of these three layers. It can be observed that the highest temperature now occurs at layer 3 that is away from the heat-sink.

Moreover, alternative materials with higher thermal conductivities can also improve heat removal of 3D ICs. While, thermal and interlayer vias are shown to mitigate thermal problems in 3D ICs [61], employing metallic carbon nanotube (CNT) bundle vias to replace copper at different locations in the interconnect stack shows substantial benefits in controlling the back-end temperature [62]. In practice, CNTs have been fabricated as thermal and source bumps for flip-chip high-power amplifiers [63]. Also, it has been experimentally shown that the thermal conductivity of TIMs can be improved by employing free-standing CNT arrays or combinations of CNT arrays and existing TIMs [64].

14.5 Summary

Three-dimensional integration technology with multiple active layers has been considered as a promising candidate to alleviate the interconnect delay problems in nanoscale VLSI circuits and to realize heterogeneous integration in the same chip. As heat and thermal effects already significantly impact reliability and performance in high-performance planar (2D) ICs, obviously, heat and thermal problems in 3D ICs are worse since 3D ICs are stacked by, and thus inherited from 2D ICs. In this scenario, accurate thermal profile estimation is critical in the early design stage (before the 3D chip is fabricated).

It is shown that the first-order analysis simply employs a 1D equivalent thermal circuit with constant and average power dissipation at each active layer and results in higher temperature at the layer which is away from the heat-sink that misleads the estimation of temperature in a 3D IC. On the other hand, the proposed self-consistent 3D temperature profile estimation methodology incorporates the electrothermal couplings that are increasingly prominent as technology scales. In addition, a realistic package thermal model is considered to improve the accuracy of the thermal profile estimation. Impact of layer stacking on the temperature profile of a 3D IC is also presented.

The 3D thermal profiles are also strongly influenced by the nature of the application running on the chip. Although various techniques for power saving or thermal management for planar (2D) ICs can be applied to 3D ICs, considerations of active layer arrangement in a 3D IC as well as 3D thermally aware placement schemes can severely influence the steady-sate temperature profile of each active layer. Furthermore, overall thermal conductivity of 3D ICs can be improved by employing higher thermal conductivity materials (e.g., CNTs) between active layers or in the packaging structure.

References

1. Moore GE (1965) Cramming more components onto integrated circuits. Electronics 114–117
2. Moore GE (1975) Progress in digital integrated electronics. IEEE International Electron Devices Meeting, pp 11–13
3. Dennard RH, Gaensslen FH, Rideout VL, Bassous E, LeBlanc AR (1974) Design of ion-implanted MOSFETs with very small physical dimensions. IEEE J Solid-State Circuits 9: 256–268
4. International Technology Roadmap for Semiconductors (ITRS), http://www.itrs.net
5. Borkar S (1999) Design challenges of technology scaling. IEEE Micro, 19:23–29
6. De V, Borkar S (1999) Technology and design challenges for low power and high performance. IEEE International Symposium on Low Power Electronics and Design, pp 163–1681
7. Gelsinger PP (2001) Microprocessors for the new millennium: Challenges, opportunities, and New Frontiers. IEEE International Solid-State Circuits Conference, pp 22–25
8. Meindl JD (2003) Beyond Moore's law: The interconnect era. Comput Sci Eng, 5:20–24
9. Banerjee K et al (2001) 3-D ICs: A novel chip design for improving deep-submicrometer interconnect performance and systems-on-chip integration. Proceedings of the IEEE, vol 89: pp 602– 633
10. Topol W et al (2006) Three-dimensional integrated circuits. IBM J Res Dev, 50:491–506
11. Rahman A, Reif R (2000) System-level performance evaluation of three-dimensional integrated circuits. IEEE Transactions on Very Large Scale Integration (VLSI) Systems, vol 8. pp 671–678
12. Loi GL, Agrawal B, Srivastava N, Lin S-C, Sherwood T, Banerjee K (2006) A thermally-aware performance analysis of vertically integrated (3-D) processor-memory hierarchy. ACM Design Automation Conference, pp 991–996
13. Im S, Banerjee K (2000) Full chip thermal analysis of planar (2-D) and vertically integrated (3-D) high performance ICs. IEEE International Electron Devices Meeting, pp 727–730
14. Kleiner MB, Kühn SA, Ramm P, Weber W (1995) Thermal analysis of vertically integrated circuits. IEEE International Electron Devices Meeting, pp 487–490
15. Rahman A, Reif R (2001) Thermal analysis of three-dimensional (3-D) integrated circuits (ICs), IEEE Interconnect Technology Conference, pp 157–159
16. Banerjee K, Amerasekera A, Dixit G, Hu C (1996) The effect of interconnect scaling and low-k dielectric on the thermal characteristics of the IC metal. IEEE International Electron Devices Meeting, pp 65–68
17. Hamann HF, Weger A, Lacey JA, Cohen E, Atherton C (2006) Power distribution measurements of the dual core powerPC™ 970 MP microprocessor. IEEE International Solid-State Circuits Conference, pp 2172–2179
18. Hamann HF, Weger A, Lacey JA, Hu Z, Bose P, Cohen PE, Wakil J (2007) Hotspot-limited microprocessors: direct temperature and power distribution measurements. IEEE J Solid-State Circuits 42:56–65
19. Tadayon P (2000) Thermal challenges during microprocessor testing, Intel Technology Journal 3rd quarter
20. Viswanath R, Wakharkar V, Watwe A, Lebonheur V (2000) Thermal performance challenges from silicon to system. Intel Technology Journal 3rd quarter
21. Prasher RS, Chang JY, Sauciuc I, Narasimhan S, Chau D, Chrysler G, Myers A, Prstic S, Hu C (2005) Nano and micro technology-based next-generation package-level cooling solutions, Intel Technology Journal 4th quarter
22. Banerjee K, Mehrotra A (2001) Global (interconnect) warming, IEEE Circuits Devices Mag, 17:16–32

23. Banerjee K, Lin S-C, Keshavarzi A, Narendra S, De V (2003) A self-consistent junction temperature estimation methodology for nanometer scale ICs with implications for performance and thermal management. IEEE International Electron Devices Meeting, pp 887–890
24. Chatterjee A, Nandakumar M, Chen IC (1996) An investigation of the impact of technology scaling on power wasted as short-circuit current in low voltage static CMOS circuits. IEEE International Symposium on Low Power Electronics and Design, pp 145–150
25. Banerjee K, Mehrotra A (2002) A power-optimal repeater insertion methodology for global interconnects in nanometer designs. IEEE Trans Electron Devices 49:2001–2007
26. Zeitzoff PM (2004) MOSFET scaling trends and challenges through the end of the roadmap. Custom Integrated Circuits Conference, pp 233–240
27. Lin Y-S, Wu C-C, Chang C-S, Yang R-P, Chen W-M, Liaw J-J, Diaz CH (2002) Leakage scaling in deep submicron CMOS for SoC, IEEE Trans Electron Devices 49:1034–1041
28. Lin S-C, Chrysler G, Mahajan R, De V, Banerjee K (2007) A self-consistent substrate thermal profile estimation technique for nanoscale ICs – Part I: electrothermal couplings and full-chip package thermal model, IEEE Trans Electron Devices 54(12):3342–3350
29. Taur T, Ning TH (1998) Fundamentals of modern VLSI devices, Cambridge Univ. Press
30. Gelsinger P (2004) Gigascale integration for teraops performance–challenges, Opportunities, and New Frontiers, 41st DAC Keynote.
31. Borkar S, Karnik T, Narendra S, Tschanz J, Keshavarzi A, De V (2003) Parameter variations and impact on circuits and microarchitecture. Design Automation Conference pp 338–342
32. Adam J, Chang C-S, Stankus JJ, Iyer MK, Chen WT (2002) Addressing packaging challenges, IEEE Circuits Devices Mag 18:40–49
33. Mahajan R, Nair R, Wakharkar V, Swan J, Tang J, Vandentop G (2002) Emerging directions for packaging technologies, Intel Technology Journal 2 nd quarter
34. Im S, Srivastava N, Banerjee K, Goodson KE (2005) Scaling analysis of multilevel interconnect temperatures for high performance ICs. IEEE Trans Electron Devices 52:2710–2719
35. Cess RD (1961) The effect of radiation upon forced-convection heat transfer. Appl Sci Res 10:430–438
36. Öziik MN (2002) Boundary value problems of heat conduction. Dover Publications
37. Haberman R (1983) Elementary applied partial differential equations with fourier series and boundary value problems, Prentice Hall.
38. Peaceman DW, Rachford HH (1955) The numerical solution of parabolic and elliptic differential equations. J Soc Ind Appl Math: 28–41
39. Douglas J, Rachford HH (1956) On the numerical solution of heat conduction problems in two or three space variables, Trans Am Math Soc 421–439
40. S-C. Lin, Chrysler G, Mahajan R, De V, Banerjee K (2007) A self-consistent substrate thermal profile estimation technique for nanoscale ICs – Part II: Implementation and implications for power estimation and thermal management, IEEE Trans Electron Devices 54:3351–3360
41. Icepak, (http://www.icepak.com/)
42. Davis WR, Wilson J, Mick S, Xu J, Hua H, Mineo C, Sule AM, Steer M, Franzon PD (2005) Demystifying 3D ICs: the pros and cons of going vertical, IEEE Design & Test of Comput, 22:498–510
43. Zeng A, Lü J, Rose K, Gutmann RJ (2005) First-order performance prediction of cache memory with wafer-level 3D integration, IEEE Design & Test of Comput 22:548–555
44. Kühn SA, Kleiner MB, Ramm P, Weber W (1996) Performance modeling of the interconnect structure of a three-dimensional integrated RISC processor/cache system, IEEE Transactions on Components, Packaging, and Manufacturing Technology-Part B, vol 19. pp 719–727
45. Wong HSP, Frank DJ, Solomon PM, Wann CHJ, Welser JJ (1999) Nanoscale CMOS, Proceedings of the IEEE, vol. 87. pp 537–570
46. De I, Osburn CM (1999) Impact of super-steep-retrograde channel doping profiles on the performance of scaled devices. IEEE Trans Elect Devices 46:1711–1717
47. Codella CF, Ogura S (1985) Halo doping effects in submicron DI-LDD device design. IEEE International Electron Devices Meeting, pp 230–233

48. Shahidi GG, Warnock J, Fischer S, McFarland PA, Acovic A, Subbanna S, Ganin E, Crabbe E, Comfort J, Sun Y-C, Ning TH, Davari B (1993) High-performance devices for a 0.15-μm CMOS technology, IEEE Electron Device Lett 14:466–468
49. Su L, Subbanna S, Crabbe E, Agnello P, Nowak E, Schulz R, Rauch S, Ng H, Newman T, Ray A, Hargrove M, Acovic A, Snare J, Crowder S, Chen B, Sun J, Davari B (1996) A high-performance 0.08-μm CMOS. IEEE Symposium on VLSI Technology, pp 12–13
50. Lo SH, Buchanan DA, Taur Y, Wang W (1997) Quantum-mechanical modeling of electron tunneling current from the inversion layer of ultra-thin-oxide nMOSFETs, IEEE Electron Device Lett 18:209–211
51. Chau R, Brask J, Datta S, Dewey G, Doczy M, Doyle B, Kavalieros J, Jin B, Metz M, Majumdar A, Radosavljevic M (2005) Application of high-κ gate dielectrics and metal gate electrodes to enable silicon and non-Silicon logic nanotechnology. Microelectron Eng 80:1–6
52. Gusev EP, Narayanan V, Frank MM (2006) Advanced high-κ dielectric stacks with polySi and metal gates: recent progress and current challenges. IBM J Res Dev 50:387–410
53. Tschanz JW, Narendra S, Nair R, De V (2003) Effectiveness of adaptive supply voltage and body bias for reducing impact of parameter variations in low power and high performance microprocessors. IEEE J Solid-State Circuits 38:826–829
54. Pedram M, Rabaey J (2002) Power aware design methodologies, Kluwer.
55. Mutoh S, Douseki T, Matsuya Y, Aoki T, Shigematsu S, Yamada J (1995) 1-V power supply high-speed digital circuit technology with multithreshold-voltage CMOS. IEEE J Solid-State Circuits 30:847–854
56. Chao KY, Wong DF (1995) Thermal placement for high-performance multichip modules. International Conference on Computer Design, pp 218–223
57. Tsai CH, Kang SM (2000) Cell-level placement for improving substrate thermal distribution. IEEE Transactions on Computer- Aided Design 19:253–266
58. Goplen B, Sapatnekar S (2003) Efficient thermal placement of standard cells in 3D ICs using a force directed approach. International Conference on Computer Aided Design, pp 86–89
59. Chandrakasan AP, Sheng S, Brodersen RW (1992) Low-power CMOS digital design. IEEE J Solid-State Circuits 27:474–484
60. Lin S-C, Banerjee K (2008) Cool Chips: Opportunities and Implications for Power and Thermal Management, IEEE Trans Electron Devices 55:245–255
61. Goplen B, Sapatnekar S (2007) Placement of 3D ICs with thermal and interlayer via considerations, ACM Design Automation Conference, pp 626–631
62. Srivastava N, Joshi RV, Banerjee K (2005) Carbon nanotube interconnects: implications for performance, power dissipation and thermal management, IEEE International Electron Devices Meeting, pp 257–260
63. Iwai T, Shioya H, Kondo D, Hirose S, Kawabata A, Sato S, Nihei M, Kikkawa T, Joshin K, Awano Y, Yokoyama N (2005) Thermal and source bumps utilizing carbon nanotubes for flip-chip high power amplifiers. IEEE International Electron Devices Meeting, pp 257–260
64. Xu J, Fisher TS (2006) Enhancement of thermal interface materials with carbon nanotube arrays. Int J Heat and Mass Transf 49:1658–1666

Chapter 15
Status and Outlook

Scott K. Pozder and Robert E. Jones

15.1 Introduction

The preceding chapters have described a wide range of three-dimensional (3D) integration technologies. There has been a very substantial level of research and development (R&D) effort underway in universities and industry. As is typical of an immature technology, there has not yet been much consolidation of the technology choices. There is also a wide range of various types of applications where 3D integration may be advantageous. While there has been limited productization of 3D integration to date, in general, there has not yet been widespread implementation. However, 3D integration appears poised to become a significant player in the market in the next few years.

In this chapter, we will look at the structure of the technology application landscape and discuss the status and challenges of various technologies to meet these requirements. We will also consider the additional challenges that 3D integration can bring to packaging. The role of economics in determining appropriate 3D technologies for various market applications will also be explored. Finally we will look at early market projections.

15.2 Technology and Applications

15.2.1 System Integration

System integration, that is, the integrating together of circuits or intellectual property (IP) blocks, is one of the major applications of 3D integration. As such, 3D integration must compete against a number of established technologies. Figure 15.1 illustrates the relative capability of several system integration methods as a function of overall form factor and interconnect density between circuit blocks. Figure 15.1 also

S.K. Pozder
Freescale Semiconductor, Inc., Austin, TX, USA
e-mail: Scott.Pozder@freescale.com

C.S. Tan et al. (eds.), *Wafer-Level 3D ICs Process Technology*,
DOI: 10.1007/978-0-387-76534-1_15, © Springer Science+Business Media, LLC 2008

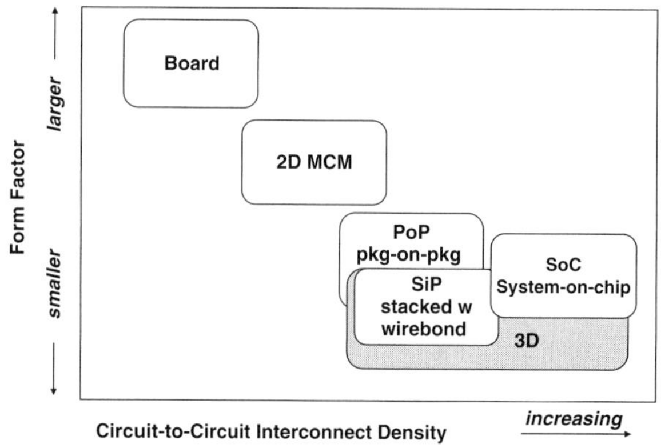

Fig. 15.1 Comparison of various system interconnect technologies in terms of form factor and circuit-to-circuit interconnect density. [1] © 2007 IEEE

illustrates two of the primary advantages of 3D interconnect: competitive circuit-to-circuit interconnect densities and the most compact form factor. Three-dimensional integration encompasses a number of different methods and occupies a relatively large region that overlaps a number of competitors, including package-on-package (PoP), the stacked chip, wired-bonded version of system-in-package (SiP), and system-on-chip (SoC). All of these have smaller form factor and greater interblock interconnect densities than separately packaged chips on a board or multichip module (MCM) with 2D layout of chips on a package substrate. Three-dimensional integration actually offers the smallest form factor, although in many cases of practical interest its advantage over SiP will be minimal, if any. In many applications, system cost will be the most important discriminator between these system integration methods and this will be discussed later in this chapter.

In comparison with other assembled systems (board, MCM, PoP, and SiP), 3D integration offers shorter circuit-to-circuit interconnect and generally higher interconnect densities. This can provide for lower power and higher-bandwidth communication. Three-dimensional integration can provide connectivity that is competitive with that of SoC. In very high performance systems, 3D integration can shorten the global interconnect compared to SoC and thus enhance system performance [2–4]. A major advantage of 3D over SoC is that it allows the use of different process platforms for the different circuits to be integrated; for example, if memory and logic functions are on separate strata, then optimized processing can be used for each stratum. This can result in substantial cost advantages over the use of an embedded memory process. Three-dimensional integration can be used to fabricate technology combinations that would otherwise be unavailable or at least delayed for monolithic or SoC integration. Other examples of heterogeneous integration include (1) combinations of analog circuits with digital logic circuits, and (2) combinations of microelectrical mechanical systems (MEMS) or sensors with either analog circuits or mixed mode analog and digital logic circuits. Heterogeneous integration also can include devices fabricated on different types of semiconductor substrates (e.g., III–V devices on Si complementary metal-oxide semiconductor (CMOS)). However, partitioning of heterogeneous technologies to different strata can result in added constraints. An obvious situation is that for a given system the die area needed for one process technology may not equal the die area for a different process technology. This is an obvious problem for wafer-to-wafer integration where the same die area must be dedicated on each stratum. Further, wafer-to-wafer integration is economically practical only if the wafer diameters are the same. A major advantage of wafer-to-wafer bonding is that post-bond processing using conventional wafer processing is greatly facilitated in comparison with the topography challenges that would be encountered in processing of a wafer with die bonded to the upper surface.

The ability to pretest and assemble only known good die (KGD) can be economically attractive. This is an option for PoP, SiP, and die-on-wafer 3D integration. It is not an option for SoC integration, and is not generally feasible for wafer-on-wafer which would require matching of yielding die patterns. Table 15.1 compares some of the major features of die-to-wafer and wafer-to-wafer technologies against SoC and wire-bonded stacked dice SiP. Die-to-die integration is also practiced and many attributes are similar to die-on-wafer.

15.2.2 Vertical Interconnect

Conceptually, there are two basic types of vertical electrical connections that are needed in 3D integration beyond the usual electrical connections in 2D integrated circuits: (1) stratum-to-stratum connections and (2) through-stratum vias (TSVs) which may also be known as through-substrate vias or through-Si vias as appropriate. Both of the connectivity functions can be provided by the same structural element in certain integration schemes. Additionally, stratum-to-stratum connections are always needed for 3D integration, while TSVs may be avoided by some integrations using face-to-face orientation. The density of interconnects on the surfaces being bonded and the TSVs is a function of the system architecture and process constraints, especially the aligning and bonding of stratum.

Table 15.1 Comparison of major attributes of different methods for integrated circuit system formation

	SiP	Die-to-wafer	Wafer-to-wafer	SoC
Partition of differentiated technologies	Yes	Yes	Yes	No
SoC-like performance	No	Yes	Yes	Yes
Economics of KGD combinations	Yes	Yes	No	No
Independent areas for different strata	Yes	Yes	No	Not an issue
Different wafer sizes for different strata	Yes	Yes	No	Does not apply
Relative form factor	Small	Smallest	Smallest	Moderate
Global interconnect latency	Moderate	Lowest	Lowest	Low
Global interconnect densities	Low	Medium	High	Highest

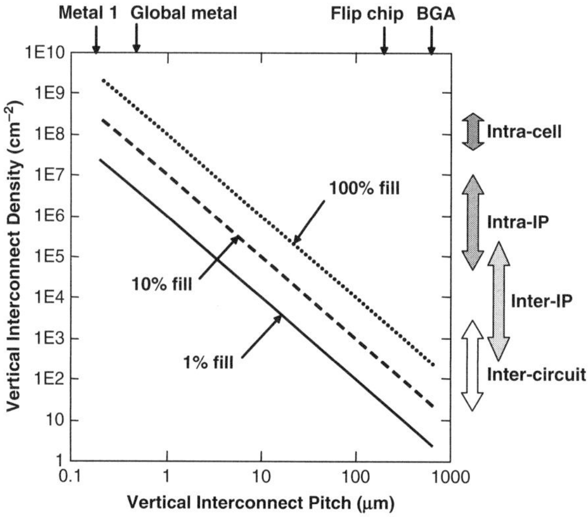

Fig. 15.2 The system interconnect architectural expectation drives vertical interconnect density and pitch. Applications range from low-density circuit-to-circuit connectivity to extremely high-density interstrata connectivity within a logic or memory cell or an imager pixel [1] © 2007 IEEE

15.2.2.1 Vertical Interconnect Density and Applications

System architecture requirements can result in vastly different vertical interconnect density requirements as illustrated in Fig. 15.2. The relationship between 3D interconnect density and the pitch for various fill factors is quite straightforward. More interesting is the mapping of this density on to the application space. If circuit-to-circuit connections are needed, the 3D interconnect densities are typically quite low. As the application requirements move to inter-IP block connections and then to intra-IP block connections, the required interconnect densities increase dramatically. This also will be associated with driving design complexity for 3D integration. The highest densities are required when 3D interconnections are employed within an individual design cell (e.g., logic gate, memory bit cell, or imager pixel).

In turn, these vertical interconnect densities tend to drive several different features of the process technology. For example, higher density requires smaller interstratum connections and TSVs. The highest vertical interconnect densities can be achieved in the case of the grown additional semiconductor device layers [5, 6] (also see Chapter 2), or in the case of transfer of unpatterned semiconductor layers [7]. In these cases there is no requirement for stratum-to-stratum alignment of previously formed structures; thus, the patterning of interstratum connections can use design rules that approximate those in a monolithic integration. A limitation of these approaches is that the devices on the previously completed strata must experience the additional thermal processing for fabricating devices in the newly added strata as well as the thermal processes for bonding or growing the additional semiconductor layer. This type of integration also can limit the interstrata bond material to inorganic bond materials which can withstand semiconductor processing temperatures. These thermal limitations can be mitigated by forming the semiconductor devices on a separate semiconductor stratum and then transferring this layer to a wafer with previously formed devices [8, 9]. Because of the reduced thermal restrictions, the strata to be bonded now have the freedom to include interconnect layers as well as the device layers. Clearly, in the case of bonding previously patterned device layers, the density of vertical interconnections can be limited by the stratum-to-stratum alignment capability. Additionally, since the aspect ratio (height-to-diameter) of TSVs is limited by process capability, higher TSV densities and smaller TSV diameters also require thinner strata. Interstratum connection density also interacts with the ability to test for KGD. At lower interstratum connection densities such testing may be quite feasible, but with increasing density a point is reached at which testing is no longer practical.

15.2.2.2 Interstratum Connections

Interstratum connections can be formed at bond using conductive bonds or a combination of conductive and dielectric bonds. The most common type of connection at bond is the metal-to-metal bond that forms the mechanical and electrical connection at the faces of two strata. Copper-to-Copper and Cu:Sn-to-Cu are demonstrated for wafer-to-wafer and die-to-wafer bond process integrations [10–12]. However, metal bonds can result in a cantilevered or unattached edge of the bonded stratum that has to be considered when sawing multiple strata and handling the edges of the wafers. Combining dielectric or adhesive bonding into a consolidated integration with metal-to-metal bonding solves the interstratum gap issue [13]. Typically the dielectric bond is initiated first and the dielectric-bonded surfaces hold the metal areas in contact or close proximity. The metal-to-metal bond is then completed in a subsequent anneal or thermal-compression process.

For wafers bonded using dielectric or adhesive bonding, the electrical connections are formed post-bond using TSVs to interconnect conductive elements on different strata.

15.2.2.3 Through-Stratum Vias

Through-stratum vias are a key feature in 3D integration. How the TSVs are implemented can interact with other key technology factors such as stratum thickness, bonded strata orientation (e.g., face-to-face or back-to-face), and interstratum connection density. Through-stratum vias are essential for most 3D integration schemes. The notable exception is the case of face-to-face, die-to-wafer bonding using micropad bonding to place a smaller die on a larger die, with subsequent connections to the package made to the uncovered portion of the larger die [14]. Through-stratum vias can be made before bonding (TSV-first) or after bonding (TSV-last). The TSV in Fig. 15.3 is a TSV-last integration where the TSV was etched and filled to form an interstratum connection of metal traces after bonding and thinning the upper wafer. In this case the upper wafer was a face-down SOI

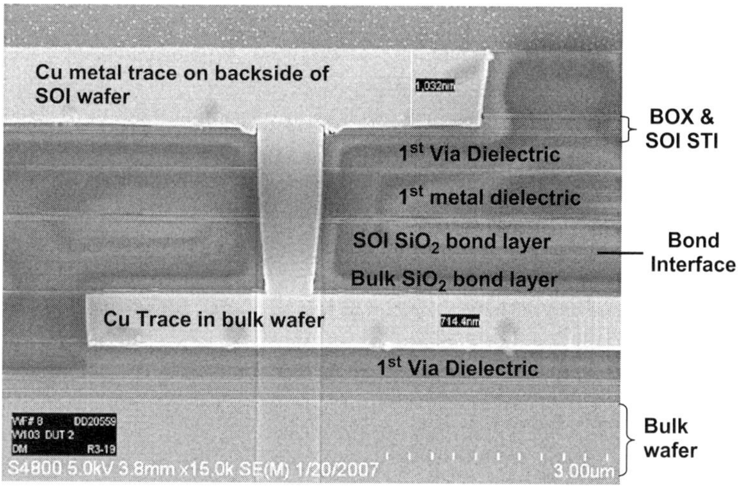

Fig. 15.3 Cross-section of an interstrata via formed after face-down dielectric bonding and thinning of a SOI wafer. From [15] © 2007 IEEE

wafer which was thinned down to the buried oxide (BOX) layer. In this integration the TSV is etched through the shallow trench isolation (STI) and various dielectric layers. Thus, no dielectric liner of the TSV is needed prior to metal fill to prevent contact to the semiconductor stratum. The metallization generally will include a diffusion barrier deposited prior to the main metal fill. Figure 15.4 is an example of a TSV-first integration where a blind TSV is etched into a Si wafer before bond. The blind TSV is exposed by backside thinning, and subsequent backside processing can

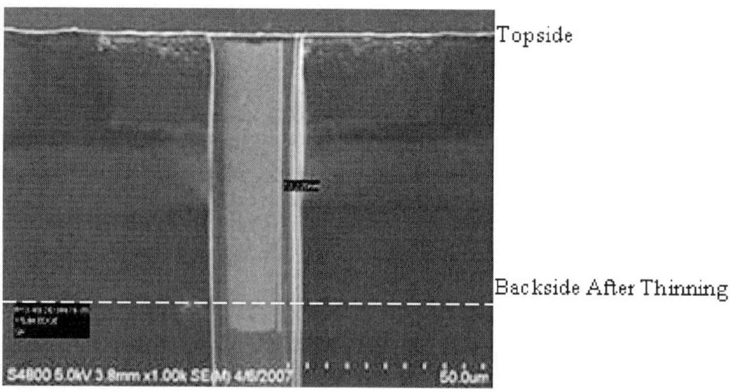

Fig. 15.4 A scanning electron microscopy (SEM) cross-section of a Cu–filled, 75-μm deep, 15-μm diameter TSV. This TSV-first integration connects a trace to be formed over the TSV on the topside to the backside using the exposed portion of the TSV after thinning as the backside connection point. (The white vertical lines adjacent to the TSV in the image are an artifact of the sample preparation using focussed ion beam (FIB) to finish cross-sectioning the TSV). From [15] © 2007 IEEE

Table 15.2 Typical 3D integration schemes used with different combinations of bonding type and TSV technology

	TSV-last	TSV-first	Microbond, no TSV
Dielectric bond	Wafer-to-wafer	No	No
Conductive bond	No	Wafer-to-wafer, die-to-wafer	Die-to-wafer
Hybrid bond	No	Wafer-to-wafer, die-to-wafer	Die-to-wafer

further reveal the TSV as a Cu stud for bonding or form a metal pad connection site. Here the TSV opening is etched through the semiconductor material, and a dielectric liner is required to isolate the metal fill from the semiconductor. The dielectric liner is generally deposited by chemical vapor deposition (CVD). This is followed by deposition of a diffusion barrier and the metal conductor fill. Both electroplated Cu and CVD W fill have been used. Chemical vapor deposition of in-situ-doped polycrystalline Si also has been used when it is desired to form the TSV early in the device integration flow.

Through-stratum via size is controlled by the TSV aspect ratio that can be etched and filled. Electroplated Cu fill typically limits the achievable aspect ratio more than the ability to etch the via opening. Aspect ratios of 5:1 or less are common, although aspect ratios of up to about 10:1 have been reported. CVD W and polycrystalline Si can achieve fill of high aspect ratios higher than 10:1.

Table 15.2 summarizes combinations of 3D technologies which are typically used for bonded strata integration. For 3D integration schemes with the additional semiconductor stratum deposited directly on the wafer, the electrical connections between strata may be formed during deposition by etching via openings prior to deposition of the semiconductor strata. In some cases, this via opening provides the crystal seed region for recrystallization of the newly deposited stratum. In other integrations of deposited stratum, the interstratum connections are provided by TSVs formed after the stratum deposition.

15.2.3 Bonded Vertical Integration Technologies

15.2.3.1 Alignment and Bonding

In any bonding of previously patterned strata, the interstratum alignment accuracy will limit the minimum stratum-to-stratum interconnect pitch. For the case of TSV-first, the alignment accuracy can limit the minimum pitch of micropads and TSVs. Wafer-to-wafer bonding is a parallel process operation which results in all of the dice on one wafer being aligned to their companion sites on another wafer in a single operation. As a practical matter the greater efficiency of alignment of the many dice on a wafer affords increased alignment accuracy. The die-to-wafer integration can be more limited in vertical interconnect densities because the value of the devices can determine the minimum affordable serial placement throughput rate which is generally a tradeoff with alignment accuracy. However, the alignment capability of existing die placement tools can meet the needs of circuit-to-circuit and, in many cases, IP-block-to-IP-block system needs. The advantage of die-to-wafer bonding is that it allows the integration of different die sizes and dice from different wafer sizes. Additionally, die-to-wafer offers the substantial economic advantage of allowing assembly of KGD to good receiver die on the wafer.

The final thickness of the bonded stratum can place a significant limitation on the TSV pitch. For TSV-first integrations, the bond can be made before or after thinning the bonded stratum. In practice, TSV-first strata have not been thinned as aggressively as strata with TSV-last, so the TSV pitch is typically not as small.

For the case of TSV-last, very aggressive thinning has been demonstrated especially with SOI wafers. Quite small TSV diameters are then feasible and the vertical interconnection density becomes limited by overlay accuracy [16]. Tool placement can align one wafer to another with an accuracy of 1 μm or less. However, as placement accuracy improves, the lens distortion and wafer bowing result in a lower bound to overlay accuracy when patterning the TSVs subsequent to bonding. Furthermore, thermally generated misalignment becomes more important for overlay requirements in the 1-μm range. The radial thermal expansion differences for a matched pair of 300-mm wafers is shown in Fig. 15.5 as a function of the temperature differences between the wafers and the average temperature of the wafers. These results were calculated using the temperature-dependent thermal expansion coefficient data of Okada and Tokumaru [17]. The temperature variation between wafers at align must be less than 0.25°C for the thermal expansion mismatch at the edge of a 300-mm wafer to approach 100 nm. It is reported that for 200-mm wafer bonding, submicron alignment accuracies are possible for room temperature bonds using tools with temperature control from 0.5°C to 1°C [18].

The die-to-wafer alignment is carried out with package pick-and-place tools and the micropad connect size and pitches are selected to allow rapid placement of the die to reduce place time and cost [19]. High-speed placement is achievable for die placement misalignment allowance of about ±10 μm (3σ). While placement

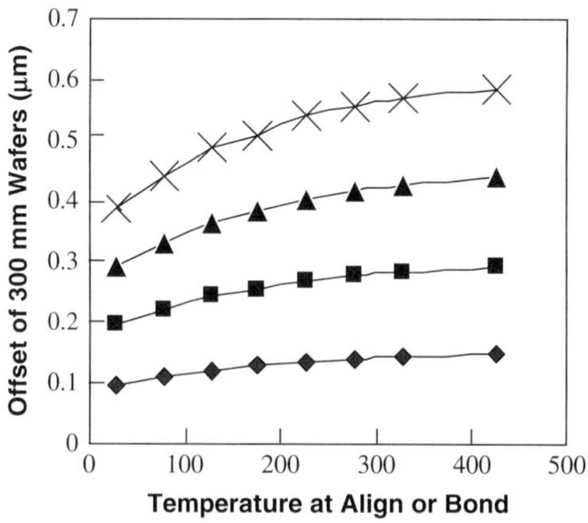

Fig. 15.5 Plot of the thermal mismatch at the edge of a 300-mm-diameter wafer pair at four temperature differences when the centers of the wafers are aligned. Wafer temperature differences of 0.25°C (diamonds), 0.5°C (squares), 0.75°C (triangles), and 1.0°C (crosses). Thermal expansion coefficients used in the calculations are from Okada and Tokumaru [17]. From [15] © 2007 IEEE

tolerances of about ±1 μm can be achieved it is presently done at substantially reduced speed. Because of the small size of dice compared to wafers, die-to-wafer has minimal misalignment due to temperature mismatch and materials differences in coefficient of expansions.

When TSV or other backside features are patterned on thinned bonded wafer strata, then analysis of the stepper lens and bonded wafer distortion have to be accounted for in stepper patterning. In silicon-on–insulator (SOI) wafers that are bonded face-down, the back silicon can be removed by grinding followed by wet etching to the BOX. The SOI is transparent and backside alignment to the frontside alignment marks is possible. The effects of lens distortion and 3D processing are reported to result in errors that limit the minimum feature size printed on a bonded wafer pair to 0.9–1.2 μm [20]. When all sources of stratum-to-stratum alignment and patterning errors are considered, a minimum practical TSV size for wafer bonding could be 1-μm TSV diameter at a pitch of ~3 μm.

15.2.3.2 Stratum Thinning

The assembly of strata can be accomplished as bond-before-thinning or bond-after-thinning. Bond-before-thinning using face-down, wafer-to-wafer bonding facilitates greater thinning of a stratum by eliminating thinned wafer handling. The extreme example is the thinning of a face-down-bonded SOI wafer to the BOX layer, leaving only the BOX, SOI layer, and interconnect layers [21, 22]. Face-down-bonded bulk Si wafers can also be thinned but not to as great a degree since the devices extend deeper into the wafer and because of the lack of BOX layer as a convenient etch-stop layer. The greater alignment accuracy and ease of wafer thinning make wafer-to-wafer preferred for very high interstratum connection densities.

For die-to-wafer bonding the dice can be thinned at wafer level prior to singulation and bonding, or the dice can be bonded face-down on the receiving wafer and then thinned. With the dice bonded to the wafer, subsequent processing is complicated by the highly nonplanar surface. Even if the planarity is corrected by the use of fill material, the pattern alignment of the dice still have greater local variations compared to the wafer. Wafer level thinning prior to singulation of the dice streamlines the thinning and backside metallization processes, but introduces the challenges associated with temporary bonds and thinned die handling. When strata to be bonded are thinned to a thickness that is not self-supporting then a temporary support wafer is required. The thinned temporarily bonded wafer is debonded, transferred to tape, and sawn into dice. The dice are picked from tape and placed on the wafer for micropad bonding. Regardless of whether a die is bonded then thinned or thinned then bonded, the placement variation has to be considered in any post-bond processing.

15.2.4 Design

In any device design the technology node, electrical performance, power requirement, reliability, and package type are optimized to meet the market needs of the finished part. Within the technology node are design rules, design for test, circuit libraries, layout specifications, and mask-generation rules that are selected based on the desired power, performance, and reliability specifications. Although stacking circuits and devices is a simple idea, it can cause increased complexity that can extend to the circuit block and transistor design level.

It is understandably attractive to design each die independently to the interface between bonded die, much as separate devices are designed. The benefit of designing to the electrical interface between the strata is that the circuits and independent die functions remain relatively unchanged for 3D. In the case of wafer-to-wafer integration with dielectric bonding, a special interface may not be needed, especially in the case of co-design of the strata. In other cases, it may be desired or necessary to provide some form of input/output (I/O) circuitry between the strata. If existing circuit designs are reused in a circuit-on-circuit 3D integration, they could use the standard I/O circuitry employed for single-chip packaging. This could however provide more output voltage and electrostatic discharge (ESD) protection than is actually needed in 3D integration. There are opportunities to reduce the I/O power and chip area. Ezaki et al. [23] have suggested smaller buffers and reduced-size ESD protection for interstratum connections. Recently, Alam et al. [24] have further suggested the use of interstratum I/O circuitry which is voltage programmable to facilitate a more independent circuit design. To allow 3D "mix and match" of independently functioning circuits, a standardized interface layout would be required. Standardization could be feasible in selected cases such as memory I/O which could facilitate integration of memory-on-logic or memory-on-memory. While KGD testing is not necessarily a requirement for standardization, the use of both can be a desirable combination for "mix and match" capability. As mention previously, if the vertical connectivity density is too great it may not be feasible to probe the separate devices. Thus KGD testing is the most feasible for the case of circuit-to-circuit or IP block-to-IP-block integration.

One of the least design-intensive integrations is die-to-wafer bonding of individual circuits because the interstrata connectivity is relatively simple. It is not restricted to like die sizes or like wafer sizes, and can use face-to-face bonding. In the integration using face-to-face, micropad bonding of two dies, TSVs are not needed if wirebonds to the larger bottom die are used for the package connections.

The 3D stacking of multiple copies of the same memory also presents one of the simpler 3D design opportunities as there are many shared address and data lines. This application makes wafer-to-wafer particularly attractive since memory repair techniques limit the yield impact of wafer-to-wafer bonding and the dice on different strata are inherently the same size. From the process aspect, this integration does require the use of TSVs but the required densities typically will be low.

The evolution of 3D design that incorporates partitioning functions vertically at a finer granularity than the circuit or major IP block level requires substantial enhancement to design tools and design methodologies. For example, design tools are required that can achieve coherent layouts by linking design data from different mask sets without data naming conflicts or unintended data over writes. This increase in design sophistication would require new rules and methodologies for design, layout, routing, circuit libraries, design for test, circuit verification, and mask generation. The leap to finer 3D design granularity increases the difficulty of testing the individual stratum prior to 3D integration because the vertical interconnect density exceeds probe card capability. This can enhance the need for built-in self-test and self-repair functionality.

15.3 Packaging of 3D Devices

The transistor density in a die and in a package is continuing to increase with scaling of 2D feature dimensions. Three-dimensional integration adds to this density increase by the addition of multilayers of active devices. Stacking die whether by stacked die in package, die-to-wafer, or wafer-to-wafer adds value by reducing form factor and reducing the packaging cost. The 3D bonded strata add additional value by increasing bandwidth between strata and reduced interconnect length. A major challenge of 3D integration is the reduced surface-to-volume ratio which restricts the available thermal and electrical connectivity to the package.

15.3.1 Signal, Power, and Ground Congestion

An important package design consideration for a 3D die is the potential for congestion of vertical interconnect. Because of the reduced footprint of a 3D chip there is an increased risk of the density of connections on the chip surface exceeding the capability of packaging technology. These limitations for connectivity to the package can occur for either perimeter connections or area array connections although the details will be different. Figure 15.6 illustrates that the vertical connections may be independent connections to the package, shared between all strata and the package, or internal to the 3D integration. In actual chips various combinations can occur. It is clear that a large number of independent connections can result in high vertical interconnect densities in certain strata. Further, this will also generate congestion in the connections to the package. On the other hand, totally internal connections can actually reduce the required connections to the package. Stacked memories can have largely shared connections. A memory-on-logic integration may

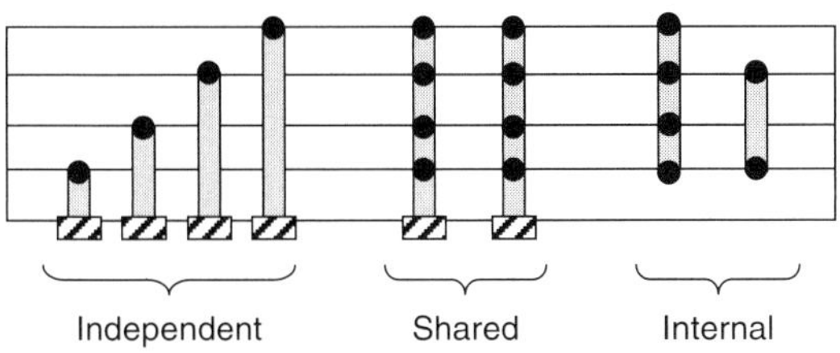

Fig. 15.6 Independent, shared, and internal vertical connections. This shows how a large number of independent connections can result in congestion in certain strata and in connections to the package [1] © 2007 IEEE

have the address and data lines as internal connections if these need not connect to the package. Even though ground and power connections potentially may be shared connections from a circuit schematic standpoint, in practice, parallel connections may be needed to provide sufficient current-carrying capability and also to provide sufficiently low resistance. A thorough accounting of the final 3D chip footprint, the required number of connections to the package, and the ability of the package and assembly operation to support this density of connectivity is required for the overall package and 3D chip design.

15.3.2 Thermal Congestion and Heat Dissipation

The decrease in die surface area for 3D integration also can be a serious issue for heat dissipation. This can make design of overall heat removal more difficult for moderate- and higher-power chips. Potential solutions can include enhanced thermal interfaces between chips and packages, better heat spreaders, packages with lower thermal resistance, and more aggressive heat sinks. Additionally, 3D design can be more complex because hot circuit blocks might be stacked on top of each other. Power-aware 3D design tools may be needed to avoid stacking "hot circuit blocks" over "hot circuit blocks" [25].

To minimize costs, designers will try to optimize the use of passive cooling within the package and chip themselves. However, in extreme cases of thermal concentration, more aggressive cooling techniques may be needed such as liquid cooling within the chip or package or thermally conductive heat pipes to dissipate heat from the stacked strata to the package heat sink. The 3D stacking of chips can be utilized in the formation of microchannels for fluid cooling.

15.3.3 Reliability of Packaged 3D Devices

The diversity of technologies that can be incorporated into a 3D die could include heterogeneous active layers, low-κ dielectric interconnect materials, Cu interconnects, Al interconnects, and basically any material found in a product die. The package types that can be used with a 3D die include flip chip or wirebond connection to any of a wide range of grid array, lead frame, and edge connect packages made with plastic or ceramic materials. Many disparate materials and processes can be included in packaged 3D devices, and reliability testing of these more complex systems will be part of any product plan.

A key parameter in die-to-package reliability is the coefficient of thermal expansion (CTE) mismatch between the 3D die and package and also any mismatch between the different strata in the 3D die. Generally the thermal conduction between the strata is high so that thermally generated stresses between strata will be minimal if the same substrate material is used or if the materials are CTE-matched. If materials with different CTEs are integrated (e.g., Si and GaAs) the thermal stresses on the bonded interface will be greater. The CTE mismatches between the semiconductor substrates and the metals and dielectrics of the interconnection also can generate substantial stresses. The advent of low-κ dielectric has been accompanied by reduced mechanical strength which will be a problem for 3D die as well as conventional die. Mechanical stress modeling will be used to design these package–chip interfaces, but reliability testing is needed to validate the success of the design. Because the technology is not yet mature only limited published reliability results arc available.

15.4 Market and Economics of 3D

The value of a die whether 2D or 3D is determined by market needs that are influenced by demands for form, function, power consumption, and fashion in consumer, industrial, and personal electronic devices. The multiple product applications and the transient nature of the market supply and demand limit the ability to estimate 3D value; however, a comparison to existing technologies can be done to identify areas where 3D has advantages.

15.4.1 Technology Maturity

The typical mature technology has process, equipment, and design infrastructure in place to enable manufacturing of the technology. When compared to this definition of a mature technology, 3D has a range of competing and complementary processes, capable equipment sets that perform individual processes, and an emerging design infrastructure. In time the various process integrations and equipment functions will coalesce to form the infrastructure for the manufacture of 3D devices. It is unclear whether the design infrastructure will lead or follow the equipment and process infrastructure. Regardless of how the infrastructure coalesces, the first introduction of 3D in any substantial scale will probably occur when 3D has demonstrated a cost advantage over an existing technology.

15.4.2 Cost

Given the large number of 3D technologies under consideration and the technological immaturity, it is difficult to accurately compare the costs against the alternative technologies. However, it is instructive to use cost modeling to investigate some of the general trends.

The cost of die-to-wafer 3D integration presents a more complex cost situation than either wafer-to-wafer or SoC. Here we introduce a hypothetical base case to investigate the impact of variations on that base case. Actual costs may vary substantially. The base case considers the situation with undifferentiated process technologies between the 3D strata or the SoC wafer. The base case assumes the cost of a fully processed 200-mm wafer is C_0, and each wafer prepared for 3-D integration has an additional 10% C_0 cost adder. A Murphy yield model is used along with a benchmark defect density D_0. It is assumed that the die-to-wafer bonding process uses a separate placement process (0.0005% C_0/die) and a thermal compression bonding process (1.25% C_0) following placement of the dice on the wafer. Only known good top die are placed on KGD on the bottom wafer. Since normalized costs are used, the calculations will be the same for other wafer sizes for all factors except for the individual die placement cost which is normalized by die rather than wafer. However, general trends should be similar. Since a wide range of technologies and technology process nodes are being considered for 3D integration, we investigate the sensitivity of the results to variation of key parameters away from the base case.

Figure 15.7 shows the calculated cost ratio of double strata 3D die to a SoC die of equivalent total area as a function of the bottom 3D die area for various top-to-bottom die area ratios. To investigate the sensitivity of the 3D/SoC cost ratio to the added test costs for the KGD, Fig. 15.7a includes zero test costs and Fig. 15.7b includes test costs for each individual wafer of 20% C_0. In both cases the die-to-wafer solution only becomes less expensive than the SoC solution for larger die sizes. (Both larger-top and larger-bottom die sizes assist in this.) The greater the test costs, the more the breakeven point of 3D integration is pushed out to larger die sizes. For the situation with undifferentiated process technology, the 3D system will have lower cost only because of the yield advantage of being able to preselect combinations of KGD. With increase in die sizes, the defect-limited die yield will fall off and 3D integration can become less expensive than SoC. To further illustrate this point, Fig. 15.8 shows the same situation as Fig. 15.7b except that D_0 is doubled from the benchmark value. While this causes the 3D/SoC cost ratio to drop below 1 more rapidly with increasing area, it has to be realized that the absolute costs will be increased for either technology by the lowered yield.

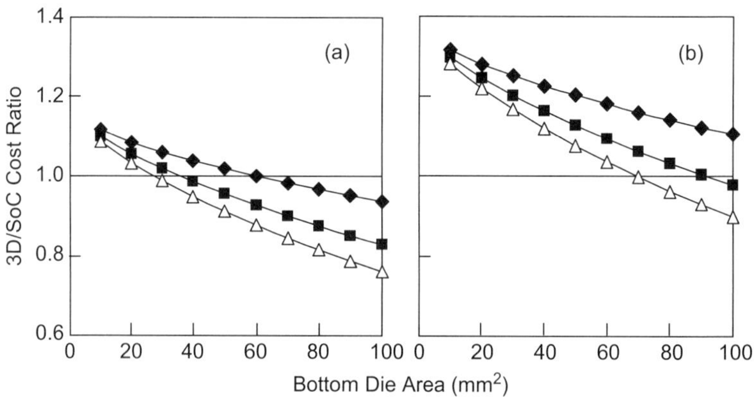

Fig. 15.7 Die-on-wafer 3D/SoC cost ratio for a hypothetical base case for undifferentiated process technologies for (**a**) no test costs and (**b**) 20% C_0/wafer KGD test costs. For relatively large die, the yield advantage of die-on-wafer for KGD can overcome the added process costs. Top–die/bottom-die area ratios of 0.25 (*diamonds*), 0.50 (*squares*), and 0.75 (*triangles*)

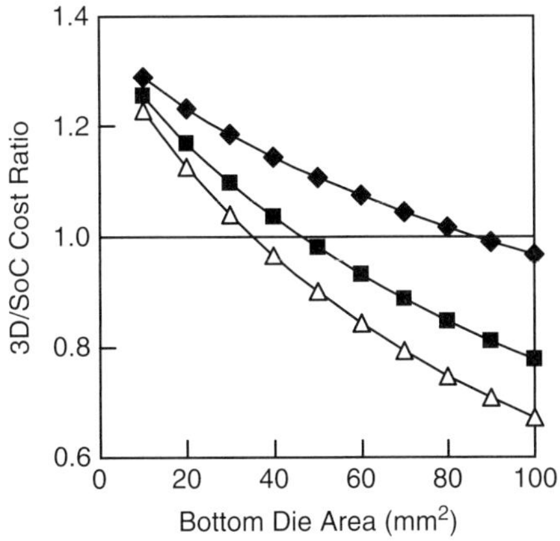

Fig. 15.8 The situation here is the same as in Fig. 15.7b except that the defect density is twice as high as the benchmark value

The advantage of 3D integration over SoC to bring together differentiated technologies has been mentioned earlier. To illustrate the impact of this effect, Fig. 15.9 shows the 3D/SoC cost ratio for the same cases as Fig. 15.7 except that the cost of the SoC wafer is increased to 125% C_0 to represent the cost of the enhanced process platform that can accommodate the features of both of the individual differentiated 3D stratum (which remain at C_0). For this scenario and no test costs, 3D integration has a cost advantage over SoC for all die sizes considered. Even with test costs of 20% C_0 per wafer, 3D integration becomes cost advantageous at relatively small die sizes.

For undifferentiated process technologies, wafer-to-wafer 3D integration will inherently be more expensive than the equivalent SoC integration. This suggests that such an integration scheme will be selected only for alternative reasons such

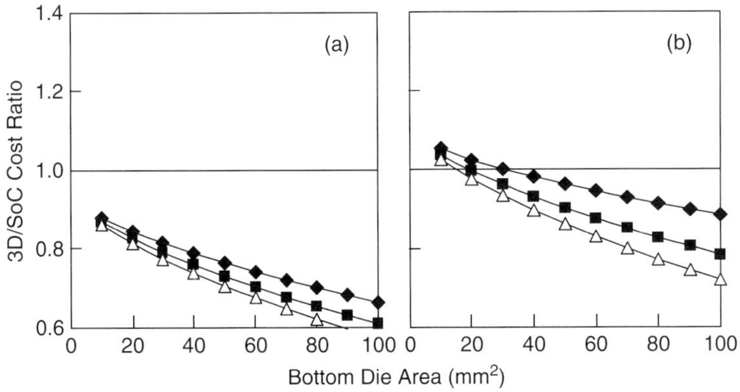

Fig. 15.9 3D/SoC cost ratio for same situations as Fig. 15.7a and b but with differentiated process technologies such that a SoC wafer costs 125% C_0. The application of 3D to differentiated process technologies can have a substantial economic advantage

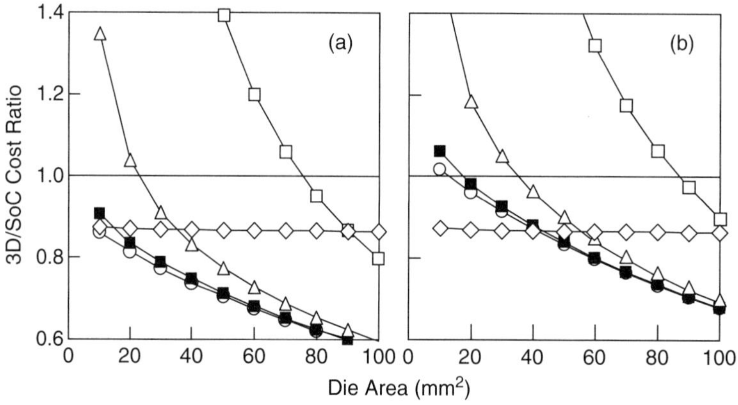

Fig. 15.10 3D/SoC cost ratio for the case where top and bottom die have the same die area. Similar situations as in Fig. 15.9a and b but with the addition of wafer-on-wafer 3D integration (*diamonds*) and a die-on-wafer integration with die placement cost of 0.0005% C_0 (*open circles*), 0.005% C_0 (*solid squares*), 0.05% C_0 (*open triangles*), and 0.5% C_0 (*open squares*)

as enhanced performance or reduced form factor. However, for differentiated process technologies, wafer-to-wafer 3D integration can be less expensive than SoC if the added process costs for the 3D integration are less than the cost added for the enhanced SoC process cost. This needs to be qualified in that the differentiated 3D strata must have equal die areas. Figure 15.10 illustrates the 3D/SoC cost ratio for wafer-to-wafer and die-on-wafer as a function of bottom die area. Here only equal top and bottom die areas are considered. The cost model parameters are as in Fig. 15.6 with the following exceptions. The wafer-to-wafer bonding process cost adder is assumed to be 15% C_0. For die-to-wafer integration, a wide range of die placement costs (0.0005% C_0 to 0.5% C_0) are investigated. For this scenario, wafer-to-wafer has a cost advantage over SoC for all die areas. Wafer-to-wafer is advantageous over die-to-wafer for smaller die areas. The breakeven point is dependent on the die placement cost and the KGD testing costs of the die-on-wafer integration.

15.5 Market Projections

Three-dimensional integration has been the subject of a rapidly expanding R&D effort. This effort is broad spread with participation by universities, government labs, consortia, process and design tool vendors, material suppliers, and semiconductor manufacturers. There is presently a strong expectation that 3D integration will move into volume manufacturing in the near future. The history of new semiconductor technologies in moving from development to volume manufacturing teaches us that a cautionary note may well be appropriate for any specific projections. While technological difficulties in manufacturability or reliability may arise, given the level of effort and the wide range of technical alternatives it is likely that such difficulties will delay the technology only as long as the economics look attractive. For many applications the cost competitiveness will be the critical factor in whether or not to use 3D.

A range of products have been announced, or at least forecast, for future market entry. These range from relatively low interstrata connection densities for die-to-wafer, face-to-face micropad bonded memory on logic without TSVs [14], to wafer-on-wafer massively parallel processors with stacked cache memory [26]. Memory products stacked as high as eight strata also have been announced [27]. Imagers with the optical detectors on one stratum overlying the pixel electronics in a second stratum have received substantial interest [28–31]. This application requires relatively high interstrata connection density since vertical connections are needed in each pixel. Motoyashi et al. [32] have suggested a 3D roadmap which is imager-centric although memory–logic combinations and stacked memory also are included. This roadmap begins with a simple imager sensor in 2006 and adds increasing functionality until an artificial retina is achieved in 2014. Others have made proposals for partitioning of N-metal-oxide semiconductor (NMOS) and PMOS transistor on to different strata which will result in extreme interstrata interconnect densities since TSVs would be required within each CMOS logic cell or SRAM bit cell. Vertical SRAMs with pass transistors stacked over p-field-effect transistor (FET) load transistors which are stacked over n-FET pull-down transistors have been reported [6]. The pass and load transistors were fabricated in single-crystal thin-film silicon layers formed over intermediate dielectrics. The pre-fabrication of n-FETs and p-FETs on separate strata and then using bonding to stack one stratum on the other also has been described [9].

Given the breadth of product requirements, several organizations are pursuing more than one technical approach to 3D integration. One of the more comprehensive 3D roadmaps is that of IMEC as has been presented by Beyne [33]. This research roadmap describes three different classifications of 3D integration. The first classification he identifies is 3D-SIP (system in package) which includes stacked and wire-bonded dice and other packaging-like techniques. The second classification is 3D-WLP (wafer-level packaging) which employs packaging-like technologies but at the wafer level. This includes two different approaches: (1) stacked and bonded die, and (2) ultra-thin chip embedding (UTCE) with built-up interconnect layers. The third group includes 3D-SIC which uses IC fabrication technologies to produce the tightest vertical interconnect pitches. Figure 15.11 shows this roadmap for the stacked and bonded die version of 3D-WLP and the 3D-SIC approaches while omitting the 3D-SIP and the various UTCE approaches. This roadmap clearly shows an increasing complexity of 3D integration technology with time; for example, 3D-WLP progresses from face-to-face-bonded die using flip-chip bumps to face-to-face bonding using micropads, then TSVs to the stacked die and finally stacking more than two dice. In the WLP technology the TSVs are relatively large and are only partially filled with Cu. The 3D-SIC technology uses wafer fab processes to produce smaller geometries. The TSVs are completely filled with Cu and the via diameters can be quite small (3–5 μm) as the final Si substrate is thinned to 10–20 μm. The 3D-SIC technology also uses Cu-to-Cu bonds rather than the microbumps of the 3D-WLP technology.

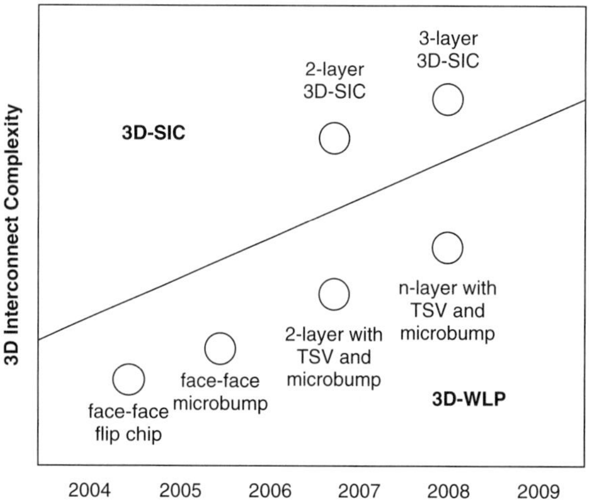

Fig. 15.11 Research roadmap for 3D wafer-level packaging and 3D SIC integration as adapted from Beyne [33] © 2006 IEEE

As the technology matures and as experience is gained in which technologies and applications are economically advantageous, roadmaps with more industry-wide acceptance will become available. While it is difficult to accurately forecast the future, 3D integration is entering a very exciting phase as it tries to break from development into volume manufacture.

References

1. Jones RE, Chatterjee R, Pozder S (2007) Technology and application of 3D interconnect. In: Proc IEEE Intern Conf Integrated Circuit Design and Techn, pp 176–179
2. Banerjee K, Souri S, Kapur P, Saraswat K (2001) 3-D ICs: A novel chip design for improving deep-submicron interconnect performance and system-on-chip integration. Proc IEEE 89:602–633
3. Meindl JD, Davis JA, Zarkesh-Ha P, Patel CS, Martin KP, Kohl PA (2002) Interconnect opportunities for gigascale integration. IBM J Res Dev 46:245–263
4. Zeng A, Lii J, Rose K, Gutmann RJ (2005) First-order performance prediction of cache memory with wafer-level 3D integration. IEEE Design Test Comput 22:548–555
5. Kunio K, Oyama K, Hayashi Y, Morimoto M (1989) Three dimensional ICs, having four stacked active device layers. In: IEDM Technical Digest, pp 837–840
6. Jung SM, Jang J, Cho W, Moon J, Kwak K, Choi B., Hwang, B, Lim H, Jeong, J, Kim J, Kim K (2004) The revolutionary and truly 3-dimensional $25F^2$ SRAM technology with the smallest S3 (stacked single-crystal Si) cell, 0.16 μm^2, and SSTFT (stacked single-crystal thin film transistor) for ultra high density SRAM. In: VLSI Technology Digest Technical Papers, pp 228–229
7. Xue L, Liu CC, Tiwari S (2001) Multi-layers with buried structures (MLBS): an approach to three-dimensional integration. In: Proc IEEE Intern SOI Conf, pp 117–118
8. Sailer PM, Singhal P, Hopwood J, Kaeli DR, Zavracky PM, Warner K, Vu DP (1997) Creating 3D circuits using transferred films. IEEE Circuits Devic Mag 13:27–30

9. Topol A, La Tulipe DC, Shi L, Alam SM, Frank DJ, Steen SE, Vichiconti J, Posillico D, Cobb M, Medd S, Patel J, Goma S, DiMilla D, Robson MT, Duch E, Farinelli M, Wang C, Conti RA, Canaperi DM, Deligianni L, Kumar A, Kwietniak KT, D'Emic C, Ott J, Young AM, Guarini KW, Ieong M (2005) Enabling SOI-based assembly technology for three-dimensional (3D) integrated circuits (ICs). In: IEDM Technical Digest, pp 352–355
10. Chen KN, Fan A, Tan CS, Reif R, Wen CY (2002) Abnormal contact resistance reduction of bonded copper interconnects in three-dimensional integration during current stressing. Appl Phys Lett 81:3774–3776
11. Morrow P, Kobrinsky MJ, Ramanathan S, Chang-Min P, Harmes M, Ramachandrarao V, Hyun-mog P, Kloster G, List S, Kim S (2004) Wafer-level 3D interconnects via Cu bonding. In: Proc Adv Metallization Conf, pp 125–130
12. Ramm P, Klumpp A, Merkel R, Weber J, Wieland R, Ostmann A, Wolf J (2003) 3D system integration technologies. In: McKerrow AJ, Leu J, Kraft O, Kikkawa T (eds) Mater Res Soc Symp Proc, vol 766, pp 3–14
13. McMahon JJ, Lu J-Q, Gutmann RJ (2005) Wafer bonding of damascene-patterned metal/adhesive redistribution layers for via-first three-dimensional (3D) interconnect. In: Proc 55th Electronic Components and Technology Conf, vol 1, pp 331–336
14. Kostner H, Huebner H (2003) New flip-chip on chip process supercedes embedded technologies. In: Proc 14th European Microelectronics and Packaging Conf & Exposition
15. Pozder S, Chatterjee R, Jain A, Huang Z, Jones RE, Acosta E (2007) Progress of 3D integration technologies and 3D interconnects. In: Proc IEEE International Interconnect Techn Conf, pp 213–215
16. Steen SE, LaTulipe D, Topol AW, Frank DJ, Belote K, Posillico D (2007) Overlay as the key to drive wafer scale 3D integration. Microelectron Eng 84:1412–1415
17. Okada Y, Tokumaru Y (1984) Precise determination of lattice parameter and thermal expansion coefficient of silicon between 300 and 1500 K. J Appl Phys 56:314–320
18. Kemevez N (2006) Innovations in 3-D circuit via direct bonding technologies. In: Proc RTI sponsored 3D Architectures for Semiconductor Integration and Packaging, conf session 7
19. Scheiring C, Kostner H, Lindner P, Pargfrieder S (2005) 3-D integration of ICs. Adv Packag 14:26–28
20. van Zeijl HW, Slabbekoorn J, Nanver LK, van Dijk PWL, Berthold A, Machielsen T (2000) Backwafer optical lithography and wafer distortion in substrate transfer technologies. In: Proc Intern Society of Optical Engineering (SPIE), vol 4181, pp 200–207
21. Guarini KW, Topol AW, Ieong M, Yu R, Shi L, Newport MR, Frank DJ, Singh DV, Cohen GM, Nitta SV, Boyd DC, O'Neil PA, Tempest SL, Pogge HB, Purushothaman S, Haensch WE (2002) Electrical integrity of state-of-the-art 0.13 μm SOI CMOS devices and circuits transferred for three-dimensional (3D) integrated circuit(IC) fabrication. In: IEDM Technical Digest, pp 943–945
22. Pozder S, Jones R, Adams V, Li H-F, Canonico M, Zollner S, Lee SH, Gutmann RJ, Lu J-Q (2006) Exploration of the scaling limits of 3D integration. In: Bower CA, Garrou PE, Ramm P, Takahashi K (eds) Mater Res Soc Symp Proc, vol. 970, 0970-Y02-01
23. Ezaki T, Kondo K, Ozaki H, Sasaki N, Yonemure H, Kitano M, Tanaka S, Hirayama, T (2004) A 160 Gb/s interface design configuration for multichip LSI. In: Proc Intern Solid-State Circuits Conf, pp 140–141
24. Alam SM, Jones RE, Rauf S, Chatterjee R (2007) Inter-strata characteristics and signal transmission in three-dimensional (3D) integration technology. In: Proc IEEE Intern Symp Quality Electronics Design, pp 580–585
25. Black B, Nelson DW, Webb C, Samra N (2004) 3D processing technology and its impact on iA32 microprocessors. In: Proc IEEE Intern Conf on Computer Design, 1063-6404/04, pp 316–318
26. Bautista J (2007) Tera-scale computing – the role of interconnects in volume compute platforms. In: IEEE Intern Interconnect Techn Conf, pp 187–189

27. Lee K (2006) The next generation package technology for higher performance and smaller systems. In: Presentations CD: 3D Architectures for Semiconductor Integration and Packaging
28. Kurino H, Lee KW, Nakumura T, Sakuma K, Park KT, Miyakawa N, Shimazutsu H, Kim KY, Inamura K, Koyanagi M (1999) Intelligent image sensor chip with three dimensional structure. In: IEDM Technical Digest, pp 879–882
29. Kurino H, Nakagawa Y, Nakamura T, Yamada Y, Lee K-W, Koyanagi M (2001) Biologically inspired vision chip with three dimensional structure. IEICE Trans Electron Jpn E84-C:1717–1722
30. Aull B, Burns J, Chenson C, Felton B, Hanson H, Keast C, Knecht J, Loomis A, Renzo M, Soares A, Suntharalingam V, Warner K, Wolfson D, Yost D, Young D (2006) Laser radar imager based on 3D integration of Geiger-mode avalanche photodiodes with two SOI timing circuit layers. In: IEEE Intern Solid-State Circuits Conf Digest of Techn Papers, paper 16.9
31. Temple D, Bower A, Malta D, Robinson JE, Coffman PR, Slokan MR, Welch TB (2006) High density 3-D integration technology for massively parallel signal processing in advanced infrared focal plane array sensors. IEDM Technical Digest, 10.1109/IEDM.2006.346980
32. Motoyoshi M, Kamibayashi K, Koyanagi M, Bonkohara M (2007) Current and future 3 dimensional LDI technologies. In: Proc 3D- System Integration Conf (Japan)
33. Beyne E (2006) 3D system integration technologies. In: Intern Symp VLSI Techn Systems Applications, pp 19–27

Index

T

U

V

W

Z

Series on Integrated Circuits and Systems

(*Continued from page ii*)

Ultra-Low Power Wireless Technologies for Sensor Networks
Brian Otis and Jan Rabaey
ISBN 978-0-387-30930-9, 2007

Sub-Threshold Design for Ultra Low-Power Systems
Alice Wang, Benton H. Calhoun and Anantha Chandrakasan
ISBN 978-0-387-33515-5, 2006

High Performance Energy Efficient Microprocessor Design
Vojin Oklibdzija and Ram Krishnamurthy (Eds.)
ISBN 978-0-387-28594-8, 2006

Abstraction Refinement for Large Scale Model Checking
Chao Wang, Gary D. Hachtel, and Fabio Somenzi
ISBN 978-0-387-28594-2, 2006

A Practical Introduction to PSL Cindy Eisner and Dana Fisman
ISBN 978-0-387-35313-5, 2006

Thermal and Power Management of Integrated Systems
Arman Vassighi and Manoj Sachdev
ISBN 978-0-387-25762-4, 2006

Leakage in Nanometer CMOS Technologies
Siva G. Narendra and Anantha Chandrakasan
ISBN 978-0-387-25737-2, 2005

Statistical Analysis and Optimization for VLSI: Timing and Power
Ashish Srivastava, Dennis Sylvester, and David Blaauw
ISBN 978-0-387-26049-9, 2005

Printed in the United States of America